The First Thousand Years of Glass-Making in the Ancient Near East

The First Thousand Years of Glass-Making in the Ancient Near East: Compositional Analyses of Late Bronze and Iron Age Glasses

Wendy Reade

Archaeopress Archaeology

Archaeopress Publishing Ltd
Summertown Pavilion
18-24 Middle Way
Summertown
Oxford OX2 7LG
www.archaeopress.com

ISBN 978-1-78969-703-2
ISBN 978-1-78969-704-9 (e-Pdf)

For my father
Peter C Reade
1930 - 2009

'... there is still merit in the simple notion that if you don't know what an object is made of, you don't really know much about it.'

Brill 1987: 1

Contents

List of Figures

Chapter 1

Chapter 2

Chapter 3

Chapter 6

Chapter 7

Chapter 8

Chapter 9

List of Tables

Chapter 2

Chapter 3

Chapter 4

Chapter 5

Chapter 6

Chapter 7

Chapter 8

Chapter 9

Appendix II

Appendix III

Appendix IV

Appendix V

Appendix VI

Preface

This project arose from my intense interest in archaeology and human technological achievement, a combination of a fascination for history and for the chemistry of archaeological materials, which forms the essence of archaeological science. Rollinson (1993: ix) as a geochemist wrote,

> 'what excuse have I as a single author for venturing into the fields of so many experts in an attempt to explain the methodologies of the main sub-disciplines of geochemistry? ...Today of course the problem is compounded, for geochemistry has a multitude of relatively new and comparatively narrow fields.'

The same can be said of a modern archaeological investigation into ancient technology, such as this one. The researcher is required to use a combination of related disciplines. There are experts in each field of scientific instrumental analysis, experts in geochemistry and its various subdivisions, those who specialise in glass chemistry, or in statistics, those whose field is archaeology, from regional and chronological studies to the classification of artefacts into typologies, the writing of history and the deciphering of ancient texts. Indeed, all of these aspects must be combined to produce a complete and well-informed work which contributes to knowledge and from which all who read it can derive useful information.

That such studies are now possible, even essential, to produce the highly detailed information required to solve archaeological problems that cannot be solved in traditional archaeological ways, is my defence for taking on such a task as an individual. Naturally, I have called on experts when possible, but take responsibility for compiling the whole into the dissertation as presented.

Moorey (1985: x) advocated the benefits of a combined approach to studying archaeological materials when he observed,

> 'In this generation of proliferating information no single scholar can hope to combine the formidable range of learning ideally required for a task of this kind. Still, ... it is much needed and, at a preliminary stage, there are distinct advantages in one scholar, rather than a team, undertaking it'.

Not only are there many aspects to the investigation of ancient glass which must be considered, but the very nature of glass itself contributes to the complexity of its study. The words of McCray and Kingery (1998: 2) describe this challenge well.

> 'Glass, in comparison to other inorganic art/archaeological materials, might be said to represent "the worst of both worlds". Glass does not have an easily characterized microstructure (except for opaque and some colored glasses) in the same manner as traditional ceramics or metals. Glass often does not retain evidence of its manufacture in the same manner that a metal's microstructure "speaks" to a scholar. And glass does not have diagnostic features such as temper or inclusions that can help in ascertaining provenance in the way that pottery does. Indeed ... the extremely complicated business of interpreting glass compositions ... should be viewed as a challenge and opportunity to glass researchers rather than a bane.'

Glass researchers include both scientists and archaeologists, and many shades in between, but it has often been the case that scientists write in order to communicate with other scientists rather than archaeologists, and not all archaeologists make use of scientific analysis. It is hoped, while the core of the present work is essentially a scientific treatise, that it will satisfy both the scientifically and historically inclined, providing a synthesis of data and ideas into which the specialist will be able to delve more deeply at will, and from which future research will spring.

There are always more questions than answers.

Acknowledgements

I owe sincere thanks to the following people without whom this study would neither have started, nor been completed. I wish to express my whole-hearted appreciation to them.

My supervisors:

Daniel T. Potts, Edwin Cuthbert Hall Professor of Middle Eastern Archaeology at the University of Sydney (now Institute for the Study of the Ancient World, New York University), for his unfailing support and belief in me, which made the journey possible and sustained me throughout.

Ian C. Freestone, Deputy Keeper of the Department of Scientific Research at the British Museum (now Professor, Department of Archaeology, University College London), for his generous and patient guidance, unflagging good humour and keen insight for which I am deeply grateful. I am indebted to him for teaching me to operate an SEM, for analysing the glasses from Pella, for generously providing samples of Egyptian glasses for analysis and for the scanning electron micrographs that appear in this study. Thanks to Pam Freestone for her kind and patient hospitality.

Associate Professor Mike Barbetti, formerly Director of the NWG Macintosh Centre for Quaternary Dating at the University of Sydney, for his support and encouragement for this project, and for providing the opportunity for teaching and research. MB introduced me to the excellent analytical facility at the Advanced Centre for Queensland University Isotope Research Excellence (ACQUIRE) (now the Centre for Microscopy and Microanalysis) at the University of Queensland.

This study would not have been possible without the financial assistance from a University Postgraduate Award, two Near Eastern Archaeology Foundation grants-in-aid, the Catherine Southwell-Keely Travel Award, two Carlyle Greenwell bequests, the NWG Macintosh Centre for Quaternary Dating and three awards from the Postgraduate Research Support Scheme, all at the University of Sydney.

The following people and institutions for their time and for the opportunity to study and sample ancient glasses from their collections. Without them, this study could not have been made:

Dr James Armstrong, Harvard Semitic Museum, Boston, for access to the glasses from Nuzi and for allowing samples to be taken;

Dr St John Simpson, Assistant Keeper, and Dr Jonathan Tubb, Assistant Keeper, Department of the Middle East, the British Museum, for facilitating access to and sampling of the museum's collection of ancient glasses, principally those from Nimrud. Thanks to SJS for ongoing assistance with the study of the material from Nimrud;

Dr Stephen J. Bourke, Director of the University of Sydney Pella Project, and the Sar As-Saraya Museum, Irbid, Jordan, for permitting the study and sampling of the glasses from Pella. I am indebted to SJB for hours of discussion about the glass and the intricacies of the contexts in which it was found, for reading chapters and for much congenial support and encouragement;

Dr Ralph-Bernhard Wartke, Deputy Director, Pergamon Museum, Staatliche Museen zu Berlin, for allowing me to study glasses from Assur and Babylon and for providing samples for analysis. Restraints of time and space have unfortunately prevented the inclusion of the results of these analyses in the present study, but they will be the subject of future publications as they provide important information on Iron Age glasses.

Dr Bernard Gratuze, of IRAMAT, Centre Ernest Babélon, C.N.R.S., Orléans, for most generously supplying me with his analytical data and reports for comparative glasses from France and thereby adding a valuable dimension to this work.

The many people who opened the collections of their institutions for me to study. The opportunity to learn about early ancient glasses by direct examination was invaluable:

Dr Ralph-Bernhard Wartke, Deputy Director, Pergamon Museum, Staatliche Museen zu Berlin, Ms Maude de Schauensee, formerly Keeper of the Near Eastern Section, the University of Pennsylvania Museum of Archaeology and Anthropology; Mr Raymond Tindel, Registrar and Senior Curator, Museum, Oriental Institute of Chicago; Dr Rachel Sparks and Mr Ian Carroll, Petrie Palestinian Collection, Institute of Archaeology, University College, London; Dr Rita Freed, Chair, Art in Ancient World Department, Boston Museum of Fine Arts; Professor Carl Lamberg-Karlovsky, Harvard University, Curator of Near Eastern Archaeology, the Peabody Museum, Dr Karin Sowada, Assistant Curator, Nicholson Museum, University of Sydney.

Dr Robert H. Brill, of the Corning Museum of Glass, for inspiration and early guidance, and for generously furnishing the Corning standard reference glasses, B, C and D.

The people who have assisted with sample preparation and who have performed analyses of ancient glasses for this project (in addition to ICF):

Dr Alan Grieg and Dr Balz Kamber, formerly of ACQUIRE, for their inductively coupled plasma mass spectrometry analyses of the glasses examined in the present study. Their instruction and advice in geochemical interpretation of the resulting highly sensitive data was invaluable; Ms Alanna Simpson, University of Queensland, for electron probe microanalyses of the glasses from Nuzi; Mr Nigel Meeks, Department of Scientific Research, the British Museum, for assistance during sample preparation and scanning electron microscope analysis at the museum; Mr Adam Sikorski, Australian Key Centre for Microscopy and Microanalysis, University of Sydney, for assistance during sample preparation.

The many fine people who have been helpful in their stimulating discussions and encouragement, in particular: Dr Andrew Shortland, Dr Caroline Jackson, Dr Paul Nicholson, Mr Mark Wypyski, Dr Colleen Stapleton, Professor Julian Henderson, Professor Michael Baxter, Professor Mike Tite, Dr Julian Reade and Professor Janet Duncan Jones.

Professor Simon Ringer, Director of the Electron Microscope Unit (EMU), Australian Key Centre for Microscopy and Microanalysis, University of Sydney, for generously supporting me while employed at the EMU during my doctoral journey. In being a part of the EMU, I have found myself immersed in a stimulating environment of scientific research, filled with people whose interest and enthusiasm have inspired me.

Dr Melanie Fillios, Australian Research Council post-doctoral fellow, University of Sydney, for performing statistical testing of the Late Bronze Age glass data, and Mr Dennis Dwarte, Senior Microscopist, the EMU, for checking the results and for assistance with use of the Sensi-Cam light microscopy system used to take light micrographs of glasses.

The efficient and helpful staff of the Fisher Library, and its branches, at the University of Sydney, who have expedited my research. Ms Beth Hylen of the Rakow Library at the Corning Museum of Glass for her equally kind assistance.

The people who work at the University of Sydney excavations at Pella in Jordan, where this project began, who each encouraged and endured my last month of writing, which I undertook on site during the 2009 excavation season. It seemed poetic that this work should end where it began.

My caring and persevering family, friends and colleagues, some of whom have already made this journey and have guided me with unending patience:

Dr Catriona Bonfiglioli, Dr Paul Donnelly, Dr Tiffany Donnelly, Mrs Fiona Wotton, Mrs Joan Downton, Ms Liz Fenner, Mr David Bridges, Dr Bernadette McCall, Dr Michael Rodriguez, Dr Liz Carter, Dr Kate da Costa, Dr Kyle Ratinac, Mr Aneal Chandra, Dr Lloyd Weeks, Dr Cameron Petrie and by no means least, Dr John Tidmarsh, and Dr Karyn Wesselingh, who have been constantly supportive, reassuring and long-suffering. KW and Mr Philip Cullan have been invaluable technical advisors for computer-related frustrations. I owe the preservation of my sanity to them.

My father, Professor Emeritus Peter C. Reade, for reading my work and patiently correcting my many sins of scientific expression, for advising me, and for being there. I will always be grateful to him, to my mother, Erica, and to my step mother, Ene, for nurturing my dream of becoming an archaeologist, and for helping me to move to Sydney after finishing school in Adelaide, to the only University in Australia which offered a degree in archaeology. Now, many years on, I hope this dissertation goes some way to thanking them.

To Will and Chloe, always.

To all of these people I am deeply grateful for sharing their vast knowledge, insight and experience from which I have tried to learn, not just about ancient glass, but about life in general. I dedicate this study to them in appreciation of the immensity of all they have done for me.

Abbreviations

Chronology

BA	Bronze Age
c.	circa
C	century
IA	Iron Age
LBA	Late Bronze Age
MBA	Middle Bronze Age
MB/LB	Middle Bronze Age to Late Bronze Age transition
n.d.	no date

Technical Terms

Av.	average
b.d.	below detection limit
EPMA	electron probe microanalysis
Fig.	figure
HCHC	high copper-high cobalt
HFS	high field strength
HMHK	high magnesia, high potash
HREE	heavy rare earth elements
ICP-AES	inductively coupled plasma atomic emission spectrometry
ICP-MS	inductively coupled plasma mass spectrometry
kV	kilo electron volt
LCLC	low copper-low cobalt
LMLK	low magnesia, low potash
LREE	light rare earth elements
MDL	minimum detection limit
mg	milligram
mL	millilitre
MUQ	mud from Queensland
nA	nano Amps (measure of current)
no.	number
O	opaque
ppb	parts per billion
ppm	parts per million
REE	rare earth elements
Ref.	reference
RSD	relative standard deviation
SBS	sedimentary background signal
SD	standard deviation
SEM	scanning electron microscope
SEM-EDS	scanning electron microscope with energy dispersive spectrometer
SRE	sediment-related elements
T	translucent
turq.	turquoise
UCC	upper continental crust
μm	micron
wt %	weight per cent
%	where % is used to indicate proportions of chemical compounds in the text, it has been calculated by weight
Z	atomic number
*	as in MgO* denotes reduced or base glass oxide normalised

Oxides

Al_2O_3	alumina
CaO	lime
CoO	cobalt oxide
CuO	copper oxide
FeO	iron oxide
Fe_2O_3	iron oxide
K_2O	potash
MgO	magnesia
MnO	manganese oxide (referred to in text as manganese)
Na_2O	soda
P_2O_5	phosphorus pentoxide (referred to in text as phosphorus)
PbO	lead oxide
Sb_2O_3	antimony oxide
SiO_2	silica
SO_2	sulphur dioxide
TiO_2	titania

Elements

(for elements in order of atomic number see Chapter 2)

Al	aluminium
As	arsenic
Ba	barium
Be	beryllium
Ca	calcium
Cd	cadmium
Ce	cerium
Co	cobalt
Cr	chromium
Cs	caesium
Cu	copper
Dy	dysprosium
Er	erbium
Eu	europium
Fe	iron
Ga	gallium
Gd	gadolinium
Hf	hafnium
Ho	holmium
K	potassium
La	lanthanum

Li	lithium
Lu	lutetium
Mg	magnesium
Mo	molybdenum
Nb	niobium
Nd	neodymium
Ni	nickel
Pb	lead
Pr	praseodymium
Rb	rubidium
Sb	antimony
Sc	scandium
Sm	samarium
Sn	tin
Sr	strontium
Ta	tantalum
Tb	terbium
Th	thorium
Ti	titanium
Tl	thallium
Tm	thulium
U	uranium
V	vanadium
W	tungsten
Y	yttrium
Yb	ytterbium
Zn	zinc
Zr	zirconium

Chapter 1

Introduction, Background and Aims

1.1 Introduction (Figure 1-1)

Until the mid-2nd millennium BC, the production of glass objects in the Near East appears to have been limited, with very few examples recovered from archaeological excavations. From the late 16th C BC, the volume of glass production increased markedly, the evidence for which is found in the beads, small mould-made pieces, core-formed vessels, inlays and ingots excavated at sites throughout the Near East and Egypt. The present study was designed to determine the compositions of a selection of the early Near Eastern 'mass-produced' glasses from the Late Bronze and Iron Ages, as a means of identifying differences and similarities in their compositions which occurred during and between each chronological period. This information has been used to define possible regional variations in composition, and to investigate the raw materials used in glass-making at this time.

The glasses analysed in the present study were excavated from the sites of Nuzi in Mesopotamia, from the neo-Assyrian capital of Nimrud (both in northern Iraq), and from Pella in the Levant (northern Jordan). The Nuzi glasses and many of the Pella glasses are dated between the 16th C to the 12th C BC, or the Late Bronze Age (LBA). The remainder of the Pella glasses and the Nimrud glasses are dated between the 11th C and the 8th C BC, or the Iron Age (IA).

Much of the early Near Eastern glass that has been studied to date, has been excavated from Mesopotamian sites, such as Nuzi, Nimrud, Assur and Tell Brak, and hence traditionally has been referred to as 'Mesopotamian', a term often used to distinguish glass found in the broader Near Eastern region from that unearthed in Egypt. Moreover, Near Eastern or Mesopotamian craftsmen are sometimes referred to as 'Asiatic'. It is necessary here to clarify the terminology used in the present study. The Near Eastern region encompasses modern Syria, south-eastern Turkey, Lebanon, Israel, Jordan, Iraq (Mesopotamia) and Iran. The glasses included in this study were found in both Jordan and Iraq, and have been described as 'Near Eastern'. This term is used in the present study to include all glasses from the Near Eastern region where the origins of glass remain obscure, because the production centres have not yet been located. The term 'Mesopotamian' is used when referring to the works of authors who have referred to glass specifically from the Mesopotamian region.

The early glasses from the LBA appear to be the product of an already complex pyrotechnological industry, the origins of which have been much debated. The products of this industry include glaze, faience and frit, besides glass. This study is concerned with the investigation of glass only.

Oppenheim (1973a: 262) contended that glass manufacturing was not a new invention in the LBA, as evidenced by the discovery of isolated glasses which preceded the *floruit* of glass-making in the mid-2nd millennium BC, the result of a change in glass-making technology that led to increased production and the broad distribution of glass. This technological advance is evident in the appearance of new production techniques, such as fusing, marvering, trailing and core-forming of vessels, and the creation of a wider range of colours than had previously been used (Moorey 1999: 193). General summaries of the nature of glass and the technology and techniques of ancient glass production, which are beyond the scope of this dissertation, have been published by Newton and Davison (1989: 1-17, 54-69), Stern and Schlick-Nolte (1994: 19-95) and Schlick-Nolte and Lierke (2002: 11-40), for example.

When studying the earliest glasses, the traditional archaeological approach of visual classification by physical type is limited by the narrow and simple range of shapes of objects, other than vessels. It is difficult to determine the origin of these objects, and hence the geographic extent of their manufacture and distribution. From physical studies alone, it is impossible to know the nature of the raw materials used to make ancient glasses (see for example Matoïan 1999: 56). Chemical studies, or compositional 'finger-printing', are developing as a result of the increased interest in the practical questions that they can help to answer. Compositional studies are providing more information because of the refined and sensitive analytical instrumentation now available, although the results obtained from such analyses can be constrained by the limited survival of glass during millennia of interment, the dependency on the rigour of excavation for the collection and recording of glass, and that much of the early glass is weathered.[1] When glass decays, or

[1] Indeed, it is possible that glass objects made in an emerging and probably experimental industry, prior to the mid-2nd millennium, have not survived due to compositional instability, although the survival of glazes from much earlier periods suggests that it was already known how to make stable glass. For the effects of weathering on glass and the problems regarding sampling, see for example

weathers, its composition is altered (Freestone 2001), but what remains of the original glass can often be sampled from a broken edge, and chemically analysed, thus providing valuable information about one of the first synthetic materials.

Until recently, the relatively undeveloped state of this kind of investigation has consisted of isolated studies of glass from several sites in the ancient Near East. The data from many of these analyses have not been systematically compared with the available data from other studies of glass compositions, and hence are not viewed in regional, chronological or technological contexts. Trace element data have largely been interpreted in very general terms, and 'beyond broad technological affiliations, meaningful compositional groupings have been difficult to establish' (Freestone 2006: 201).

Initially, it is possible to investigate glass-making groups by using the major elements which comprise the base glass compositions, determined with relative accuracy by most investigators of early glass. The data produced by different analysts may be compared, and indeed this is usually unavoidable in archaeological work because it is either impossible or undesirable to sample artefacts repeatedly (Freestone 2005: 5). While acknowledging the difficulties of sampling archaeological glasses, which has often meant the analysis of small numbers of samples of random selection, Shortland and Eremin (2006: 582) advocated the analyses of a large number of samples to reveal compositional patterns successfully, and to provide a cohesive data-set for comparative purposes for other researchers. The present study aimed to fulfil these criteria for the glasses analysed therein.

With the improvement in the availability and capability of chemical analytical techniques in recent decades, it is now possible to characterise the compositions of ancient glasses to ultra-trace element levels measured in parts per billion (ppb). The data produced by such analyses are used to refine the initial major element characterisations of the glasses, and to determine as far as is possible the raw materials used in their production. In turn, it might be possible to relate glasses to their primary production group or region, as has been done with 2nd millennium BC Egyptian glasses (Shortland *et al.* 2007) and with AD 1st millennium glasses (Freestone 2005: 1). This information enhances the interpretation of the earliest glass-making practices and the sourcing of raw materials. Glass is, by definition, an amorphous material, made from a variable combination of natural materials of variable composition, making the determination of its elemental composition a difficult task. The more detailed the chemical data available, the more enlightening, but the more complex, this task becomes.

A model which has been developed in the form of a flow chart to illustrate the processes involved in the present study appears in Appendix I.

1.2 Historical Background of Early Glass

The earliest glass objects appear throughout the Near East in quantity as the products of an already complex pyrotechnology[2] in the mid-2nd millennium BC, with the date generally agreed to be *c.*1500 BC or slightly earlier (Lilyquist *et al.* 1993; Nicholson 1993; Shortland and Eremin 2006: 581). Before this time, vitreous technology centred on faience production and the glazing of stone, with rare exception, for example a piece of raw glass from Eridu (Barag 1985: no. 179) and other sporadic finds described by Moorey (1999: 190-191).

During the LBA, the large territorial states of Egypt and the kingdom of Mitanni interacted with one another (Kuhrt 1995: 283) and, located between them, especially in the Syro-Levantine region, was a set of smaller states that owed allegiance to their more powerful neighbours. Egypt at that time exercised northward territorial expansion. To the east lay Kassite Babylonia and to the north was Hittite Anatolia. When the Mitanni state declined in the mid-14th C BC, it was followed by the rising domination of Assyria. To the east of Assyria was the kingdom of Elam, while the city-states of Syria and Palestine persisted, but were dependent on the great powers around them (van de Mieroop 2004: 121) (Figure 1-2).

The development and proliferation of glass-making occurred in the cosmopolitan cultural framework of the Hurrian dominated northern Syro-Levant and Mesopotamia, and the powerful Egyptian state and its vassals. In the mid-2nd millennium BC, these cultures were connected by trade routes which led north-south along the Levantine coast and the Jordan valley from Egypt to Syria, and east along the arcs of the Tigris and Euphrates Rivers which formed the Fertile Crescent (Moorey 1989; 1999: 5-6, 10; Van De Mieroop 2004: 127-135).

At this time, Cyprus, the Aegean and the coastal regions of the eastern Mediterranean, including Egypt, were connected by regular maritime trade. The inland regions, such as Mesopotamia, were not directly involved but were connected to the coastal traders by inland trade routes. The quantity of goods that entered the Near East from the west increased

Turner (1963: 37).

[2] It is beyond the scope of the present study to discuss the origins of glass in earlier pyrotechnologies.

in the second half of the 2nd millennium BC to unprecedented levels, and included ceramics, copper, tin, ivory and other products in a complex exchange network, vividly recalled by LBA shipwrecks such as those of Cape Gelidonya and Ulu Burun (Bass *et al.* 1989; Knapp 1993; Akkermans and Schwarz 2003: 352; van de Mieroop 2004: 117). Textual evidence and arts and crafts demonstrate the internationalism of this period and the movement and exchange not only of goods, but of ideas and of people (Akkermans and Schwarz 2003: 354). Cross-cultural contacts have an impact on the development of technology through trade, itinerant craftsmen, written records and native craftsmen being educated abroad, with the consequent spread of ideas, styles or the techniques of crafts (McGovern 1995: 30).

The end of the LBA, *c.*1200 BC, and the commencement of the IA belongs to a period which is termed a 'Dark Age' that affected the whole of the eastern Mediterranean region. From this time, written records were scarce and there has been much debate regarding the various events that are thought to have contributed to the collapse of the LBA civilisations. The large urban centres and political systems of the eastern Mediterranean world experienced a period of crisis and collapse which saw the abandonment of several cities including Ugarit on the Syrian coast, Alalakh in south-eastern Turkey and Tell Brak in northern Syria. The Hittite state to the north, and Mycenaean Greece and Cyprus to the west, were destroyed, while Egyptian imperial involvement in the Near East was curtailed. The Middle Assyrian empire in Syria, which had emerged after the defeat of Mitanni in the late 14th C BC, held out until its territories were finally reduced by the middle of the 11th C BC. The once flourishing maritime trade of the LBA ended with the widespread destruction and political instability of the region (McGovern 1987a: 271; Liverani 1987; Dever 1992: 18-19; Drews 1993; Akkermans and Schwarz 2003: 358-359).

McGovern (1987a: 267) argued that the LBA cosmopolitan city-state system was transformed into a much more insular society in the early IA, with few foreign contacts and a lower standard of living, in part due to the disruption of trade. Elements of early IA material culture superficially appear to be different to that of the LBA, but cultural and technological continuity was also evident in the uninterrupted use of the same cemeteries and settlement sites and in elements of material culture (McGovern 1987a: 267-269).

The economic decline of the 'Dark Age' is reflected in the apparent reduction in the production of glass at this time. There is an almost total absence of glass finds dating from the end of the 2nd and the beginning of the 1st millennia BC, until the 9th to 8th C BC when glass vessels and other glass objects, such as inlays and beads,[3] were being made again in the Near East (von Saldern 1965; 1970: 205; 2004: 5-67; Harden 1968: 49, 53; Barag 1970: 131; 1985: 51; Webb 1987: 145; Moorey 1999: 198-199). Harden (1968: 53) proposed that from the 9th C BC onwards, during the IA, there were probably two centres of glass vessel production, one on the Syrian coast and one in Mesopotamia. Both made core-formed vessels and, more rarely mosaic glass, and they developed casting and cold-cutting of glass, as found at Nimrud, for example.

At the end of the 10th C BC, the Assyrian kings began a series of successful campaigns against the peoples to the west, establishing the Neo-Assyrian empire (*c.*900-609 BC), which they maintained by continued expansion through annual military campaigns to raid and subjugate their neighbours (Akkermans and Schwarz 2003: 377). In the IA, trade developed into a more commercialised and geographically widespread venture, no longer exclusively controlled by elite powers. The use of camels, after their domestication in *c.*1000 BC, meant that goods could be transported over long distances through formidable desert, such as that between Mesopotamia and the Levant, complementing the traditional LBA trade routes along the rivers of these regions. At the same time, Phoenician inhabitants of the Levantine coast sailed further west than the LBA traders had, opening up new sources of raw materials in a far reaching trading network and establishing colonies in Cyprus, North Africa, Sicily, Sardinia and Spain (Akkermans and Schwarz 2003: 9-10, 360-361, 386).

By the end of the 8th C BC, the Assyrians had conquered lands as far as the Phoenician states of western Syria and the Lebanese coast, and Babylonia was reduced to subject status, making the Assyrian empire the largest political entity the Near East had ever seen (Akkermans and Schwarz 2003: 378-379).

Before the fall of the Neo-Assyrian empire in the late 7th C BC, the sites of Nimrud, in Mesopotamia, destroyed in the late 7th C BC, and Hasanlu, in Iran, destroyed in the 8th C BC, are known for the quantity and variety of glass finds, which are securely dated to this period of the resurgence of glass-making in the IA (Moorey 1999: 198). At Nimrud, vessels are predominantly monochrome (many are colourless) mould-made and cut glass bowls (von Saldern 1965), rather than the polychrome core-formed vessels of the preceding LBA, which is indicative of significant technological change.

[3] Harden (1968: 49) assumed, without evidence, that beads and other small objects probably continued to be made through the 'Dark Age', especially in Syria.

This change can also be seen in the typological distinctions between the LBA core-formed vessels and those of the IA, which are smaller, and of poorer quality glass and workmanship, and which exhibit no direct links with the earlier LBA vessels (Moorey 1999: 200). Barag (1970: 135, 171-174) remarked on the limited distribution of core-formed vessels from the LBA by contrast with the diffusion of vessels to remote and very different regions in the IA. The LBA vessels have a smaller variety of shapes, which were frequently made with a blue base colour, whereas the IA vessels have a much wider range of shapes usually made with very dark colours, such as dark brown, green-black (often referred to as 'black'), and sometimes with blue, turquoise, brown, yellow, and rarely white.

Glass all but disappeared from Egypt between the 11th/ 10th C BC and the 7th/ 6th C BC, the time of the troubled Third Intermediate Period. A small number of glass samples have been found, dated from the 26th Dynasty, but glass remains rare until the 4th C BC (Moorey 1999: 200; Nicholson 1993: 61). The lack of evidence for IA glass-making from Egypt, has led to the suggestion of a Phoenician or Assyrian origin for the Nimrud mould-formed bowls (Moorey 1999: 199). Von Saldern (1970: 209) was undecided as to whether the origins of IA glass were Phoenician, Syrian or Assyrian, and Barag (1985: 52-55) later argued for manufacture by Phoenician craftsmen perhaps working in Assyrian court workshops. Barag's opinion was based on circumstantial evidence which links the glass from Nimrud with Phoenician production, including glass inlays attached to 'Phoenician' carved ivories, and similar glass inlays on a cut glass bowl (von Saldern 1970: nos 29-30), a proposition which has been supported by Orchard (1978) and Moorey (1999: 200). The Phoenicians have been recognised as the manufacturers and exporters of well-crafted metal vessels, jewellery, ivory furniture components and glass objects (Akkermans and Schwarz 2003: 387-388).

1.3 The Early Glass Industry

Many writers have commented that glass was made to imitate precious stones, primarily lapis lazuli, turquoise and later, rock crystal, amongst other stones (see for example Pliny in Eichholz 1962: 155-157, 185, 255; Oppenheim *et al.* 1970: 9-15; Moorey 1999: 77). The status of glass as a highly prized material in its own right is evident, not only from the predominance of glass in the archaeological record in temples, palaces and elite burials, but from the ancient texts, such as the Amarna Letters, which draw comparisons between stones and glass, and which list glass as a prized commodity exchanged between the elite. It could be said that glass, inspired by stones, came to compete with them in value, and surpass them in the variety of its colours and applications. The Assyrian glass-making texts contain recipes for the production of glass, but are not technically explicit, and hence the technology and raw materials of ancient glass-making remain incompletely understood (Oppenheim *et al.* 1970: 22-68; Moran 1992; see also Nicholson 1993: 49; Nicholson and Henderson 2000: 195-196; Nicholson 2007: 2; Shortland 2007: 261-262).

Little is known about where glass was produced in the LBA and IA Near East. The trade in glass ingots, such as those found in the Ulu Burun shipwreck of the late 14th or early 13th C BC close to the coast of Turkey (Pulak 1988: 14; Bass 1986; Bass *et al.* 1989), indicates that glass-making and glass-working could be separate industries, and that the formation of glass into objects could be carried out in locations far from the raw glass production site. The discovery of glass ingots at LBA and IA sites such as Ugarit, Nuzi, Tell Brak and Nimrud suggests that glass was worked at these places, if not made there (Moorey 1999: 202-203). That the division between glass manufacture and object production was widespread in antiquity has been widely discussed (see for example Baxter *et al.* 1995; Freestone *et al.* 2000: 66-67; Gorin-Rosen 2000; Nenna *et al.* 2000; Matoïan 2000a; 2000b; Rehren *et al.* 2001; Shortland 2007: 262-267).

Glass-making remains are rare and, to date, are documented only in Egypt at Tell el-Amarna (Nicholson 1995a; 1995b; 1996; Shortland and Tite 1998; Shortland 2000; Nicholson 2007 for example), Malkata (Mass *et al.* 2002: 68; Nicholson 2007: 21) and at Qantir-Piramesses (Rehren and Pusch 1997; 1999; 2005; 2008; Rehren *et al.* 1998; Schoer and Rehren 2007). It is believed that glass was made in the Near East at this time, possibly in the Syro-Mesopotamian region in several production centres, but no definitive glass-making remains have been identified (Moorey 1999: 201-202). Although Pliny's account in his *Natural History* (Eichholz 1962: 151) of the origins of glass-making in the sands at the mouth of the Belus River in modern Israel is considered to be apocryphal[4] (Wedepohl 1997: 247), other strands of circumstantial evidence point tantalisingly to the production, if not the origin, of glass-making being in the north Syrian or Mesopotamian regions. Such evidence includes the use of two key words in the textual references to glass; *mekku*, of West Semitic origin, found in texts from Anatolia, Syria and Nuzi in Mesopotamia, is synonymous with the word *ehlipakku* which is used in the Assyrian glass-making texts and is possibly of Hurrian origin. Both refer to 'raw glass' which the Egyptian king sought from the local rulers of Palestine, such as Tyre, recorded in the Amarna Letters of the 14th C BC. The Assyrian glass-making texts from the library of Ashurbanipal, of the 7th C BC, are

[4] Turner (1956c: 277T-279T) summarised the ancient references to glass-making ingredients and the connection of the Belus River sands with glass-making along the Syrian coast for many centuries.

believed to have had their origins earlier in the 12th C BC (Oppenheim 1970; 1973a; Moorey 1999: 195).

While the regular production of glass in quantity, and particularly of vessels, is considered to have occurred in northern Syria or Mesopotamia because of the geographic distribution of excavated glass artefacts at sites such as Nuzi, Tel al-Fakhar, Tel al-Rimah, Tel Brak, Assur and Nineveh, all sites which were under the domination or influence of the Hurrians (Petrie 1926: 230; Beck 1934: 8, 19; Harden 1956: 319; 1968: 46-47; Barag 1970: 134; Grose 1989: 45-48; Tite *et al.* 2002: 589), glass manufacturing in Egypt is now generally believed to have begun shortly afterwards (Oppenheim 1973a: 263; and see Nicholson 1995a: 11-19, Nicholson *et al.* 1997: 143, 147; Jackson *et al.* 1998: 11-12; Nicholson and Jackson 2000: 11-21; Shortland 2000c; Shortland and Eremin 2006: 581 for example).

Bryan (2000: 247) proposed, from her study of Egyptian tombs from the time of Thutmosis III and Amenhotep II, that 'the prestige of things Syrian began to soar'. Not least amongst these luxury items was glass, for example the glass vessels made to imitate marble, found in the tomb of Kenamun (TT162). That the military aspect of the Egyptian encounters with the Mitanni was directly responsible for the introduction of glass to Egypt is strongly suggested by Bryan (2000: 247), who believed that the production of glass at this time could have developed in Mitanni centres, such as Tell Brak and Tell Rimah.

Egyptian representations of Syrian workmen being brought to Egypt, carrying vases of metal or glass they had made, had also suggested to Petrie (1926: 230) that glass originated in Syria or Mesopotamia, although glass from this time is much better preserved in Egyptian soils than in those of the regions to the north, and provides a larger corpus for study (Harden 1968: 48; Moorey 1985: x; Newton and Davison 1989: 19). The Syrian origins of glass were discussed by Oppenheim (Oppenheim *et al.* 1970: 57, 84-6; 1973a: 9-11; 1973b: 261) Moorey (1985: 197, 201-2) and McGovern *et al.* (1991: 401).

Nicholson (1993: 45, 47) and Nicholson and Jackson (2000: 11-12), through their work at the pivotal Egyptian glass production site of Amarna, acknowledged that the Mesopotamian region was the 'cradle' of glass-making, and that the first glass to be found in Egypt was either imported from Mesopotamia or was made in Egypt by Mesopotamian craftsmen.[5] They reviewed current opinion which held that Egyptians of the early New Kingdom could not make glass (for example Newton 1980: 176; Newton and Davison 1989: 62, 107), whereas Lilyquist *et al.* (1993: 43) determined that there was evidence for the importation of Near Eastern glasses and that 'Asiatic' craftsmen were present and making glass in Egypt, as Oppenheim had suggested earlier from philological evidence (1973a: 263). That early glass from the reign of Thutmosis III was imported into Egypt from the Near East as tribute from the chiefs of Syria was discussed by Shortland (2000c) with reference to descriptions of glass in two Theban tombs of this date. Shortland (2001) continued his examination of the earliest glasses found in Egypt concluding that while some of these glasses were likely to be imported from the Near East, the cobalt blue glass found from the time of the reign of Thutmosis III must have been a local Egyptian product.

Nicholson and Jackson (1998) showed, by making glass in a replicated furnace, that the excavated furnaces at Amarna could have been used for making glass, a conclusion that was supported by the work on Amarna glasses by Shortland and Tite (2000). Rehren and Pusch (2005: 1757) and Schoer and Rehren (2007: 173) have showed convincingly that the finds from Qantir-Piramesses, including cylindrical vessels filled with partly fused glasses, are evidence that glass was made at this location in Egypt. In his detailed archaeological analysis of the origins of glass found in Egypt, Nicholson (2007: 1-4, 8) highlighted the important relationship of Mitanni with Egypt when he observed that the Amarna letters recorded Mitanni as a source of glass which was exploited into the Amarna period, but noted this did not mean that the New Kingdom Egyptians could not make glass themselves (Nicholson 2007: 4). The Egyptian origin of the glass ingots from the Ulu Burun shipwreck was established by Nicholson *et al.* (1997) and, considered with the remains from Amarna and Qantir-Piramesses, provided convincing evidence for local Egyptian glass-making (Jackson 2005a: 1751).

The organisation of LBA glass workshops is thought to have been centred in palaces and temples of major cities, where much of the glass and glass-working remains have been found, although Moorey (1999: 197, 202) did not discount the possibility of small-scale production outside the sphere of palace and temple workshops. McGovern *et al.* (1991: 401) thought that the different chemical compositions of the glasses from sites such as Tell al-Rimah, Nuzi, Nippur and from the Baq'ah Valley demonstrated that glasses were largely produced by well-developed, local industries, while their similarities indicated that they shared a common technological tradition.

Sparks (2001: 96, 109) explored the practice of locating workshops which produced valuable goods under palace control, with reference to a luxury obsidian workshop at Tell Atchana, commenting that 'a high degree of organisation and wealth would have been

[5] See section 1.1 for explanation of the usage of the terms 'Mesopotamian' and 'Near Eastern' in this study.

essential in maintaining a craft requiring imported luxury raw materials'. Glass-making from raw materials and the importation and subsequent working of glass ingots might be considered in the same light. Glass ingots and many pieces of cullet were found within the LBA palace at Tell Brak in northern Syria which Oates *et al.* (1997: 81, 85-86) interpreted as evidence for glass manufacture, or at least glass-working, from as early as the 15th C BC. There was no evidence of a workshop at LBA Ugarit (Ras Shamra) on the Syrian coast, but it has been hypothesised by Matoïan (1999: 56) that vitreous materials were made there because of the 'exceptional and original' objects recovered, amongst the small chunks of glass, Egyptian blue and cobalt, and hundreds of vitreous objects (Matoïan 2000b: 41-42).

Evidence from two known glass-making sites in Egypt provide insight into the organisation of the glass industry in that region during the 2nd millennium BC. Rehren *et al.* (1998) described the industrial complex of Qantir-Piramesses in the Egyptian delta where evidence suggested that bronze casting and vitreous materials production were integrated in a highly specialised organisation of shared pyrotechnologies and skills. The interdependence of glass, faience, frit and pottery production at Amarna was described by Shortland and Tite (1998) who examined and chemically analysed the material evidence.

The study of the organisation of glass- and faience-making at Amarna was pursued by Shortland *et al.* (2001) who believed that although faience and glass were made using the same facilities, the raw materials used in their production were different. Glass was linked to royalty, partly by the connection evident in textual evidence, and also, the authors suggested, by the possible temple or state control of special colourants like cobalt.

Faience appeared to have a lower socio-economic distribution because it was a common material found in many houses (Shortland *et al.* 2001: 151, 154-155). Shortland (2007) concluded from philological and textual evidence and from the remains of glass-working premises at Amarna, that the glass-workers there were of lowly status in a hierarchy of possibly three ranks that encompassed a glass-making and glass-working industry that was closely controlled by the royal court.

Ritual and divination techniques were applied to the more highly regarded glass production process (Oppenheim 1970: 32-33, 44-47), which was probably carried out within the temple grounds. Robson (2001: 54) made the observation that the 'boundaries between science and religion, medicine and magic were always blurred in the ancient Near East' and hence it is not surprising that the production of glass was a combination of practical and ritual procedures. Shortland (2008: 67) made the significant comment that the ritual nature of glass-making had the 'effect of stifling technological choice and prohibiting experimentation and innovation'.

Robson (2001: 51-52) argued that different colours of glass were believed to have magical properties including curative powers. That the magical and medical fields lay within the sphere of priests explains why the recipes for making artificial stones were included in the corpus of literature to which the glass texts belong. It is believed that these recipes were handed down from the mid-2nd millennium BC as part of a codified tradition that might have been outdated by the 1st millennium BC when, as Robson believed, it was no longer fashionable to make coloured glass (Robson 2001: 53). While different colour preferences are evident in IA glasses, there is no reason to conclude that they did not continue to imitate, and indeed even rival, stones. Dark and turquoise blues, red and purple, known from the LBA, are found also at IA sites, such as Nimrud, and the emergence of larger numbers of colourless glasses might be the result of an advance in technology that enabled the imitation of rock crystal.

Rehren *et al.* (2001), in view of material evidence from Qantir and other sites, proposed that the Egyptian glass industry consisted of primary and secondary workshops, where glass-making and glass-working were carried out separately, a theory not incompatible with that proposed by Shortland (2007). Rehren and Pusch (1997; 2007: 231-233) and Rehren (2000b; 2005), from their research at Qantir-Piramesses, proposed that in the LBA different workshops specialised in the production of particular colours, for example red glass was made at Qantir and cobalt blue glass at Amarna (Jackson and Nicholson 2007: 114-115). The manufacture of these colours required special ingredients or specialist knowledge, but it has not yet been possible to attribute other colours, such as yellow and white, to particular workshops due to the lack of evidence regarding primary production sites. Petrie (1926: 230), in noting that early glass was coloured in imitation of precious stones, found it interesting that the earliest glasses were in fact coloured rather than left in their natural 'colourless' state, because the ability to colour glass required a high degree of specialisation. Nicholson and Henderson (2000: 197, 214) commented that it was easier to colour than to decolour glass and this might more readily explain the lack of truly colourless glass at this time, but does not explain the lack of naturally uncoloured glass, which would have been easier to produce than either intentionally coloured or decoloured glass.

Rehren (2000b: 22) suggested that copper blue glass could have been produced at a number of different manufacturing locations in LBA Egypt and the Near East because copper was readily available. Copper concentrations in blue glasses have a wider variability of composition, which suggests that the glasses were the products of different workshops. Glass working, on the other hand, encompassed the entire range of colours, which could have been traded from their places of specialist manufacture.

The compositional homogeneity of LBA Egyptian plant ash glasses[6] for approximately 500 years was seen by Rehren (2008: 1346) as a possible result of a closely controlled raw material supply and workshop practice, as could be expected from an elite material under strict control, in the same way that LBA Near Eastern glass production was described by Robson (2001) and Shortland (2008). Glass working remains in the royal palace at Nimrud suggested that royal control continued to be exerted over this industry in the IA (von Saldern 1965: 241.2; Moorey 1999: 202-203).

1.4 Object Studies

The typological studies of LBA and IA Near Eastern glasses, in which vessels and moulded objects have been assembled and classified with regard to shape, decoration, chronology and origin, have been central to our understanding of early glass-making in this region, providing the archaeological context for subsequent research. This information was established largely by the works of Fossing (1940), Harden (1968),[7] Barag (1962; 1970; 1985) and von Saldern (1965; 1966a; 1966b; 1970). Fossing (1940) collected the limited material available at the time, although his dating of the early glass industry in Mesopotamia has now been extended earlier into the LBA. Von Saldern traced the history of Mesopotamian cut glass vessels of the 8th C BC onwards from the site of Gordion in central Turkey (1959), and mosaic glass from Hasanlu and Marlik in Iran, Tell al-Rimah (1966) and from Nimrud (1965; 1966) in Iraq.

Barag (1970: 131) surveyed the archaeological evidence for the production of core-formed glass vessels made in Mesopotamia from the mid-2nd millennium to the mid-1st millennium BC, for which there is no textual evidence. He identified two distinct groups, corresponding to the LBA, 15th-13th C BC, and the IA after the 'Dark Age', 8th -6th C BC. From his examination of the archaeological evidence from Palestine and Syria, Barag (1970: 185) was in no doubt that contemporary glass vessels from Mesopotamia had a different, probably local, origin. He surveyed the connections between the Aegean, the Syro-Levantine region and Mesopotamia, considering nude female pendants, star and plain disk pendants, and spacer beads, such as those found at Nuzi. Barag concluded that they were produced in Mesopotamia, but noted that they also could have been made in the Syro-Levantine region (Barag 1970: 188-193). The work of von Saldern (1970: 203-212) complemented that of Barag with a chronological study of Near Eastern glass, other than core-formed glass, from 1500 to 600 BC. He observed the move from opacity to translucency in IA glass, and that the bowl, a shape found at Nimrud, became the most popular form at this time. After Nimrud fell in 612 BC, the manufacture of transparent luxury glass continued in the same region and further west into Achaemenid and Hellenistic times.

Catalogues of glasses from the Near East, Egypt and the Mediterranean from the mid-2nd millennium BC to the early Roman period were compiled by Grose (1989) and Stern and Schlick-Nolte (1994). Grose's study focussed on the typology and chronology of objects, and Stern and Schlick-Nolte added a summary of manufacturing techniques.[8] The addition of chemical 'typologies' to the physical object data continues to add immensely to the current state of knowledge of LBA and IA glass production and is summarised in the following sections.

1.5 Early Chemical Investigations

Early chemical studies of BA glasses from the Near East, from the early sporadic finds prior to the mid-2nd millennium BC through the increased production during the LBA, are summarised well by Moorey who commented that, although the compositional database was slowly increasing, there remained a remarkable lack of progress in the chemical study of the earliest Near Eastern glass (1999: 190-198, 206-210; see also Henderson 2000: 48-51).

Although only small numbers of glasses were analysed in the early compositional studies, they were important, nevertheless, for providing the basis of current knowledge and the foundations for subsequent research. The work of Turner (1930; 1954; 1955; 1956a-d; 1963), Turner and Rooksby (1959; 1961; 1963), Sayre (1963; 1964; 1965), Sayre and Smith (1961; 1967; 1974), Smith (1963) and Brill (for example 1999a; 1999b, amongst others) made important contributions to our understanding of the compositions of early glasses. Their work is summarised briefly in the following sections and recent research is reviewed to provide the background, context and impetus for the present study.

For ancient Egyptian glass technology, works by scholars such as Lucas (Lucas and Harris 1989: 179-194), and Nicholson and Henderson (2000: 195-226) have

6 Early base glass compositions are reviewed in sections 1.5 to 1.9.

7 Harden (1984) published an interesting review of previous, and sometimes forgotten, early studies of glass which fall outside the focus of the current work.

8 Earlier, Stern (1993) published a brief description of glass-working techniques before the advent of glass blowing.

made their own valuable contributions to this field, combining chemistry and history in a more holistic approach. Henderson's textbook on the science and archaeology of inorganic materials is an informative, and general work, in which the glass technology of several periods and regions is summarised (2000: 24-109).

Turner (1930), in his study of ancient glass, reported that soda acted as a flux to lower the melting temperature of silica, and lime served as a stabiliser. Ash of the plant *Salicornia* was thought to have been used to provide the soda for the flux (Turner 1956a). The composition and use of plant ashes in glass-making have been studied intensively in recent years in an attempt to understand which plants were used and how they were processed (section 1.8).

Turner's (1956c: 282T-291T) findings provided the essential characterisation of ancient glasses which are now termed 'soda-lime-silica' glasses after their main components, which were silica from quartz or sand, soda as the dominant alkali from plant ash, or mineral soda, and lime which was introduced as part of the plant ash or sand component. To this base glass was added colourant, opacifying or decolourant agents. The remaining components, such as potash, magnesia, alumina and iron oxide, were included in the glasses by association with the main ingredients. Each ingredient could be complex in character and variable in composition (Turner 1956c: 296T), which adds to the difficulty of identifying raw materials from glass compositions. These early chemical descriptions of ancient glasses were important in forming the basis of our understanding of ancient glass compositions and the raw materials used to make them, and have been the inspiration for subsequent studies.

Sayre and Smith (1961) looked at a similarly wide chronological and geographical range of glasses and, building on Turner's work, identified five main categories of glass composition. The group designated as glass of the 2nd millennium BC (15th C to *c.*7th C BC) from Egypt, Mesopotamia, Mycenae and Elam, was found to be of the soda-lime-silica type with a high magnesia content. The other four categories are comprised of glasses of later dates than are relevant to the present study, but with compositions that served to distinguish them from the earliest glasses.

Smith (1963) refined the classification of the five categories of ancient glass by identifying high- and low-magnesia glasses, of which the 2nd millennium glasses were classed as high-magnesia while the later 'antimony-rich' and 'Roman' groups were of low-magnesia content. Potash content was found to follow the same pattern as magnesia, but with greater overlap of concentration between the high- and low-magnesia groups. Sayre and Smith (1974: 56, 58) observed that there were two groups of soda-lime-silica glasses: high-magnesia, high-potash (HMHK) and low-magnesia, low-potash (LMLK), which depended on the use of different types of alkali. Brill (1999a: 277) has more recently defined HMHK glasses as having contents of these oxides in excess of approximately 1.5%, whereas LMLK, or natron-type, glasses have contents of these oxides between 0.5% to 1.5%.

The HMHK glasses were known from the 2nd millennium BC Near East until approximately the mid-1st millennium BC when HMHK glass from sites west of the Euphrates all but disappeared. The discovery of a few high magnesium glasses from the last half of the 1st millennium BC suggested that there was some continuation of the high magnesium tradition at the relatively eastern sites of Nippur and Persepolis. The HMHK technology was reintroduced widely in Islamic times (Sayre 1964: 9-11; 1965: 151; Sayre and Smith 1967: 283). The results of Brill's (1999b: 43-44, 46-47) analyses of glasses from Nimrud and Hasanlu show that the high magnesia tradition also continued at these eastern sites.

The LMLK glasses, on the other hand, were believed to have appeared from the mid-1st millennium BC when the HMHK tradition declined (Sayre and Smith 1974: 58). This proposition was not challenged until the discovery of the 10th C BC LMLK glasses from the burial of Nesikhons in Egypt (Schlick-Nolte and Werthmann 2003; section 1.5), and the analysis by Brill (1999b: 46-47) of LMLK (non-cobalt) glasses from Nimrud, the results of which were compiled in a publication without interpretation, show that these glasses pre-dated the appearance of other documented LMLK glass by up to half a millennium.

The translation of the neo-Assyrian glass-making texts published by Oppenheim, with the accompanying chemical interpretation by Brill, and the surveys of core-formed and other vessels from the mid 2nd millennium BC to the mid 1st millennium BC by Barag and von Saldern (section 1.4), made a highly significant contribution to our knowledge of the technological history of Mesopotamian glasses (Oppenheim *et al.* 1970). Brill (1970a) interpreted the chemical and technical information of the texts, which were of a literary nature rather than explicit instructions for glass-makers, and he and Oppenheim identified quartz pebbles and plant ash as the chief ingredients, although many other ancient terms remain unclear (Brill 1970a: 109-110). To show that the recipes in the texts were indeed for glass, Brill sourced appropriate raw materials and replicated the production of ancient glass, and found it to be of a high quality. Brill (1970a:

113) commented that different ashes and silica sources could be used to achieve similar results.

After the early definitive studies of glass compositions and their general observations on the nature of raw materials, scholars have often pursued research into specific and increasingly complex aspects of glass-making and raw materials studies. The availability of sensitive analytical techniques on which this kind of research relies has already been mentioned, and it is these techniques that allow researchers to undertake detailed compositional studies of larger numbers of ancient glasses from a range of sites. The data obtained can be used to investigate a range of archaeological and technological questions about ancient glass production. Research concerning the various components of ancient glass and compositional studies of glass from the Near East and Egypt are considered in the following sections.

1.6 Silica

The major ingredient of glass is silica in the range of approximately 60% to 70%, sourced primarily from quartz or sand. In early studies, Turner (1956c: 279T, 281T) analysed the compositions of sands and showed how their alumina and iron contents, components of clay and other minerals which occur as impurities in sands and in the refractory containers used in melting glass (Matson 1951: 84; Henderson 1985a: 270-271; Lucas and Harris 1989: 481), would have contributed to the glass compositions. More recently, Rehren and Pusch (2007: 223-225) have experimented with grinding stones from Qantir to show how they contributed contaminating elements, such as aluminium, potassium, calcium, titanium, strontium, zirconium, barium and transition elements, to the silica when quartz pebbles were being crushed.

The composition of possible glass-making sands and quartz pebbles from the Near East, Egypt and southern Turkey have since been analysed by Bezborodov (1975: 57-60), Brill (1999b: 474-479) and Degryse *et al.* (2005: Tables 5 and 6, 294-295) amongst others, identifying the range of contaminants that would have been incorporated into glasses made from them, for example alumina, iron, lime, soda and potash. Tite *et al.* (1998: 118) proposed that iron oxide content in glass of less than 0.6% indicates the use of quartz as the source of silica, which introduces very low concentrations of impurities to the final glass. Nicholson and Henderson (2000: 197) noted that a sand source of silica must have been used to make Egyptian glasses because of their relatively high alumina and iron oxide contents (see also Shortland and Eremin 2006: 589).

Freestone (2006: 205-206) listed potential sources of silica as quarried siliceous minerals and rocks, such as vein quartz, chert and quartzite, as well as pebbles of these materials and coastal or inland sands. Freestone investigated the nature of the coastal sands of Israel, including those from the Bay of Haifa into which the Belus River flows. Emery and Neev (1960) noted that beach sands of Israel are derived mainly from the Nile River with mineral contributions from other sources. Their research showed that sands along the coast of Israel have variable composition, including that of lime derived from sea shells and the remains of other forms of marine life. The variable composition of these coastal sands is reflected in the glasses made from them as demonstrated by Freestone *et al.* (2002: 265) in their study of Byzantine glass from Maroni Petrera in Cyprus.

If lime-bearing plant ash was used to flux glass made from sand, that sand would have to have a low lime content to avoid the incomplete melting or devitrification of the resulting glass that is caused by an excessive concentration of lime (Freestone and Gorin-Rosen 1999; section 1.7). For this problem to have been avoided, HMHK or plant ash glasses are assumed to have been made with pure quartz pebbles or relatively pure sand, and LMLK glasses with a lime-bearing sand and mineral soda mix. When Egyptian mineral soda was first used to make glasses in the 10th C BC, for example, silica was derived from quartz or low-lime sand in the HMHK tradition. The resulting glasses, such as those belonging to Nesikhons (Schlick-Nolte and Werthmann 2003), were chemically unstable due to low lime levels, showing that the selection of a glass-making sand with sufficient lime would have been critical for the preservation of mineral soda fluxed glass. It is inevitable that other early unstable LMLK glasses would not have survived without favourable burial conditions, explaining the limited evidence of this important technological change (Freestone 2006: 208).

Nicholson and Jackson (2000: 20; Jackson and Nicholson 2007: 98) found that they could make a good quality glass of similar composition to those analysed from New Kingdom Egyptian contexts using a soda-rich plant ash and sand, although it must be assumed that the total lime content of the sand and ash mix was low enough to achieve an enduring glass that did not vitrify (see also Nicholson and Jackson 1998; Freestone and Gorin-Rosen 1999). Shortland and Eremin (2006: 589) argued that the apparent wide availability of reasonably pure sands in Egypt meant that sand could possibly have been used to make plant ash fluxed glass. Rehren and Pusch (2008: 30) have more recently confirmed that crushed quartz was the source of silica in glasses made at Qantir Piramesses, thus it would appear that either a low lime, or non-calcareous, sand or quartz could be used to make plant ash fluxed glasses.

Similar work has not been done for glasses from the Near East for this period, although the research of

Freestone and co-workers (Freestone and Gorin-Rosen 1999; Freestone *et al.* 2000) on the compositions of Byzantine and Early Islamic glasses from Israel showed that coastal sands near the mouth of the Belus River were too rich in lime to have been combined with plant ash to produce enduring glass (section 1.7).

1.7 Lime

Lime acted as a stabiliser to reduce weathering of the silica-soda glass network (Henderson 1985a: 277). While elements such as aluminium or iron also had preservative qualities at elevated levels, lime was a key component of ancient glasses, hence their designation as 'soda-lime-silica'.

The question as to whether lime was deliberately added to ancient glasses has been much debated. According to the two-component model of glass-making, silica and a soda-rich flux were the chief ingredients, with lime entering the glass with either or both of them (section 1.5).

Pliny (Eichholz 1962: 153) made an enigmatic reference to the addition of shells to the glass batch, presumably as a source of lime, but the otherwise scant mention of lime in ancient texts, and no mention of lime or magnesia in later texts until the 18th C AD, led Turner (1956a; 1956b: 175T-176T) to conclude that until this time glass-makers did not know that their glasses contained these substances, and yet lime has been an important and substantial component of most of the glasses that have survived from early times, acting as a stabiliser for the soda-silica network.[9] The anomaly between what has been written in past times and the actual composition of the ancient glasses has led to much discussion as to how the glass-makers could not know about lime, and in what form it must have been added to the glasses.

Henderson (1985a: 272) cited the greater durability of prehistoric soda-lime-silica glasses compared with that of medieval potash glasses as being an example of a long-term technical regression in glass-making. It could be extrapolated from this that the medieval texts concerning glass-making are not necessarily reflective of the knowledge of ancient glass-makers, and could not be an accurate indication of their knowledge of the use of lime. Smedley *et al.* (1998: 149) suggested that if the ratio of alkalis was monitored by taste, this knowledge in time would be taken for granted, and not recorded. Shortland (2007: 263), in considering the lowly depictions of medieval glass craftsmen in contemporary and later manuscripts, questioned the validity of using what we know of medieval glass craftsmen to represent ancient glass-workers, and concluded from archaeological and philological/textual evidence that 'the more spiritually complex and challenging world of the second millennium BC' could not be represented by more recent glass production models (Shortland 2007: 271-272) (section 1.3).

Turner (1956c) noted the high lime content of sands from Tell el-Amarna and Karnak, which are adjacent to long limestone bluffs, and from Gizeh and the mouth of the Belus River. Sayre (1965: 146) agreed with Turner's argument that lime was likely to have been present in the sand. Sayre and Smith (1967: 280) found a correlation between magnesia and potash, and other alkali metals, and a lack of correlation between lime and magnesia, which seemed to discount the idea that dolomitic limestone was used to make ancient glasses, as had been proposed by Matson (1951: 84). To enable the addition of measured quantities of lime to the glass batch, Nicholson and Henderson (2000: 216) proposed that shell fragments were separated from sand and used as an individual lime source, although Brill (1970: 109) had suggested that the proportion of shell in a sand would be constant. Deliberate incorporation of lime or limestone as a separate component in Egyptian glasses was supported by Tite *et al.* (1998: 118) and Shortland and Eremin (2006: 590-591), while Henderson (1985a: 288) argued against the addition of limestone and Rehren (2008: 1346) also disagreed, maintaining that neither calcareous sand nor lime were deliberately added by the glass-makers. Cosyns and Hurt (2007: 6) were also of the opinion that lime was unintentionally added to the ancient glass batch as a component of sand or plant ash.

Rehren and Pusch (2007: 231-233), through experimental work on the significance of the melting process and melting temperature on the alkali content of ancient glass, concluded that the lime parting layer applied to the inside of the melting crucibles added lime to the glass batch as the furnace temperature increased (see also Rehren and Pusch 2005: 1756; Smirniou and Rehren 2011). It is not known whether the glass-makers realised that when they used lime in this way that it was also a key ingredient of the glass. Rehren and Pusch (2007) did not consider whether a lime-rich wood ash could have been used as fuel in the furnace and thereby become incorporated in the glass batch. Contamination of the glass batch by fuel ash was suggested by Jackson *et al.* (2003: 451) when explaining the composition of Roman glasses from Coppergate, York. Shugar and Rehren (2002: 147) proposed that the melting temperature of the glass batch directly controlled the lime content of the glass melt, but this must be dependent on the quantity of lime added to the batch before firing.

[9] If glasses were made with insufficient lime, they might not have survived unless under extraordinary circumstances, such as the low-lime glasses found in the protective environment of the tomb of Nesikhons in Egypt (Schlick-Nolte and Werthmann 2003), section 1.5. McGovern (1995: 30) wrote, 'Archaeological data represent a very small, highly selective fraction of the original technology'.

When investigating Byzantine and Early Islamic glasses from Israel, Freestone *et al.* (2000: 70) found that a mixture of sand and natron, or of quartz and plant ash must have been selected by the glass-makers to produce glasses with similar ratios of silica: lime: soda. If lime-rich sand were mixed with plant ash, the resulting lime content would be too high causing the incomplete melting or devitrification (section 1.6), as was observed in the Bet She'arim slab from Israel of the 9th C AD, which contained approximately 16% lime (Brill and Wosinski 1965; Brill 1967: 92; Freestone and Gorin-Rosen 1999; and see also Matson 1951: 83). Brill (1988) demonstrated that the composition of the glasses from the Late Roman site of Jalame in Israel were consistent with manufacture from natron and beach sand from the mouth of the Belus River. Lime, in this instance, was introduced as shell occurring naturally in the sand. When Freestone (2006) examined Near Eastern glass production in Late Antiquity and the early Islamic period, he concluded that lime was rarely deliberately added to glasses of the 1st millennium AD (2006: 207), agreeing with Brill's similar observation that early Indian glasses were made from a two-component batch (1987: 7-8). Brill, however, commented that if a third basic ingredient became necessary as a stabiliser because the first two were too pure, then the third ingredient must have been intended to introduce lime, implying that these glass-makers had knowledge of the role of lime in glass-making.

The question of whether lime was added deliberately by the ancient glass-makers has not been resolved, although the weight of evidence suggests that lime was usually added either as part of the plant ash flux, or in a natural combination with silica sand. It is probable, however, that the glass-makers learnt from experimentation which combinations of particular raw materials produced an enduring glass. Henderson (1985a: 286) commented that the glass artisans must have understood their materials because they were able to achieve close control of the amount of colourant and presumably of the kiln conditions. It might be said, however, that cause and effect in colouring is readily observed, while presence or absence of lime is not visually identified. Research and argument continue.

1.8 Plant Ash

The role of plant ash in glass-making is to provide the alkali, particularly soda, as a flux to reduce the melting temperature of the silica component from 1710° Celsius (Henderson 1985a: 270-271). It is present typically between approximately 13% to 20% in early glasses. The continuing study of plant ash compositions has provided an understanding of the complex nature of this raw material, and how its composition might be related to the final glass it was used to make, although many questions remain to be answered.

In his commentary on the chemical aspects of the ancient Assyrian glass texts, Brill (1970a: 110) noted that the ashes of the '*Naga* plant' could have been of the *Salsola* or *Salicornia* genera, as suggested previously by Turner, who recorded that plants grown near the sea would have had a relatively high soda content, desirable for glass-making, and that plant ashes contain many other constituents besides soda and potash (Turner 1956a: 42T;1956c: 282T-291T).

Turner's (1956c: 282T-291T) basic analyses of 'Keli'[10] ash from Syria, amongst other ashes, showed the great variability in the relative proportions of the constituents of ashes from different plants, and from the same plants depending on the locality in which they were grown. The variability of ash composition was confirmed by Brill's (1970a: 110, 124) analyses of Near Eastern and Mediterranean saline coastal and arid region plant ashes, and by Bezborodov's (1975: Table 5) analyses of different parts of the same plants which exhibited compositional variation. Brill (1970a: 110) commented that the soda to potash ratio in ancient glass of approximately 10:1 was an important indicator of the source of the alkali, as did Matson (1951: 84) and Turner (1956c: 282T-291T) before him.

The study of plant ashes as raw materials was advanced by Sanderson and Hunter (1981: 27-30) when they too examined the compositional variability of plant ashes (seaweed and wood ash) and of glasses made with plant ashes, and noted that there must have been many combinations of natural materials which resulted in the observed compositions of ancient glasses. Sanderson and Hunter were concerned that the conclusions of previous studies had been of a comparatively general nature and, like Smith before them (1963), urged further quantitative analysis of raw materials. The work of Smedley *et al.* (1998) on the glass-making practices recorded by Theophilus, highlighted again the variability of plant ash composition and that all of the oxides found in the ash would not necessarily be incorporated into the final glass. The variability of ash composition was reiterated by Canti (2003: 347) who added that soil adhering to the plants would affect the mineral content of the ash.

Brill (1970a: 110) noted that the way the ash was burned was also responsible for causing variation in its composition, a point made by Smedley *et al.* (1998: 152) who found that not all of the oxides present in wood ashes would be incorporated in the final glass because of decomposition or reaction during heating. Although wood ash compositions are high in potash[11] and hence

[10] The word 'Keli' probably derives from the Arabic *al-qali* meaning plant ash (Biek and Bayley 1979: 6).

[11] The high-lime, high-potash nature of wood ash, used to make European glasses, is discussed by Henderson (1988) and Wedepohl

distinct from the high soda ashes used to make ancient Near Eastern glasses, the results of their study are worth considering, because they highlight issues which could apply to the study of other plant ashes. In view of this, it is interesting to note that in their study of wood ash, Misra *et al.* (1993) found that the content of potash and soda decreased as a function of temperature during ashing, while the concentration of lime was assumed to remain constant. This might give an indication of the effects that ashing could have on the original elemental content of the plants, and the complexities that this might impose on the interpretation of the relationship between the original composition of the plants and the final composition of glasses made from the plant ashes.

The proposition that alkali earth oxide concentrations in LBA plant ash glasses were controlled largely by the melting temperatures and the melting behaviour of the system was put forward by Tanimoto and Rehren (2008: 2567) from experimental evidence.[12] Chemical processes which occurred during melting affected the ratio of soda to potash, but the total molar content of these alkalis in the glass remained similar to their content as carbonates in the batch. The final concentrations of magnesia and lime were unchanged (Tanimoto and Rehren 2008: 2571).

Another possible variable in the nature of plant ashes was demonstrated by Smedley and Jackson (2006; Jackson and Smedley 2008) when they measured the changes in yield weight and in the composition of bracken during its growth cycle. The compositional differences in glasses made with ashes from plants harvested at different times during the growth cycle has implications for the interpretation of archaeological glass compositions, and although bracken was not used to make Near Eastern glasses, the principle of these findings should be considered in the study of Near Eastern plant ashes.

Further work on plant ash compositions was published by Tite *et al.* (2006) in which they analysed plant ashes, mostly from the *Salsola kali* plant, from different locations in the Middle East and beyond, but the ash compositions did not match the composition of Egyptian glasses. The suitability of ashes for the production of ancient glass was assessed by calculating and comparing soda to potash ratios (see also Brill 1970a: 110), lime and magnesia contents against total alkali $(CaO + MgO)/(Na_2O + K_2O)$, and the percentage of alkali present as carbonates (Tite *et al.* 2006: 1286) for both ashes and selected ancient glasses. Similarly, Rehren (2008: 1345) advocated the characterisation of the plant ash component of glasses by using the ratios of soda to potash, lime to magnesia, and total alkali oxides to total alkali earth oxides.

A variety of plant ashes specifically from Syria was studied by Barkoudah and Henderson (2006) who provided new chemical data for ash compositions, and canvassed the issues of variability of composition and of relating ash composition to final glass compositions. They found a number of sodium-rich plant ashes, and claimed that the positive correlation between magnesium, potassium and calcium in plants of the *Salsola* genus helped to confirm its use in ancient glass production on the grounds that such correlations are found in some ancient plant-ash glasses, although the absolute concentrations of alkali in the ashes of these plants could be argued to be insufficient to have been used to make LBA glass. The results of Barkoudah and Henderson's study indicate that the same plants grown in different locations have different compositions (2006: 306-310), supporting Brill's earlier finding (1970a: 110).

Despite the variability of plant ash composition, the ancient glass produced from it was of a relatively consistent composition, within geographical and chronological limits (Rehren 2000a), although how this was achieved remains unclear, as do the relationships between the compositions of plants, their ashes and the glasses produced from them (Freestone 2006: 205; see also Shugar and Rehren 2002; Jackson *et al.* 2005; Tite *et al.* 2006). Nicholson and Henderson (2000: 197) argued that if the first glass-makers were brought to Egypt from Mitanni, they might have used plants for ashing that were similar to those in their native land, presumably ensuring a certain similarity of composition to those glasses made in the Near East.

The issues of compositional variability of plants and their ashes, which occur for a number of reasons, and the translation of elemental concentrations from ashes to final glass, continue to occupy researchers and to inform our understanding of the complexities of raw material selection and the sophistication of the art of ancient glass-making.

1.9 Mineral Soda

Mineral soda was used as an alternative to a plant ash source of soda flux to make ancient glasses in the IA. Glasses of the LMLK type are considered to have been made using a mineral soda source, most widely believed to be natron,[13] or trona, the naturally occurring sodium

(1997).

[12] For recent experimental work on the melting characteristics of glasses made from vegetable ash and sand mixtures see Jackson and Smedley (2004). Although these were potash rich ashes, the study is interesting because it showed how the choice of alkali could affect the melting temperature of the batch.

[13] The term 'natron' refers to a particular mineral, and so the non-specific terms 'mineral soda' or 'natron-type' are often used in preference (see Freestone 2006: 204) and are interchangeable with it in the present study.

carbonate/ sodium bicarbonate (Na_2CO_3, $NaHCO_3$) found in the Western Desert oases of Egypt, such as the Wadi Natrun (Shortland 2004).[14] Natron was used also in making soap and in embalming (Nicholson and Henderson 2000: 216). Turner (1956c: 282T-291T) investigated the sources of alkali salts, and natron in particular, as the most likely mineral soda source used in ancient glasses of low alkali content, noting its complex and variable composition.

Sayre and Smith (1967: 290) analysed the composition of natron salts from 11th and 18th Dynasty tombs to show that it had a low concentration of both potash and magnesia (in the order of 0.1%) and hence would not have contributed these oxides to the glasses. They concluded that HMHK glasses could not have been made with natron, but rather with plant ashes or, less likely, evaporated river water (Sayre and Smith 1967: 289-291), as suggested by Turner who returned similar findings from his analyses of natron from 18th Dynasty tombs and from the Wadi Natrun (1956c: 283T-284T). The analyses of natron by Bezborodov (1975: 56), Brill (1999b: 480), Shortland (2004) and Shortland *et al.* (2006: 525) have confirmed the composition and relatively uncontaminated nature of natron as characterised in the previous studies.[15]

Until recently, it has been thought that glass of the natron-type was dominant in the Mediterranean and surrounding regions from the mid-1st millennium BC through to the late 1st millennium AD (Smith 1963; Sayre and Smith 1974: 58). Shortland *et al.* (2006) traced the history of the use of natron as a flux in ancient vitreous materials noting that there was no conclusive evidence for the use of natron as the main flux in 2nd millennium BC glasses, but it appears to have been introduced as early as the 10th C BC in the glasses from the Egyptian tomb of Nesikhons (Schlick-Nolte and Werthmann 2003; section 1.5).

1.10 Colourants, Opacifiers and Decolourants

Ancient glasses were modified with a wide range of additives which imparted colour, made naturally translucent glass opaque or decoloured naturally tinted glass to achieve a colourless product.

Early studies of ancient Near Eastern and Egyptian glasses recognised copper as a common colouring agent which produced blue, green and opaque red glasses,[16] and cobalt was identified as a dark blue colouring agent in both Egyptian and Babylonian glasses, although it was used less frequently than copper and its source was unknown. Manganese was identified as a purple or black colouring agent,[17] and antimony as a refining agent in glasses (Farnsworth and Ritchie 1938: 159-164; Turner 1956a;1956b: 179T).

Studies since then have focussed on the chemistry of colourants, decolourants and opacifiers, which because of the availability of more sensitive analytical techniques, are able to be measured in increasingly small quantities, thereby allowing detailed characterisation of the colourant components of ancient glasses. This in turn helps to relate the analysed glass compositions to possible raw materials. Recently, the nature and uses of individual colourants and opacifiers have been conveniently summarised by Nicholson and Henderson (2000: 217-218) and Shortland (2002: 517-519), amongst others. Although these summaries were based on Egyptian LBA glasses, they are equally applicable to LBA Near Eastern glasses. In addition to those colours mentioned above, glass was made in the following colours; calcium antimonate which produced white, lead antimonate which produced yellow, and a mixture of copper blue and calcium antimonate which produced a turquoise opaque colour, cobalt blue and calcium antimonate made a dark opaque blue and lead antimonate yellow and copper together created an opaque green, and iron oxide present from 0.3% to 1.5% would have imparted a greenish hue to glass (Schreurs and Brill 1984: 199). Dark brown or amber could also be produced by the presence of iron oxide in combination with sulphur (Schreurs and Brill 1984: 199, 206-207; Nicholson and Henderson 2000: 217).[18] In the IA, antimony was used for both decolouring and opacifying glass (Brill 1999b: 47), while iron in sufficient quantity (usually in excess of 5% iron oxide) produced a dark green to black coloured glass (Stapleton and Swanson 2002b: Table 1).

The presence of tin in some blue glasses from Malkata and Lisht suggested the use of the by-products of bronze-making and working to Mass *et al.* (2002: 67, 69). Shortland (2005: 2) commented that the ratio of tin to copper, in the order of 1:10, indicated the derivation of these elements from bronze. Egyptian copper blue

[14] The locations of possible natron sources around the Dakhla and Kharga oases were considered by Shortland (2004) and Shortland *et al.* (2006), who analysed specimens for composition. Trona was also sourced in antiquity from the Beheira province in Lower Egypt (Nicholson and Henderson 2000: 216).

[15] Degryse *et al.* (2005: Table 6) analysed lacustrine salt patches near Sagalassos in southern Turkey as possible sources of mineral soda for the Roman and Byzantine glasses they were studying, and found them to be highly variable in their contents of soda, magnesia, potash and lime.

[16] Red glasses are not analysed in the present study and will not be discussed here.

[17] The use of manganese as a decolourant in glasses occurred from a later period and thus is beyond the scope of this study, but information on the chemistry of decolouring with manganese can be found in the publications by Weyl (1953: 164-165) and Henderson (1985a: 284).

[18] For further information see Weyl (1953: 91) and Bamford (1982: 6) who explain the chemistry of iron as a colourant.

glasses may contain copper derived from bronze, but this is not the case in copper blue glasses from the Near East, such as in those from Nuzi. This suggests that copper sources were different in the two regions.

1.10.1 Cobalt

Early investigations into the nature of cobalt as a colourant began to recognise elemental relationships which led eventually to the identification of two possible sources of this colourant in the Near East and Egypt. Early work by Farnsworth and Ritchie (1938: 159-164) attributed the dark blue colour of Egyptian 18th Dynasty glasses to the presence of cobalt with associated manganese and copper. They summarised previously published results which reported cobalt to be present in ancient Egyptian glass and glaze. At this time, cobalt was considered to have been deliberately added from an unidentifiable source. Garner (1956a; 1956b) proposed that an Iranian arsenical cobalt ore with low manganese content had been used in a very early non-Egyptian glass from Eridu, dated *c.*2000 BC. Sayre (1963: 267-268) noted that not only manganese, but nickel and zinc concentrations were enhanced in cobalt blue glasses from Egypt and Mycenae, but that the elemental content was different in two Western Asiatic cobalt blue glasses, which Sayre and Smith concluded must have been because a different source of cobalt was used (Sayre and Smith 1974: 51, 54).

Kaczmarczyk and Hedges surveyed the composition of Egyptian faience from Predynastic to Roman times (1983) and established that the cobalt colourant used in ancient Egyptian New Kingdom vitreous materials was probably derived from the alum[19] deposits of the Western Oases of Egypt, especially the Kharga and Dakhla Oases (1983: 41-55). They noted that the 'striking feature of the occurrence of cobalt in New Kingdom faience is its association with manganese, zinc, nickel and aluminium' (Kaczmarczyk and Hedges 1983: 46). These findings were confirmed by Kaczmarczyk (1986), who analysed Egyptian blue pigments from the 2nd millennium BC, establishing that cobalt was invariably associated with elevated concentrations of aluminium, magnesium, manganese, iron, nickel and zinc in cobalt blue Egyptian glasses, and that the raw material was derived from the Great Western Oases in Egypt.

Kaczmarczyk and Hedges (1983: 45-47) noted the virtual disappearance of cobalt from post-New Kingdom vitreous materials, from the 12th to the 7th C BC, concluding that when it reappeared during the 26th Dynasty in the Late Period (7th C BC), it must have originated from a different ore source, because the concentrations of manganese, nickel and zinc returned to the average level of impurities, no longer being found at the elevated levels associated with cobalt alum, and there were no clear correlations with other elements. Cobalt in vitreous materials from this time resembled cobalt detected in Mesopotamian pigments of all time periods, which Kaczmarczyk and Hedges assumed were from Iran (1983: 53; Kaczmarczyk 1986: 373-374). The distinction between an Egyptian alum source of cobalt and an eastern, possibly Iranian ore source of cobalt which was used in Near Eastern glasses, was clarified. Their results highlighted to Kaczmarczyk and Hedges just 'how fragmentary and inadequate is the present state of knowledge of 1st millennium B.C. Mesopotamian faience and glass' (1983: 294).[20]

Analyses of alum from the Western Desert Oases in Egypt proved that it contained the same trace elements, in approximately the same relative amounts as those found in cobalt blue glasses, including relatively high levels of alumina (Kaczmarczyk 1986: Table 34.3; Shortland and Tite 2000: 145-146). Further chemical analyses of Egyptian cobalt alums were published by Shortland *et al.* (2006: 157) and Gratuze and Picon (2005: 272-273) confirming the results of earlier compositional studies, and showing that the levels of soda and lime were comparatively low and that compositions of alums from around the Kharga and Dakhla oases were variable. The unpredictable composition of cobalt alum was noted also by Jackson and Nicholson (2007: 104).

The work of Lilyquist *et al.* (1993: 36, 41-43; Lilyquist and Brill 1996) showed that the Egyptian cobalt blue glasses of the 2nd millennium BC, from as early as the reign of Thutmosis III, were compositionally distinct from other contemporary glasses, having high alumina and magnesia contents from the addition of the alum colourant, but with low potash levels. This anomalous composition does not fit the accepted models of HMHK (plant ash) or LMLK (natron) glass and has provoked discussion about the alkali source for the Egyptian cobalt glasses. Shortland and Tite (2000: 146-147) suggested that they were made with natron, which explained the low potash levels, while the high magnesia content was due to magnesia in the alum.

Rehren (2001) believed that the cobalt blue glasses were made using a low-potash plant ash of a different type to that used in Near Eastern plant ash glasses, questioning why, if it were a natron-based glass, this technology would have been restricted to the one colour. Rehren preferred the view that LBA glass-making on the whole was based on plant ash alkali sources. Jackson and Nicholson (2007: 112-113) similarly concluded that copper and cobalt coloured glasses were made using

[19] 'Alum' is defined by Shortland *et al.* (2006: 153) as the hydrated double sulphate of aluminium with or without an alkali, alkali earth or transition metal.

[20] Henderson (1985a: 278-281; 1989c: 33-36; 2000: 30-32) has summarised the early work on cobalt as a colourant.

different types of plants, one of which was lower in potash than the other. Alternatively, Tite and Shortland (2003: 302) tentatively proposed that cobalt blue glass was made with a mixture of cobalt coloured frit, plant ash and quartz, which explained the lower potash level of the cobalt glass compared with contemporary copper blue glasses, and repeated Shortland's (2000b) finding that the cobalt contents of New Kingdom glasses correlated with alumina, manganese, nickel and zinc, but not magnesia, because of the addition of plant ash containing magnesia. The experiments of Shortland *et al.* (2006: 164) showed that the magnesia content of the alums could vary widely when cobalt, alumina and magnesia were precipitated from the alum.

The results of analyses of Egyptian cobalt coloured glasses and cobalt alums from the only known source in the oases of the Egyptian western desert, show that the cobalt colourant in these high alumina glasses was derived from the cobalt alums, and that the glasses are of an unusual high-magnesia, low-potash type. Brill's analyses of IA cobalt blue glasses from Nimrud also indicate a cobalt colourant associated with elevated levels of the same trace elements (Brill 1999b: 47-48). Various reasons for this composition have been proposed but not satisfactorily resolved.

1.10.2 Antimony as an Opacifier and a Decolourant

Turner and Rooksby published several studies concerning opalising or opacifying agents used in early glasses, for example from Thebes (*c.*1450 BC) and Nimrud (8th to 6th C BC) (Turner and Rooksby 1959; 1961; 1963; Rooksby 1959; 1962). In these studies, Turner and Rooksby identified antimony in the form of calcium antimonate ($Ca_2Sb_2O_7$ or $CaSb_2O_6$) as the opacifier for white opaque glasses, and when mixed with translucent copper blue, produced turquoise opaque glasses.[21] The intentional inclusion of a material such as stibnite (antimony sulphide, Sb_2S_3) was suggested for this purpose, because there are no known mineral forms of calcium antimonate in Egypt or the Near East (Mass *et al.* 2001: 41; Mass *et al.* 2002: 70, 78).[22] Opacity resulted from the presence of antimony pentoxide (Sb_2O_5) in the glass (Sayre and Smith 1974: 59).[23]

Sayre (1963; 1965) discussed the presence of antimony and manganese in early glasses and concluded that these components were added deliberately either as decolourants or as opacifiers, with colourless glass being rare in the 2nd millennium BC. He noted that of five colourless Egyptian specimens and three colourless Mesopotamian specimens from 13th C BC Tchoga Zanbil in Iran, none contained large concentrations of either antimony or manganese, indicating that these elements were not used as decolourants in LBA glasses (Sayre 1963: 269). Colourless glasses of the 2nd millennium BC from Egypt, analysed by Bimson and Freestone (two samples) (1988: 12), Wypyski (one sample) (in Lilyquist *et al.* 1993: 36), Shortland and Eremin (ten samples) (2006: 584) and Jackson and Nicholson (one sample) (2007: 182) appear to contain little or no antimony or manganese, supporting Sayre's finding. Decolouring of the glass might instead have been achieved by careful control of the furnace atmospheres (Jackson and Nicholson 2007: 106), or could have been the result of using pure batch materials (Nicholson 2007: 1).

Fossing (1940: 42) noted the consistent appearance of high concentrations of antimony in colourless glasses from approximately the mid-6th C BC, until antimony as a decolourant was largely replaced by manganese at the end of the 1st millennium BC (see also Biek and Bayley 1979: 6). Sayre (1963: 272) found that all 12 specimens of colourless 7th to 1st C BC glasses he analysed contained elevated levels of antimony similar to the levels found in opaque glasses. He explained that the ability of antimony to both opacify and decolour resulted from its presence in the glasses in different states of chemical valence. The tradition of opacifying with antimony was evident from the earliest glasses and into Roman times (Brill and Moll 1963: 299), and the use of antimony as an opacifying agent appears to have led to the discovery that antimony could, under the appropriate conditions, decolour glass (Sayre 1963: 272). Brill (1970a: 116) summarised the use of antimony in ancient times for three distinctly different purposes as either a decolourant, by oxidising iron to render the glass colourless, as a fining agent to remove small seeds or bubbles from the glass, or as an opacifying agent at the levels of approximately 1% to 2% antimony oxide in the glass.

The property of calcium antimonate to act as an opacifier was discussed by Shortland (2002: 519-523) who suggested that during cooling of the melt, antimony combined with the lime thus drawing it out of the batch to form calcium antimonate. This proposal was based on Shortland's understanding at the time that lime levels were lower in opacified LBA Egyptian glasses than in their translucent counterparts. Shortland considered that if calcium antimonate had been added as a preformed material, the lime content would be expected to have risen in opaque glasses and not be reduced.

21 Although no yellow opaque glasses were analysed in the present study, it is noted that lead antimonate ($Pb_2Sb_2O_7$) was used to create opaque yellow, and opaque green (Mass *et al.* 2002: 70-71).

22 Mass *et al.* (2002: 70) suggested stibnite as a lead-free antimony source, found throughout the Mediterranean and the Near East, for Roman glasses (see also Mass *et al.* 1998).

23 Unintentional opacity could be due also to devitrification, incomplete vitrification, many gas bubbles or impurities in the glass (Henderson 1985a: 286).

Nicholson and Henderson (2000: 208) proposed that antimony was added as an opacifier to translucent blue Egyptian glass to make turquoise opaque glass. Shortland suggested that to make opaque blue glasses, antimony could have been added to a translucent blue glass by mixing translucent blue and opaque white glasses (Shortland 2002: 522) at the same time that Mass *et al.* (2002: 78) suggested the addition of ready-made white glass to achieve opacity in Egyptian glasses. Further research by Shortland and Eremin (2006: 591) found that, apart from the antimony levels, the compositions of white and colourless Egyptian glasses were closely similar, including the average lime concentrations, which suggested to them that antimony alone was added to the colourless glass to create white glass.[24] Whether antimony or white glass was added to a translucent blue glass to achieve opacity remains unanswered.

1.11 Trace Element Analyses

After early work on ancient glass had 'systematised' knowledge of basic glass compositions and colourants, Brill (1969: 48) looked forward to the 'long hoped-for useful pattern of trace-element concentrations, for example the rare earths'. He encouraged the analysis of

'large numbers of well authenticated specimens of fragments and vessels encompassing a wide range of places and periods. The analyses must be accurate where necessary and comprehensive in terms of elements determined. They must include a large number of specimens to allow statistically significant inferences to be drawn. It is only in this way that a useful catalogue of ancient compositions can be compiled. ... The meaningful variability appears only to be in minor, or trace, or 'trace-trace' elements.'

The difficulty of establishing the origin of ancient glass was noted by Velde and Gendron (1980: 183) who, when referring to previous compositional analyses of glass, commented that 'Chemical analyses of major elements composing the glass itself has not given any great hope of distinguishing one area of production from another', although general compositional groups could be distinguished by a broad chronological period or geographical region, but composition and stylistic criteria were not sufficiently detailed to provide more refined classifications.

Compositional studies which include trace element analyses have become more frequent with the availability of, and improvement in analytical instrumentation, and have allowed more detailed characterisation and comparison of groups of glasses.[25] Freestone *et al.* (2000: 65-82), in their study of glasses of Byzantine and Early Islamic date from Israel, employed ICP-MS to analyse a selection of glasses from Apollonia and Bet Eli'ezer, confirming the grouping of glass based on major element composition, and allowing further discrimination between the glass groups.

Freestone *et al.* (2002) applied ICP-MS and ICP-AES trace element analyses to AD 6th to 7th C glasses from Maroni Petrera in Cyprus, and normalised the data for a selection of elements to the mean composition of the continental crust. This showed which elements were enriched or depleted relative to the crustal values, and indicated the use of a mineralogically pure sand to make the glass. Normalisation of glass data to mean crustal values allowed compositional anomalies to be recognised.

Shortland *et al.* (2007) analysed LBA glasses from Nuzi, Tell Brak and Egypt for 32 trace elements by laser ablation ICP-MS. They advocated the potential of trace element compositional fingerprinting to resolve difficult debates about provenance and details of manufacturing technology. The calculation of the ratios of zirconium to titanium, and of chromium to lanthanum, enabled the authors to discriminate between Egyptian and Near Eastern glass production in the 2nd millennium BC. They showed that elevated chromium contents and high chromium to lanthanum ratios are characteristic of Near Eastern glasses, while Egyptian glasses contain elevated levels of zirconium (Shortland *et al.* 2007: 781, 787-789). This important development has advanced our ability to distinguish the consistent differences between the two regional groups of glass. These differences indicate the use of different raw materials and/or manufacturing processes, because these trace elements provide distinctive geological signatures that vary between regions.

Solution ICP-MS was used by Mirti *et al.* (2008) to analyse 40 major, minor and trace elements, including the REE, of Parthian and Sasanian glasses from Iraq, but their discussion was concerned principally with the base glass and colourant composition with limited reference to the REE and associated elements, the levels of which indicated the use of an impure sand silica source (Mirti *et al.* 2008: 441, 443). Similarly, Santagostino Barbone *et al.* (2008) have studied late 1st millennium BC to

[24] Recent studies of colourless glasses have involved those predominantly from the Roman period when colourless glass was common (Jackson *et al.* 2003; Jackson 2005b). In some Roman glasses from Britain, Jackson (2005b: 771) has found that both antimony and manganese were present, but that antimony was probably the intended decolourant for the majority of glasses analysed. See also Sayre (1963: 279-280) for the mixed use of manganese and antimony in Rhenish glasses of the 1st millennium AD.

[25] Much has been written about the scientific fingerprinting of materials and of artefacts to trace the source of raw materials or places of manufacture, for example, Pollard and Heron (1996), Pollard *et al.*(2007), Brothwell and Pollard (2001: Section 6, 441-518).

AD 1st millennium glasses from southern Italy using ICP-MS, amongst other techniques, to produce a suite of 54 oxides and elements, including the REE. Their discussion centred on the main colourants with only brief reference to their trace element data (Santagostino Barbone *et al.* 2008: Table 2, 460).

The collection of detailed compositional data to answer archaeological questions about the technology, organisation and recipes for glass production, including the nature, sources and craftsmen's choices of raw materials, and the geographical and chronological distribution of raw materials, raw glass and finished glass objects, is being made possible by the development of sensitive analytical techniques such as ICP-MS, with 'interpretation...becoming the central challenge' (Lilyquist and Brill 1996: 8).

The published chemical analyses (as at 2009) of selections of glasses from the LBA and IA Near East and LBA Egypt are compiled in the following sections. These studies form part of the growing data-base of compositional information that scientists, technologists and archaeologists alike use to try to unravel the details of ancient glass production.

1.12 Analyses of LBA Near Eastern Glasses

The early chemical investigations of glass composition should be viewed with caution because these analyses, assembled by Turner (1956b), involved small numbers of samples, often of uncertain context and date, and were performed using techniques that were subject to error and were inadequate by comparison with current analytical instrumental methods (Bimson 1987: 165-166). Smith (1963: 286) made the significant observation that well-dated samples are of 'extreme importance to a research program', but acknowledged that for various reasons this was not always possible. Archaeological contexts have not always been well recorded and relatively few Near Eastern glasses have been chemically analysed. Moreover, the data from the published studies have not been brought together and examined as a whole and consequently our understanding of the LBA glass industry remains fragmentary. Investigating the technological history of glass is, as Moorey (1985: ix) put it, 'not so much a matter of absent evidence, a problem not to be underestimated, as of scattered and incoherent source material which has never been examined'. Moorey compiled available information from early Near Eastern glass studies in an historical survey of early glass-making which did much to clarify the context and chronology of excavated glasses (Moorey 1985; 1999).

Recent analyses of LBA Near Eastern glass compositions have been made by several researchers in an attempt to learn more of the nature of these early glasses and the raw materials used in their manufacture. Unusually early glasses from Dinkha Tepe in Iran, *c.*1800 to 1600 BC, were analysed by McGovern *et al.* (1991) and found to be coloured blue with copper. Glasses from 14th C BC Tell Brak in northern Syria have been analysed for composition by Brill and Shirohata, by Velde and by Henderson (in Oates *et al.* 1997: 89-94, 94-100; Brill 1999b: 39), by Shortland and Eremin (2006: 595) and by Shortland *et al.* (2007: 784). Brill (1999b: 40-41) and Vandiver (1982: 84; 1983: 239-247) have analysed glasses from 15th to 14th C BC Nuzi, as have Shortland and Eremin (2006: 595) and Shortland *et al.* (2007: 784) since the present study of glasses from Nuzi was begun. The chief purposes of the latter two studies were, first, to provide a comparison for Egyptian glass compositions rather than to investigate in any detail the compositions of the Mesopotamian glasses themselves and, second, to answer questions about where and how widespread glass-making was in the 2nd millennium BC, whether workshops were specialising in particular colours and whether local or imported raw materials were being used. Brill has analysed a selection of glasses from Mesopotamian Tell al-Rimah of the 15th to 13th C BC (1999b: 42-43), and 13th C BC Tchoga Zanbil in Iran (1999b: 45). Brill's (1999a; 1999b) collected data from the analyses of a variety of glasses were published, but without interpretation, nevertheless providing an invaluable tool for future compositional work.

McGovern and co-workers examined silicate industry development, the origins of the glass industry in the region of Syria-Palestine in *c.*1600 BC, and craft interaction between Egypt and the Levant, especially with reference to the vitreous materials found at the site of Beth Shan in Israel, and in the Baq'ah Valley in Jordan (McGovern 1980; 1985; 1986; 1987b; 1995; Swann *et al.* 1989; McGovern *et al.* 1993; James and McGovern 1993; 157). They used the PIXE-PIGME technique to analyse the compositions of the 13th C BC vitreous materials from Beth Shan, but an inter-laboratory comparison of the results for the PIXE-PIGME they used with results from the SEM-EDS system at the British Museum suggested that serious errors occurred in the PIXE-PIGME analysis (Freestone, personal communication).

In general, LBA Near Eastern glasses are remarkably homogeneous in base glass composition being HMHK glasses, many of which are coloured copper blue. The location and number of production centres is unknown, but the industry appears to have used well-controlled raw materials of similar composition.

1.13 Analyses of LBA Egyptian Glasses

Much research has been undertaken on Egyptian glasses, which provides comparative material for the present study and will be reviewed briefly in this section. Studies of Egyptian glasses by many of the early researchers have been referred to in previous sections of this chapter. These studies provided general information on the composition of 2nd millennium BC Egyptian glasses and contributed to the current understanding of their chemistry. An archaeological overview of Egyptian glass from the earliest occurrences to the Graeco-Roman period has been published by Nicholson (1993).

Bimson and Freestone (1988: 11-15) analysed the base glass and colourant components of several early glasses from the reign of Queen Hatshepsut (1473-1458 BC) and the Amarna period (1353-1333 BC). The use of cobalt as a colourant was discussed, and the colourless glasses were found to contain no manganese or antimony, indicating that these elements had not been used for decolouring ancient glasses (section 1.10.2). Analyses of glasses from Amarna, Malkata, Lisht, Timna and Wadi Qirud have been analysed and collectively published by Brill (1999b: 27-37; see also Brill and Barnes 1988). The chemical analyses of 2nd millennium BC glasses performed by Wypyski (Lilyquist *et al.* 1993: 36-39) categorised early Egyptian glasses from the reigns of Hatshepsut to Thutmosis IV, and were grouped as 'pre-Malkata' glasses. This chemical archaeological study was prompted by the authors' observation that previous compositional analyses of Egyptian glasses had been concerned with raw materials, opacifiers and colourants and were based on few samples broadly dated to the Malkata and Amarna periods of the late 18th Dynasty. Wypyski and Brill (Lilyquist *et al.* 1993: 36-39) filled in the 150-year gap prior to this time and included Brill's, at that time, unpublished analyses of Amarna, Malkata and Nuzi glasses. Both Brill's and Wypyski's analyses, and their regional comparative approach to the study of early glasses have remained as a standard for compositional comparisons with other glasses from Egypt and the Near East. Much of the glass that they analysed was of the HMHK type and provides interesting comparative material for contemporary Near Eastern glasses.

Egyptian LBA glasses were analysed by Shortland and Tite (2000) who determined that there were both copper blue plant ash glasses and cobalt coloured 'natron' glasses used in Egypt at this time. The focus of the work by Mass *et al.* (2002) was on the colourant elements and opacifying agents of the Malkata and Lisht glasses, which included copper blue glasses, and were of a soda-lime-silica type made with plant ash.

Analysis of 226 LBA glasses from Egypt and Mesopotamia was carried out by Shortland and co-workers (Shortland and Eremin 2006: 581-603; Shortland *et al.* 2007: 781-789) in a two-part study which focussed primarily on the identification of elements brought in with each of the raw materials, and the characterisation of those raw materials. They followed the proposition that different glass-making sites in Egypt and the Near East would have used local raw materials with identifiable chemical variations that provided the opportunity for provenancing the glass.

Shortland and Eremin (2006: 591-593) summarised the nature of the colourants used in the glasses they analysed, and for opacified glasses concluded that antimony must have been added as an individual ingredient because the base glass compositions of white and colourless glasses were otherwise closely similar. Manganese was added as a relatively pure ingredient to colourless glass to create purple (section 1.10). They observed that none of the early copper blue glasses from the reigns of Thutmosis III to Amenophis II contained a significant level of tin, whereas tin in the copper blue glasses from Amarna was usual at levels that indicated the use of bronze scrap as the colourant source. Based on this finding, they have suggested that the early low-tin, copper blue glasses originated in Mesopotamia, a region from which glasses were revealed by chemical analysis to contain insignificant levels of tin (Shortland 2005: 1-5). Mass *et al.* (2002: 67, 69, 75) found the presence of tin in some blue glasses from Malkata and Lisht which suggested the use of a bronze by-product.

Shortland and Eremin (2006: 596-598) observed inconsistency in the concentrations of lime between the glasses they analysed from the early or 'pre-Malkata' and the Malkata/Amarna periods, by comparison with those from mid-13th C BC Lisht. The similarity of magnesia and potash content in all glasses, combined with the variation in lime levels, led the authors to propose that the same source of plant ash was used for all three groups of glass and that the lime levels were adjusted by the deliberate addition of lime (section 1.7).

The general similarity of Mesopotamian glasses from Nuzi and Tell Brak to those from Egypt were noted, although ICP-MS analyses showed that they were probably of distinct compositions. The value of the low detection limits of ICP-MS were acknowledged as the most sensitive method for provenancing glass by examining the transition and trace elements (Shortland *et al.* 2007). Shortland *et al.* (2007) recognised that trace elements and their ratios could be used to discriminate between Egyptian and Near Eastern glasses of the LBA (see also Shortland and Eremin 2006: 596-598).

The publication by Nicholson (2007) of comprehensive investigations at the 14th C BC site of Amarna provided new compositional analyses of glasses (Jackson and Nicholson 2007: 100-116; Appendix 5) and a detailed and informative discussion of Amarna's important role in early Egyptian glass production, drawing together the results of excavation, experiment, previous research and materials analysis in the wider context of the New

Kingdom vitreous materials production to provide convincing evidence that Amarna was indeed a glass-making site.

The 2nd millennium BC glass production site of Qantir-Piramesses in the Egyptian delta has been investigated by Rehren and Pusch (1997; 1999; 2005; Rehren *et al.* 1998; 2001; Pusch and Rehren 2007) and, most recently, a discussion of the glasses from the LBA site of Qantir-Piramesses, including a small selection of compositional analyses of glasses associated with glass production materials, have been published by Schoer and Rehren (2007). Rehren and Pusch (2007) experimented with the pounding and grinding of quartz stones to determine the extent to which the grinding tools contaminated the silica by leaving a trace element signature in experimental glasses. They found that the pattern of contamination in the ground material was the same as in the grinding tools, a consideration when interpreting the trace element data from analyses of ancient glasses. The factors that affected the final glass composition, including raw material types and sources, contamination of the batch throughout the glass production process, the choice of batch recipe and firing regime and chemical reactions which occurred during firing, have been summarised by Rehren (2008: 1350).

From their work on the ancient Egyptian glass industry at Qantir, Rehren *et al.* (2001) proposed a model for LBA glass production and working which stated that while common colours, such as copper blue, were produced at a number of sites in the Near East and Egypt, specific colours were produced only at specialised sites which were dependent on particular raw materials. Long distance trade allowed these colours to be used in many different centres, along with the commonly produced glass colours.

1.14 Analyses of IA Glasses

Few IA glasses have been analysed and no work has been done to compile and interpret the Near Eastern IA assemblage as a whole in its chemical, geographical and chronological contexts. Previous chemical analyses of IA glasses include those of glasses from Hasanlu by Stapleton and Swanson (2002a; 2002b), Stapleton (2003) and Brill (1999b: 43-44), 12th to 11th C BC glasses from Marlik Tepe in Iran,[26] one 8th C BC sample from Altin Tepe in Iran, and 9th to 8th C BC glasses from Nimrud by Brill (1978; 1999b: 45, 47-49, 52) and two by Turner and Plenderlieth (Turner 1954: 449T, 455T; 1955: 61, 67).

A brief examination of the limited published data reveals that IA object forms and colours exhibit some differences to those of LBA glasses, and although more widespread were still limited. There was an increase in the use of strong dark colours, which could appear black, and in the number of colourless and transparent glasses, indicative of the significant variations in chemical composition that were beginning to occur (section 1.2).

1.15 Glass Raw Materials in Brief

Ancient Near Eastern glasses were typically of a soda-lime-silica or high alkali base composition. Soda was the chief alkali flux for the silica, being derived from either plant ashes, which brought with them magnesia, potash and lime, or mineral soda such as natron, which was of relatively pure composition. It is generally believed that pure sand or quartz would have been combined with plant ash, or mineral soda with a lime-rich sand, in the widely accepted two-component model of ancient glass-making. It has long been argued whether the addition of lime, a key constituent which stabilised glasses to prevent dissolution, was intentional or accidental.

Colours were produced by the addition of metal minerals such as cobalt, copper, manganese and iron, and the transparency of the glass was reduced by adding antimony as an opacifying agent, which combined with lead or calcium in the glasses to form opaque yellow or opaque white glasses respectively. Opaque turquoise glass was probably made by adding antimony to translucent blue glass. Copper blue glasses were ubiquitous, having been excavated throughout the Near East and Egypt, while colours such as cobalt blue, black, red, yellow, purple, brown and white are much rarer, perhaps having been produced at specialist workshops.

Current knowledge suggests that antimony was used to decolour glasses from the mid-1st millennium BC, whereas analyses of the limited number of colourless glasses made before this time show that decolouring was achieved without the addition of a decolourant.

HMHK glasses were prevalent in the LBA in Egypt and the Near East, and LMLK glasses have their earliest known occurrence in Egypt in the 10th C BC. The Egyptian cobalt blue glasses have an anomalous composition of high magnesia and low potash but whether this can be attributed to an alum plus plant ash mix, or an alum plus natron mix, remains unclear.

1.16 Aims

While considerable advances had been made in the knowledge of ancient glass-making at the beginning of this project, information on the nature and extent of LBA and IA Near Eastern glass production was still

[26] While this is strictly a 2nd millennium date, it falls into the Near Eastern IA.

disjointed and limited, and the distinctions between LBA Egyptian and Near Eastern glass production were rudimentary and not well defined. There had been considerably more work done on Egyptian glasses than on their LBA Near Eastern counterparts, and little work had been done on Near Eastern glass from the IA. This study was designed to address questions regarding the nature of glass from the Near Eastern region from the mid-2nd to mid-1st millennia BC through the acquisition and interpretation of a substantial and unprecedented collection of chemical compositional data. This data can be used to make detailed, systematic comparisons of glasses from the LBA and from the IA, and between the LBA and IA glasses. Comparisons between the data from the present study and from published data can be made to improve our understanding of early glass-making.

The following questions have been considered. What is the chemical nature of LBA and IA glass from the Near East? How do LBA and IA Near Eastern glasses compare chemically with New Kingdom Egyptian glasses? How does IA glass production compare with that of the preceding LBA? Did LBA glass-making practices continue into the IA, or were there innovations in glass production which might be reflected in glass compositions and the choice of raw materials? How did the use of plant ash and mineral soda for fluxing the glass batch develop at this time? How were additives such as colourants and opacifiers used in the IA compared with the LBA? Can long distance trade or the movement of raw materials and/ or finished glasses be identified by the characterisation and comparison of the compositions of glasses from different regions, and by the identification of glass raw ingredients?

LBA glasses from Pella and Nuzi, analysed in this study, typify the earlier glass compositions from the 2nd millennium BC, and early 1st millennium BC glass beads from Pella provide information on glass-making during a time when the decline in glass production has been noted (section 1.2). The 9th to 8th C BC glasses from Nimrud provide information on the later, more wide-ranging IA glass compositions available in the Near Eastern region. In this way, a continuum of production can be traced from the earliest glasses of the mid-2nd millennium BC through to the fall of Nimrud in the late 7th C BC.

The present study aimed to draw together the threads of LBA and IA glass production through examination and interpretation of the compositions of the glasses with regard to the possible raw materials used in their manufacture. In this way, the analyses undertaken in the current study complement and build upon those already published, and provide new insights into LBA to IA glass manufacture in the Near East. It is intended by this process of synthesis of new and existing information to define the differences and similarities between identifiable glass chemical types from both periods, and between the two periods, in an attempt to provide a better understanding of the development of the earliest glass production in this region and to identify, where possible, trade in raw materials and glass objects.

This study was designed to provide detailed information from the compositional analyses of LBA and IA glasses of the Near East to allow for comparison with other studies, to add to our understanding of the development of the technology of glass-making, and to identify continuity and change in this key period of the history of glass production leading up to the established traditions of the later and far-reaching Hellenistic and Roman glass-making industries. This study represents the first attempt to provide detailed compositional data for a large number of IA Near Eastern glasses, and to compare these data with those from the analyses of other IA and LBA glasses of the Near East.

To answer the questions posed about the development of glass-making during these times, chemical characterisation of the compositions of selected Near Eastern glasses was undertaken using a combination of techniques which provide the most reliable, precise and accurate analyses of the widest range of elements possible at the time. Scanning electron microscopy with energy dispersive X-ray analysis (SEM-EDS), electron probe micro-analysis (EPMA) and solution inductively coupled plasma mass spectrometry (ICP-MS) were selected because of their sensitivity and availability (Chapter 2).

Baxter *et al.* (1995: 140) commented that many glass data-sets that had been studied previously contained small numbers of samples with few measured elements or variables. Instrumental methods such as ICP-MS are capable of producing large data-sets containing a large number of variables for many samples. Baxter *et al.* (1995: 140) suggested that 'the analysis of compositional variation within typological groups is a potentially illuminating way of looking at the data that complements, and may even be more informative than the inspection of mean compositions'. It was anticipated that, in the present study, the recording and handling of the chemical data of a large number of oxides and elements analysed from 132 glass samples would require the development of an innovative system for organising this data in a comprehensive, systematic way which facilitated its study and interpretation. Such a large collection of data from the analyses of eight oxides and 44 elements, including the rare earth elements, from different glass types over a broad chronological period has not been presented in one study previously.

It is anticipated that this compilation of data will enhance and facilitate comparisons with glass compositions from Egypt and from other historical periods to refine and deepen our knowledge of the growth, development and spread of glass-making, the raw materials of glass production and the implications of this information for the appreciation of the sophistication and organisation of the craftsmen who produced a technologically complex, valuable material.

The specific aims of the present study were:

1. to develop a method for reporting and manipulating the compositional data of in excess of 50 elements and oxides analysed from ancient glasses that should facilitate the meaningful grouping and interpretation of the data;
2. to obtain detailed and sensitive characterisation of the chemical composition of a range of LBA and IA glasses from the Near East using SEM-EDS, EPMA and solution ICP-MS as a means to define the glass compositions of each period;
3. to compare the compositions of LBA and IA glasses analysed in the present study to describe the nature and development of glass production and the use of raw materials from the earliest recorded industry, through change and innovation into the mid-1st millennium BC;
4. to assess the LBA and IA glass compositions in the context of recent research into raw materials with particular reference to the nature of the fluxes, colourants and decolourants and, where possible, to further current understanding of the types of raw materials chosen by the early glass manufacturers;
5. to examine and interpret the chemical data obtained from the present study together with comparable data from published studies of glasses of the 2nd and 1st millennia BC, to gain an improved understanding of the regional and chronological characteristics of the earliest glass production, and
6. to identify the possible distribution of glass-making ingredients and/ or finished glass objects.

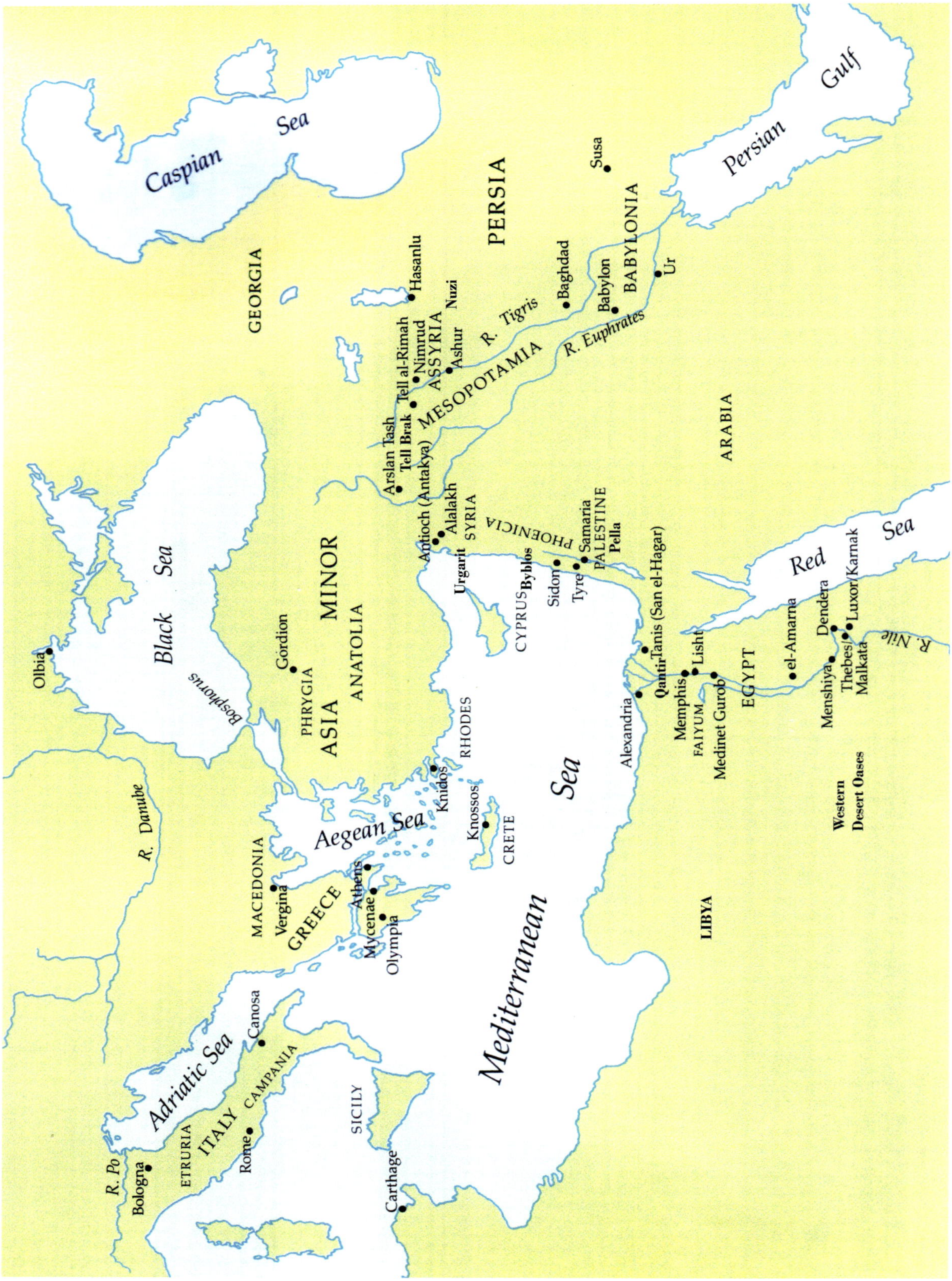

Figure 1-1 Map of the eastern Mediterranean region. Adapted from Tatton-Brown and Andrews 1991: 2

Chapter 2

Methods for Compositional Analyses and Data Interpretation

2.1 Introduction

In this study, the term 'analysis' refers to quantitative measurements that provide information about the chemical composition of ancient glass samples. The aim was to identify or characterise unknown glasses against recognised standards of known composition. It was intended that the acquired data will constitute a reference tool that can be used to answer archaeological and chemical questions related to ancient glass-making technology, raw materials, trade and production. Scanning electron microscopy with electron dispersive x-ray analysis (SEM-EDS) and electron probe microanalysis (EPMA) were used to analyse the glasses sampled to determine the concentrations of nine oxides (Table 2-6). Solution inductively coupled plasma mass spectrometry (ICP-MS) was used to detect and quantitatively analyse 44 elements, including the 14 rare earth elements (REE) (Table 2-4).

Analyses of base glass oxides with EPMA or SEM-EDS provide new information concerning the raw materials and recipes used by ancient glass-makers. These analytical methods have been used in the present study in an attempt to determine the nature and diversity of glass compositions. Trace elements can provide valuable insights into geological and geographical sources of raw materials and to some degree the nature or identity of those raw materials. As EPMA and SEM-EDS do not have sufficiently low detection limits, the more sensitive technique of solution ICP-MS has been used to investigate further the concentration of trace elements in the glasses analysed here.

The often unstable composition of the earliest Near Eastern glasses means that little remains to us for physical, let alone chemical study. The choice of archaeological glass samples for this investigation was dictated by the scarcity and limited available amount of unweathered ancient glass from this date and region, and was dependent on the kind permission of its custodians to take small samples. The sampling of unweathered glass may have biased the data toward that range of compositions which are most stable and durable.

Sampling is a process that inevitably causes a degree of destruction to the glass and/or object. These unavoidable practicalities have defined the population analysed here and demanded that the analytical techniques used be of proven reliability, requiring the smallest possible amount of sample, and providing a broad suite of elements measured as accurately and precisely as possible.

The geochemical definition of major, minor and trace constituents as given by Gill (1997: 2), is adopted in this research. Major elements are those present at concentrations greater than 1% by mass of oxide, making up the main constituents of the glass. Minor elements are those with concentrations between 0.1% and 1%, and trace elements are defined as occurring in concentrations less than 0.1%, that is less than 1000 parts per million, expressed as ppm or more rarely parts per billion (1 billion = 10^9) of the element. Parts per billion (ppb) defines the unit of trace element concentration per unit mass of sample, i.e. micrograms of element per kilogram of sample ($\mu g\ kg^{-1}$). Minor and trace element studies have become a vital part of modern archaeological material investigation and increase greatly our capability of differentiating between chemical compositions of ancient materials above relying on data from the major elements alone.

For major and minor element analyses, SEM-EDS was carried out at the British Museum, London, by the author with Professor Ian Freestone, for samples from all sites excepting Nuzi. Nuzi glasses were analysed using EPMA at the Centre for Microscopy and Microanalysis at the University of Queensland, Australia, by Ms Alanna Simpson. Trace and ultra-trace element analyses of glasses from all sites were carried out by Dr Alan Grieg of the Advanced Centre for Queensland University Isotope Research Excellence, ACQUIRE, now the Centre for Microscopy and Microanalysis, by solution ICP-MS. The sample preparation and analytical protocol for each instrument is described individually in sections 2.3 and 2.4.

2.2 Sampling

The quantity of each sample was dictated by the minimum requirements of the analytical methods chosen. The number of samples was restricted by availability of glass for sampling and the state of preservation of that glass.

Many collections of glass were investigated and a selection of 132 glass artefacts was chosen for sampling from three major LBA to IA Near Eastern sites: 16

samples from LBA Nuzi in northern Iraq, 43 LBA and 23 IA samples from Pella in the north Jordan valley, and 50 samples from IA Nimrud in northern Iraq.[27]

When taking samples, care was exercised to ensure that they were as representative of the original glass composition as is possible.[28] Weathering or corrosion of the glass influenced the choice of sample from each site with the inclusion of such glass being minimised as far as was possible. The initial selection of the least weathered glass was made visually and by light microscope examination during sample preparation and by using the imaging capabilities of SEM and EPMA at the time of microanalysis, as discussed below. Glasses made before the mid 1st millennium BC are more heterogeneous and more prone to extensive corrosion than many later glasses (Freestone 2001). Weathering or corrosion results from the often fatal combination of chemical composition, burial environment, water and time. The process of weathering causes leaching of elements including potassium, sodium, calcium and magnesium and the enrichment of others, including titanium, aluminium, iron, manganese and silicon, while others remain unchanged (Turner 1963: 37; unpublished analyses of author). Hydration and partial crystallisation of the original amorphous structure of glass occur. The result, if not complete loss, is a change in chemical composition and physical appearance, and the limited survival of the earliest glass. Much of what has survived is partially if not completely weathered.

Samples were selected to best represent the variety of colour and form, and the range of context and date of glass object found at each of the sites included in the present study. Colours sampled from the LBA glasses include opaque turquoise blue, translucent blue and translucent green. Colours sampled from the IA glasses include dark green (almost black), colourless, purple, dark blue, and both transparent and opaque light blue. Object forms are dominated by beads, with vessels, pendants (variously termed 'plaques'), inlay pieces, statuary, ingots, and unidentified fragments occurring in smaller numbers.

For catalogues of samples from each site see Table 3-1 for glasses from Nuzi, Table 4-2 for glasses from LBA Pella, Table 5-1 for glasses from IA Pella and Table 6-1 for glasses from IA Nimrud. The contexts in which the glasses were found are tombs and predominantly official structures such as palaces and temples.[29] The dates range from mid 2nd millennium BC to the early 1st millennium BC, i.e. from the LBA to IA, forming a temporal sequence that allows the investigation of glass-making during the earliest stages of its regular and deliberate production.

2.3 Scanning Electron Microscopy and Electron Probe Microanalysis (SEM and EPMA)

2.3.1 Introduction

Ancient glass often exhibits non-uniform composition due to chemical and mineralogical heterogeneity, the presence of air bubbles, or seeds (Figures 2-3, 2-4, 2-6), and the presence of weathered glass (Figure 2-1). Chemical and mineral heterogeneity can result from the inclusion of stones, that is incompletely melted and unmixed raw material (Figures 2-4, 2-5), or from the presence of crystals of opacifying compounds, such as calcium antimonate, for example (Figures 2-1, 2-2, 2-3, 2-5). Figure 2-1 illustrates clearly the demarcation between weathered and unweathered glass found in a single sample.

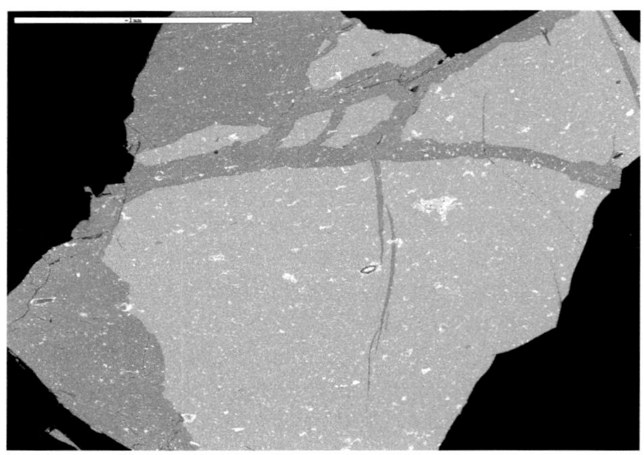

Figure 2-1 Scanning electron micrograph of Pella sample 188, a turquoise opaque bead. Dark grey areas are weathered glass which form along microscopic cracks in the glass. White grains are the opacifying compound calcium antimonate. The scale bar represents ~1mm.

For analysis of the bulk composition of glasses, it is difficult to avoid incompletely melted inclusions and weathered glass that are not representative of the glass matrix, and are invisible without the aid of the imaging capabilities of the SEM and EPMA. The back-scattered electron optical imaging and scanning capability of the SEM and EPMA, combined with their analytical facility, allows the operator to observe clearly the components of the sample being analysed, and to produce spatially-

[27] Samples which proved to be weathered and therefore unsuitable for analysis are not included in sample totals given here.

[28] It is of course possible to obtain valuable compositional information from weathered samples (unpublished data of author), and physical study of form may still be possible, but colour and size may change due to the chemical and physical effects of weathering.

[29] While Akkermans and Schwartz (2003: 12) urged archaeologists to 'guard against the facile use of present-day functional categories like "temple", "palace", "kitchen", "storeroom" etc since the ancient use of space is likely to have been frequently multifunctional', it is nevertheless how different built spaces are defined in all archaeological literature. To avoid confusion the terms applied by the excavators in each case are used here.

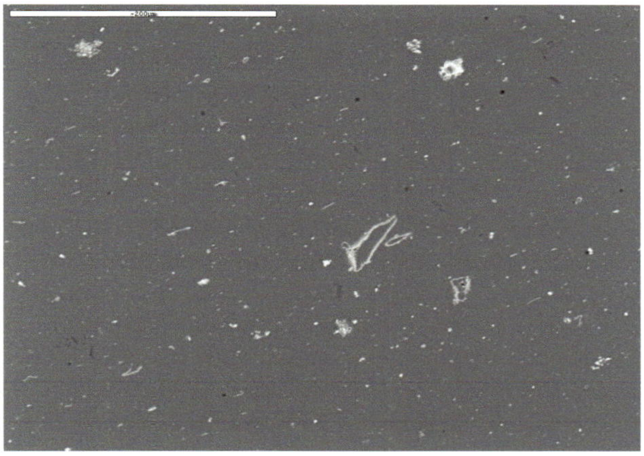

Figure 2-2 Scanning electron micrograph of Pella sample 79, from a turquoise bead opacified with calcium antimonate, visible as the white crystalline phase. The scale bar represents ~200μm.

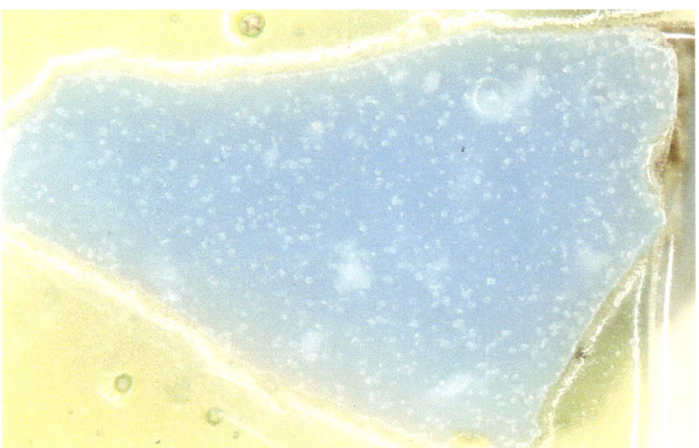

Figure 2-3 Light micrograph of Pella sample 195 from a turquoise bead opacified with calcium antimonate, visible as white inclusions. Sample size is ~ 3mm across.

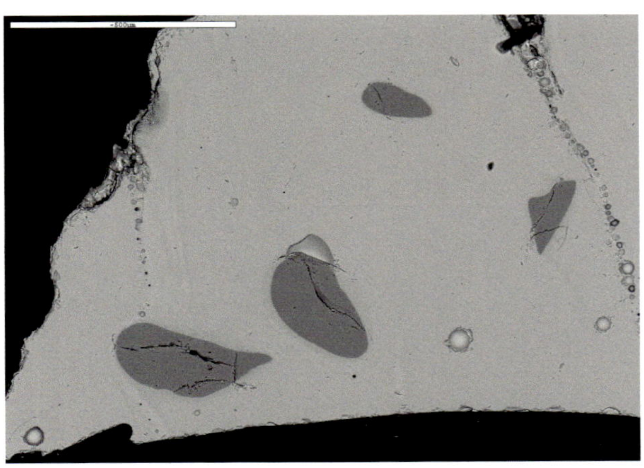

Figure 2-4 Scanning electron micrograph of Pella sample 55, a dark green glass, showing streams of small bubbles, several larger individual bubbles along the lower region, and four large dark grey silica grains which are incompletely melted and exhibit differential cooling cracks. The scale bar represents ~ 500 μm.

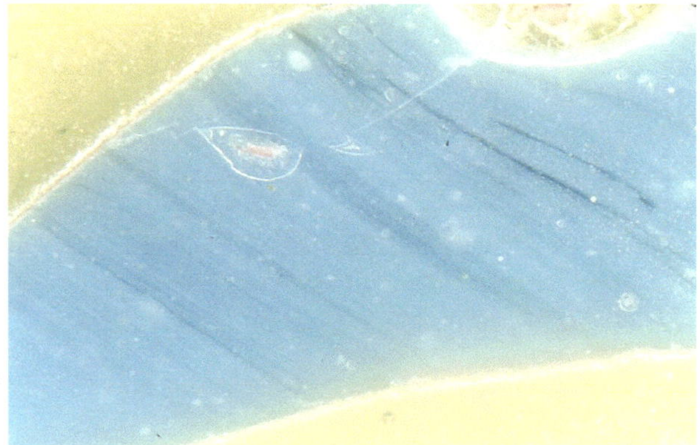

Figure 2-5 Light micrograph of Pella sample 199 from a turquoise bead opacified with calcium antimonate. Streaks of less well opacified glass due to incomplete mixing or formation of the opacifier can be seen. Sample size is ~ 2.5 mm across.

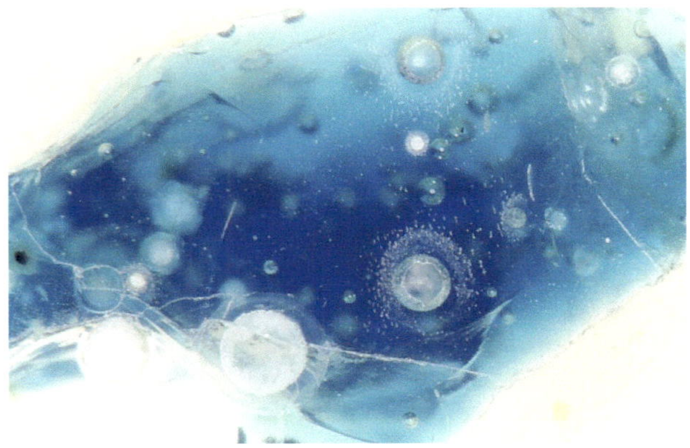

Figure 2-6 Light micrograph Pella 189 from a blue translucent bead. Bubbles are dispersed throughout the glass. Sample size is ~ 3mm across.

resolved analyses. This imaging capability is invaluable for locating heterogeneous features like individual mineral inclusions and for distinguishing weathered from unweathered glass on a micron scale. The delineation between weathered and unweathered glass is sharp allowing the analyst to avoid weathered areas of glass. Weathering products can form along fissures, for example, which are invisible to the naked eye (Figure 2-1) (see Freestone 1991: 56; Stapleton and Swanson 2002a).[30] The ability to target areas of fresh glass for analysis ensures that the volume analysed is representative of the bulk composition of the glass matrix. Hence, the imaging capability of electron microscopy is the important for the initial chemical investigation of ancient glasses.

[30] Freestone discussed the nature of weathering in a black glaze from Nimrud, describing dark areas leached of most components except silica and alumina, light grey areas which are the result of re-deposition of material leached from elsewhere in the glaze, and two small areas or 'islands' which are essentially unaltered relics of the original glaze.

Another highly desirable advantage of electron microscopy for the analysis of archaeological samples is found in the ability to characterise non-destructively very small samples which can, therefore, be reanalysed at a later date, if required. Microanalysis of ancient glass using electron microscopy is well attested by many researchers including Brill and Moll (1963), Bimson and Freestone (1983), Henderson (1988a), Wypyski (Lilyquist *et al.*, 1993) and Verita *et al.* (1994).[31]

2.3.2 Sample Preparation for SEM-EDS and EPMA

SEM samples were prepared in the Department of Conservation, Documentation and Science at the British Museum, London, and the Electron Microscope Unit, the Australian Key Centre for Microscopy and Microanalysis at the University of Sydney, Australia by the author. EPMA samples were prepared at the Centre for Microscopy and Microanalysis, University of Queensland.

Samples of typically 1mm to 3mm were removed from the edges of fragments. The sampling of already broken glass was carried out by scoring the edge of the fragment with a steel stylus. A group of complete (unbroken) beads from Pella was sampled, at the Electron Microscope Unit, University of Sydney, by adhering each bead on the edge of a glass microscope slide.[32] Each slide in turn was mounted in a Struers Accutom-50 computer-controlled diamond saw lubricated with distilled water. A sliver of approximately 1mm in width was cut from each bead, leaving it intact apart from the missing slice. The traditional method of scoring the glass with a sharp point to break off a piece would have destroyed the objects. The objects were demounted from the glass slides in baths of acetone.

All samples were mounted in epoxy resin blocks 2.5cm or 3cm in diameter (Figures 2-7, 2-8). The SEM sample blocks each contained two soda-lime-silica glass standards (either Corning A, B or RM01; see below) and 99.99% cobalt wire, used to calibrate the energy dispersive spectrometer for instrumental drift. The SEM samples were adhered to the base of a container with Araldite Five Minute epoxy resin and covered with a slow setting epoxy resin. After hardening in an oven, the blocks were freed from the containers. The surface to be analysed was ground with increasingly fine grades of wet and dry silicon carbide sandpapers (to 1200 grit), lubricated with water. The bases of the blocks were turned on a lathe to exactly 1cm in depth to ensure a uniform working distance in the SEM. Milling was followed by wheel polishing of the surface to be analysed with successively fine diamond pastes to 1μm grade, exposing a surface of fresh glass. This helps to eliminate compositional changes due to weathering, and ensures a flat surface with a known X-ray take-off angle, thereby allowing the application of standard correction procedures to the raw X-ray data. The polished sections were examined with a light microscope and each sample was photographed. The sample blocks were vacuum coated with a thin conductive layer of carbon to avoid charging during analysis.

Figure 2-7 Light micrograph of Nimrud glass samples, cobalt wire (centre top) with a standard reference glass on either side, set in an epoxy resin stub, 25mm in diameter, and polished for SEM-EDS analysis.

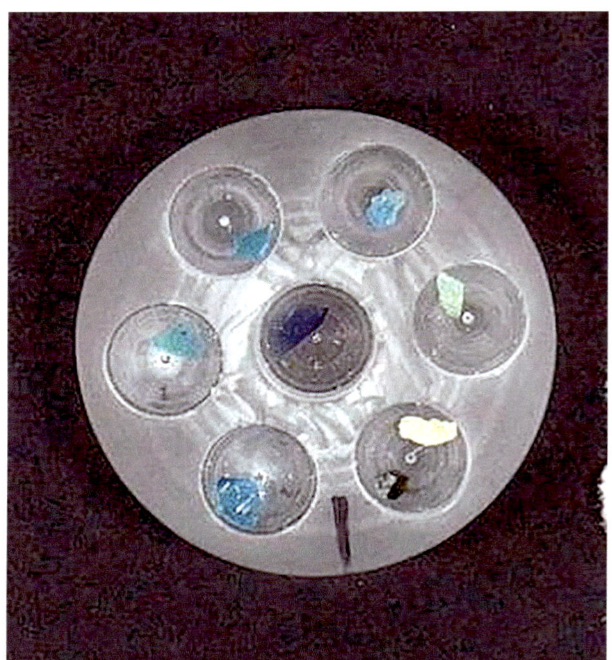

Figure 2-8 Photograph of Nuzi glass samples with standard reference glass (centre) set in a pre-prepared epoxy resin stub, 20mm in diameter, and polished for EPMA analysis.

[31] For further information regarding the principles of the operation of the SEM and microprobe, which will not be repeated here, the reader is referred to Goldstein (2003).

[32] Prior to sampling, each object was recorded by photograph and drawing.

For EPMA analysis, small holes were drilled into blocks of epoxy resin. The glass samples were placed in the holes and secured with epoxy resin. Grinding and polishing followed the same procedure that was used to prepare the sample blocks for SEM.

2.3.3 SEM-EDS Instrument and Analytical Protocol

Major and minor elements were analysed by energy-dispersive X-ray analysis in a JEOL JSM840 scanning electron microscope (SEM-EDS), using an Oxford Instruments ISIS spectrometer with a GEM germanium X-ray detector at the British Museum. The operating conditions were a 20 kV accelerating potential, 35° take-off angle, and a 1.0 nA beam current, giving a count rate of approximately 3700 counts per second on metallic cobalt, for a counting livetime of 200 seconds.

2.3.3.1 Standards

Turner (1963: 37-8) wrote that 'examination of the records of analyses of ancient glasses made during the past 150 years affords many examples of the need to apply stringent standards'. Looking back through analyses up to that time, he found that they were 'by no means free from errors of technique', concluding that 'it is abundantly clear ... that many pitfalls exist for the analyst of ancient glasses'. As a result of such observations, standard methods of analysis have evolved. Well characterised standard glasses are used to check the reliability and accuracy of results of different analyses.

The X-ray spectrometer was calibrated with primary standards of pure metals, synthetic oxides of known composition and well-characterised simple minerals. Society of Glass Technology standard glass No. 1 was used for calcium.

Results were checked against secondary standards: Corning Museum of Glass standard reference glasses A, B and D (Brill 1999b: 539-544), Society of Glass Technology standards 1 and 11, DGG (Germany) soda-lime-silica glass no. 1 and Glen Spectra soda-lime-silica RM01. Two of these standards were included in each sample block of typically eight samples. The mean analyses of selected standards are compared with their given compositions in Table 2-1. The Corning glasses, replicates of ancient glasses, are commonly used by analysts as comparative standards for archaeological samples.

2.3.3.2 Precision and Error

Every measurement taken in quantitative analysis of composition is associated with a degree of uncertainty. This may be reduced but not eliminated by repeat measurements. A statement of the uncertainty associated with the values quoted is usual. This uncertainty consists of both random and systematic error. Random error relates to the precision of the analysis or how close the agreement is between results obtained by repeated analysis under the same conditions. Standard deviation is used to quantify how precise an analysis is from replicate measurements. The smaller the standard deviation, the more precise is the analysis (Gill 1997: 7-8, 18).

Analytical totals, which fell typically between 98% and 101%, were re-summed (normalised) to 100%, a procedure that improved precision and comparability with reference standards. On the basis of the standard analyses, accuracy and precision are believed to be better than 2% relative for silica, 3% for soda and lime and 10% for other components over 1% absolute Table 2-1.

Detection limits (two sigma on the background) are 0.1 wt% for the oxides of magnesium, aluminium, phosphorus, sulphur, chlorine, potassium, calcium, titanium, manganese, and iron, 0.2 wt% for sodium, silicon, cobalt, zinc, barium and lead, 0.3 wt% for copper and antimony, and 0.4 wt% for tin. Precision deteriorated as the detection limits were approached.

2.3.4 EPMA Instrument and Analytical Protocol

Major and minor element analyses of 16 glass samples from Nuzi were conducted on a JEOL 8800L four-spectrometer electron microprobe, with a JEOL EDS detector, at the Centre of Microscopy and Microanalysis, University of Queensland under standard operating conditions of 15kV accelerating voltage, 15nA probe current and a 30µm beam diameter. To minimise volatilisation of the alkalis, potassium and sodium were the first elements measured at each spot (see Stapleton 2003: 70). All elements were analysed for 20 seconds with a 20 second background measurement. Results are reported as oxides of the element.

2.3.4.1 Primary Standards

The X-ray spectrometer was calibrated with standard natural minerals from the JEOL Taylor Block including albite, MgO, kyanite, wollastonite, orthoclase, rutile, spessartine-garnet, hematite, and apatite. In addition, pure copper, cobalt and antimony metals were used for calibration (Table 2-2). The data were corrected using the oxide ZAF correction program.

Results were checked against Corning Museum of Glass standard reference glasses A and B, and RM01. Each sample was analysed three times and the averaged results are used to represent the glass in discussions and in graphs.

Table 2-1 Analyses of standard glasses by SEM*

RM01	Na_2O	MgO	Al_2O_3	SiO_2	P_2O_5	SO_3	Cl	K_2O	CaO	TiO_2	MnO	FeO	CoO	CuO	SnO_2	Sb_2O_3	BaO	PbO	Anal Total
Analysed mean	13.35	3.91	0.68	72.84		0.35		0.29	8.77										100.19
Given	*13.50*	*3.99*	*0.61*	*72.40*		*0.24*		*0.30*	*8.61*										*99.65*
given/measured	**1.01**	**1.02**	**0.90**	**0.99**		**0.69**		**1.05**	**0.98**										
RM01 SD	0.12	0.04	0.03	0.16		0.03		0.01	0.03										
RM01 % SD	0.88	1.07	4.14	0.22		9.24		4.84	0.33										
n=6																			
Corning A	14.08	2.60	1.00	66.66	0.09	0.27	bd	2.91	5.10	0.80	1.08	1.06	0.14	1.33	0.31	1.87	0.58	bd	99.89
Corning A given	*14.30*	*2.66*	*1.00*	*66.56*	*0.13*	*0.16*	*0.10*	*2.87*	*5.03*	*0.79*	*1.00*	*0.98*	*0.17*	*1.17*	*0.19*	*1.56*	*0.56*	*0.12*	*99.35*
given/measured	**1.02**	**1.02**	**1.00**	**1.00**	**1.44**	**0.59**		**0.99**	**0.99**	**0.99**	**0.93**	**0.92**	**1.21**	**0.88**	**0.61**	**0.83**	**0.97**		
n=1																			
Corning B	16.63	1.03	4.25	61.69	0.90	0.66	bd	1.17	8.85	0.18	0.24	0.28	0.05	2.75	bd	0.35	0.04	0.42	99.86
Corning B given	*17.00*	*1.03*	*4.36*	*61.55*	*0.82*	*0.54*	*0.20*	*1.00*	*8.56*	*0.09*	*0.25*	*0.31*	*0.05*	*2.66*	*0.04*	*0.41*	*0.12*	*0.61*	*99.60*
given/measured	**1.02**	**1.00**	**1.03**	**1.00**	**0.91**	**0.82**		**0.85**	**0.97**	**0.50**	**1.04**	**1.11**	**1.00**	**0.97**		**1.17**	**3.00**	**1.45**	
n=1																			

*Results are recorded as wt% of oxide, except chlorine which is always expressed as an element.

Table 2-2 EPMA Analytical Protocol. All standards are from the JEOL Taylor Block

Elements	Standard for calibrating machine	Crystal used	Minimum Detection Limit
SiO_2	wollastonite	TAP	0.50%
TiO_2	rutile	LIF(H)	0.50%
Al_2O_3	kyanite	TAP	0.50%
FeO	hematite	LIF(H)	0.50%
MnO	spessartine garnet	LIF(H)	0.50%
MgO	MgO	TAP	0.50%
CaO	wollastonite	PET	0.50%
Na_2O	albite	TAP	0.50%
K_2O	orthoclase 2b	PET	0.50%
P_2O_5	apatite	PET	0.50%
CoO	Co pure	LIF(H)	1000 ppm
CuO	Cu pure	LIF(H)	1000 ppm
Sb_2O_5	Sb pure	PET	1000 ppm

The results of analyses of standards RM01, and Corning A and B, presented in Table 2-3, indicate that the accuracy and precision of EPMA was similar to SEM-EDS for major elements, but was considerably improved for elements below 1% due to the superior sensitivity of the wavelength dispersive analytical technique.

2.4 Solution ICP-MS

A broad range of elements can be determined by solution ICP-MS. Forty-four elements were analysed in the present study (see Table 2-4). The ICP-MS instrument provides the possibility for the simultaneous determination of most elements in the periodic table at minor, trace and ultra-trace levels (Jarvis 1992; Gill 1997: 182, 211).

2.4.1 Sample Preparation ICP-MS

ICP-MS analysis does not have the advantage of the imaging capabilities of SEM and EPMA. The glass samples were broken to reveal the freshest pieces thereby reducing the possibility that weathered glass was analysed. Indications that glass might be weathered are reflected in unusually low concentrations of the elements, such as Li, which are likely to be depleted during the process of weathering (unpublished analyses of the author). Monitoring of such elemental concentrations helps to confirm that glass unaffected by weathering was analysed.

Samples were prepared for ICP-MS analysis in ACQUIRE's clean-room based laboratories. For most glasses, a fragment or multiple fragments were obtained by gentle crushing in an agate mortar. Total sample weights were generally between 5mg and 10mg. All teflon-ware was boiled in aqua-regia, followed by 8N nitric acid, and then Milli-Q water. Because of the potential of tantalum contamination in teflon, 1mL of 24N distilled hydrofluoric acid was added to the teflon in a further cleaning step. The samples were digested in 28mL Savillex screw-top beakers in a mixture of double sub-boiling distilled hydrofluoric and nitric acids at 120°C on a hotplate until required and PTFE bombs were placed in an oven at 180°C for one to two days. The hydrofluoric acid was then discarded.

2.4.2 ICP-MS Instrument and Analytical Protocol

Trace element data for samples from all sites were obtained by solution ICP-MS on a Fisons PQ2+ and a Thermo Electron XSeries instrument at ACQUIRE. This instrument has the ability to obtain extremely accurate trace element concentration data for an extensive range of elements at the sub-ppb level (Isotope Research Centre information brochure). Solution-based sample introduction ICP-MS has been successfully used to analyse trace elements in other glass studies (Heyworth *et al.* 1989; Lambert *et al.* 1996; Hartmann *et al.* 1997; Rising 1999; Freestone *et al.* 2002).[33]

The analytical protocol and calibration followed in general that described by Kamber *et al.* (2003: 292-3; 2005). Analytical and drift correction procedures as well as Fisons PQ2+ICP-MS operating parameters are comprehensively described in Eggins *et al.* (1997), with minor modifications and were essentially those used in the present analyses (Kamber, personal communication; Niu and Batiza 1997; McGinnis *et al.* 1997). Long term averages and relative standard deviations for BHVC-1, BCR-2 and JA-2 produced at ACQUIRE exhibit the high precision of the standard analyses (see Kamber *et al.* 2005: 1044, Table 2).

Instrumental drift monitors used were a mixture of a portion of each sample to be run spiked with a standard such as W-2 such that tantalum concentrations were neither excessively high nor low relative to the unknown values in the glasses. Because only a single calibration standard was used, data were easily normalised to other sets of preferred values for standards.

Geologically unusually high concentrations of antimony, copper and cobalt were calibrated against synthetic pure element solutions. Many samples with high concentrations of these elements were analysed at two dilution factors to bring the ion response of

[33] Vicenzi *et al.* (2002) reported ICP-MS analyses of Corning Standard reference glasses, adding new data to the Smithsonian microbeam standard collection.

Table 2-3 Analyses of standard glasses by EPMA

Av. 3 analyses	Na_2O	MgO	Al_2O_3	SiO_2	P_2O_5	K_2O	CaO	TiO_2	MnO	FeO	CoO	CuO	Sb_2O_3	Analytical Total
Corning A														
analysed mean	13.52	2.67	0.92	66.74	0.12	3.00	5.23	0.78	1.03	0.97	0.17	1.23	1.77	98.15
given	*14.30*	*2.66*	*1.00*	*66.56*	*0.13*	*2.87*	*5.03*	*0.79*	*1.00*	*0.98*	*0.17*	*1.17*	*1.56*	*98.22*
given/mean	**1.06**	**1.00**	**1.09**	**1.00**	**1.08**	**0.96**	**0.96**	**1.01**	**0.97**	**1.01**	**1.00**	**0.95**	**0.88**	
SD	0.57	0.08	0.03	0.92	0.02	0.07	0.13	0.07	0.03	0.04	0.03	0.06	0.06	
% SD	4.22	3.00	3.26	1.38	16.67	2.33	2.49	8.97	2.91	4.12	17.65	4.88	3.39	
Corning B														
analysed mean	16.85	0.99	4.26	61.81	0.85	1.06	8.88	0.10	0.24	0.30	0.05	2.92	0.46	98.77
given	*17.00*	*1.03*	*4.36*	*61.55*	*0.82*	*1.00*	*8.56*	*0.09*	*0.25*	*0.31*	*0.05*	*2.66*	*0.41*	*98.09*
given/mean	**1.01**	**1.04**	**1.02**	**1.00**	**0.96**	**0.94**	**0.96**	**0.90**	**1.04**	**1.03**	**1.00**	**0.91**	**0.89**	
SD	0.24	0.02	0.07	1.10	0.06	0.04	0.11	0.02	0.03	0.03	0.02	0.08	0.03	
% SD	1.42	2.02	1.64	1.78	7.06	3.77	1.24	20.00	12.50	10.00	40.00	2.74	6.52	
RM01														
analysed mean	13.66	3.99	0.59	72.01	0.02	0.30	8.80	0.03	0.01	0.04	0.01	0.01	0.02	99.49
given	*13.50*	*3.99*	*0.61*	*72.40*		*0.30*	*8.61*							*99.41*
given/mean	**0.99**	**1.00**	**1.03**	**1.01**		**1.00**	**0.98**							
SD	0.36	0.09	0.03	0.36	0.02	0.01	0.13	0.02	0.01	0.02	0.02	0.02	0.02	
% SD	2.61	2.14	4.89	0.51	77.80	4.33	1.42	54.54	184.96	47.29	154.02	167.79	92.04	

Table 2-4 Elements measured by ICP-MS in the present study*

Element Symbol	Name	Element Symbol	Name	Element Symbol	Name
Li	Lithium	Zr	Zirconium	*Tb*	*Terbium*
Be	Beryllium	Nb	Niobium	*Dy*	*Dysprosium*
Sc	Scandium	Mo	Molybdenum	*Ho*	*Holmium*
Ti	Titanium	Cd	Cadmium	*Er*	*Erbium*
V	Vanadium	Sn	Tin	*Tm*	*Thulium*
Cr	Chromium	Sb	Antimony	*Yb*	*Ytterbium*
Co	Cobalt	Cs	Caesium	*Lu*	*Lutetium*
Ni	Nickel	Ba	Barium	Hf	Hafnium
Cu	Copper	*La*	*Lanthanum*	Ta	Tantalum
Zn	Zinc	*Ce*	*Cerium*	W	Tungsten
As	Arsenic	*Pr*	*Praesodymium*	Tl	Thallium
Ga	Gallium	*Nd*	*Neodymium*	Pb	Lead
Rb	Rubidium	*Sm*	*Samarium*	Th	Thorium
Sr	Strontium	*Eu*	*Europium*	U	Uranium
Y	*Yttrium*	*Gd*	*Gadolinium*		

*n = 44. Lanthanides (REE) appear in italics. Promethium is excluded as it does not occur naturally.

Table 2-5 Minimum Detection Levels for ICP-MS analyses of glasses*

Element	Detection Limit (ppb)	Element	Detection Limit (ppb)
7Li	*79.17*	135Ba	*31.36*
9Be	*1.66*	139La	*1.06*
45Sc	*10.68*	140Ce	*1.22*
49Ti	*163.79*	141Pr	*0.15*
51V	*3.87*	146Nd	*1.27*
53Cr	*48.53*	149Sm	*22.43*
59Co	*1.62*	151Eu	*0.04*
60Ni	*39.39*	153Eu	*0.09*
65 Cu	*969.63*	159Tb	*0.04*
66Zn	*1301.07*	160Gd	*0.16*
71Ga	*1.03*	161Dy	*0.71*
75As	*10.47*	165Ho	*0.09*
85Rb	*3.72*	167Er	*0.14*
86Sr	*259.15*	169Tm	*0.01*
89Y	*0.45*	172Yb	*0.13*
90Zr	*36.95*	175Lu	*0.01*
93Nb	*0.56*	178Hf	*0.77*
98Mo	*1.40*	181Ta	*0.11*
111Cd	*1.28*	184W	*0.69*
120Sn	*55.32*	205Tl	*0.17*
121Sb	*136.97*	avPb	*11.76*
133Cs	*0.22*	232Th	*0.41*
		238U	*0.26*

*Dilution factor of 2000. Data provided by ACQUIRE, University of Queensland.

these analytes within the linear range of the detector. Particular care was taken to correct for oxides of these elements (by estimating their oxide formation rates from pure elemental solutions analysed at the end of the run), but in some samples $^{123}Sb^{16}O$ interference on ^{139}La could not be completely removed, resulting in artificial lanthanum anomalies.

Internal precisions were always better than 2% (except for single digit ppb concentrations) and external reproducibility is of similar magnitude (see standard deviation of BHVO-1 in Table 2, Kamber *et al.* 2005: 1044).

Minimum detection limits for each element analysed with ICP-MS are presented in Table 2-5. Data for all analyses are above these limits.

2.5 Organisation and Presentation of Data

2.5.1 Introduction

The data from the analyses of the oxides reported in this study consist of the eight oxides of the elements not assayed for by ICP-MS, and which are reported in the widely accepted manner described in section 2.5.4. Where elements were detected by all analytical techniques, the ICP-MS data have been used in preference (Table 2-6).

The EPMA and SEM-EDS analyses of the oxides of silica, soda, lime, magnesium, potassium, phosphorus, aluminium and iron provide information about the

Table 2-6 Elements detected by SEM-EDS and EPMA, giving oxide names used in the text, and indicating those that were also detected with ICP-MS*

Oxide Symbol	Element Name	Oxide Name	SEM/ EPMA (as wt % oxide)	ICP-MS (as element)
SiO_2	Silicon	silica	X	
Na_2O	Sodium	soda	X	
Al_2O_3	Aluminium	alumina	X	
MgO	Magnesium	magnesia	X	
K_2O	Potassium	potash	X	
CaO	Calcium	lime	X	
FeO	Iron	iron oxide	X	
TiO_2	Titanium		X	X
P_2O_5	Phosphorus	phosphorus	X	
SO_2	Sulphur		X	
MnO	Manganese		X	
CoO	Cobalt		X	X
CuO	Copper		X	X
Sb_2O_3	Antimony		X	X
PbO	Lead		X	X

*Data from ICP-MS is used in preference when available.

primary ingredients of the glasses, referred to as the base glass composition. The base glass oxide compositions have been used to identify and classify the glasses as being of the soda-lime-silica type with either high-magnesia and high-potash contents (HMHK), or low-magnesia low-potash contents (LMLK). Manganese oxide is reported with the colourants.[34] The oxides are referred to by their names, rather than by chemical symbols.

To this initial characterisation of the glasses is added the large amount of complex information provided by the trace element analyses. It has been necessary to develop an innovative method for the organisation, presentation and interpretation of the trace element data, which is described in the following section.

2.5.2 Organisation of Trace Element Data

The grouping of behaviourally similar trace elements helps to simplify what is otherwise a large and unwieldy data-set. Trace elements are normally grouped on the basis of their position in the periodic table, for example, the lanthanides or REE (atomic numbers 57 to 71) form a chemical group. The elements in each group have similar chemical properties and for this reason are expected to show similar geochemical behaviour (Henderson 1984b).

In the present study, the trace element results are reported in groups that seem most appropriate for and which facilitate the interpretation of the glass chemical composition. Elements will be referred to by their chemical symbols (not including isotope number) as listed in Table 2-5. The REE will be reported first, and are discussed in more detail below (section 2.5.3). The elements Li, Rb, Cs, Be, Sr and Ba and those reported as oxides of sodium, potassium, magnesium and calcium are grouped on the periodic table as alkali and alkaline earth elements, and are known to be present in plant ashes.[35]

The next group to be reported consists of the sediment-related elements (SRE):[36] the transition metals Sc, Ti, V, Cr and the elements Y, Zr, Nb, Mo, Hf, Ta and W, the actinides Th and U, and Tl. Elements with the same ionic charge and size are expected to show very similar geochemical behaviour, for example, the high field strength (HFS) elements[37] included in the SRE group are Sc, Ti, Y, Zr, Nb, Hf, Ta, Th, and U. The SRE group contains elements not associated with plant ash and/ or colouration, excluding the REE and alumina and iron. The SRE are likely to be the elements most closely associated with sedimentary material, such as river alluvium. The REE are HFS elements and they exhibit similar geochemical behaviour to each other (for further explanation of element grouping see Rollinson 1993: 102-5, Fig. 4.2).

[34] Elemental manganese was not assayed by ICP-MS.

[35] Barkoudah and Henderson (2006) report these elements/oxides as plant ash components with the exception of Be.

[36] Iron and alumina are reported as base glass oxides.

[37] HFS elements are those with a high charge relative to ionic radius.

The final group to be reported is that of the elements responsible for colouring and opacifying, and those elements related to them. They include the transition metals, Co, Ni, Zn, Cu, Cd and Mn, and Sb, Sn and Pb. This group does not include the transition metal iron, which may influence colour, because it is dealt with in the base glass oxide results sections. Manganese was measured only by EPMA and SEM-EDS as an oxide. Cu and Co are traditionally recognised as blue colourants. Sb combined with either Pb or lime, is responsible for yellow or white opaque colours respectively. Calcium antimonate opacifies translucent blue to give a light turquoise blue opaque colour. Ni, Zn and manganese are known to be associated with Co from Egyptian alum sources, and Sn can be associated with Cu, for example if both are derived from bronze. Initially, Sb and later manganese could also act as decolourants (section 1.10).

This novel approach to the grouping and examination of the elemental data from the compositional analyses of ancient glasses has resulted in a new method for the organisation and interpretation of large and complex glass data-sets. Many of these elements, including the REE, have not traditionally been reported for ancient glasses, and it is anticipated that the larger number of elements assayed with the highly sensitive technique of solution ICP-MS will assist in furthering our understanding of ancient glass compositions.

2.5.3 Rare Earth Elements (REE)

The REE are a group of 14 naturally occurring elements from La (atomic number or Z = 57) to Lu (Z = 71) with very similar chemical and physical properties. They form a particularly interesting elemental group which have proved in this study to have considerable potential for the elucidation of ancient glass compositions.

Although, in terms of geochemical behaviour, Y mirrors the heavy lanthanides Dy to Ho, it is not typically included with them when reported. The REE are a useful geochemical group, as they are relatively stable during fractionation processes, including weathering. The relative abundance of these elements is most likely to be an accurate estimate of their geological origin (Kamber *et al.* 2005: 1047), and consequently can indicate whether glasses are derived from raw materials of the same or different geochemistry. The REE are divided into two sub-groups: the light REE (LREE) from La to Sm, and the heavy REE (HREE) from Gd to Lu. The REE occur naturally as a group rather than singly or as a combination of a few of them. For detailed discussion of REE geochemistry, see for example, Fleet (1984: 345, 365), Henderson (1984a), McDonough and Frey (1989: 99,136), and Rollinson (1993: 133).

During the processes of soilformation and sedimentation, the REE are found in the clay and shale components, the bauxite minerals, and the oxidates of Fe and Mn. The REE contents of sediments and sedimentary rocks naturally reflect the mineral contents of these deposits, and hence the processes by which the minerals formed and were incorporated into the deposits. The bulk of the REE in eroded material, or sediments, is contained in clays, while low concentrations of REE are associated with quartz, which is the other major silicate present in eroded material. This suggests that the silica from quartz or sand is less likely to be the source of REE in ancient glass, although other components of the sand may contribute, as may clay or soil possibly attached to or incorporated with the plants at the time of ashing, and the ash itself. During growth, plants take up the REE from soil and the REE are concentrated when the plants are ashed (Barkoudah and Henderson 2006: 308-311; Tite *et al.* 2006: 1285, 1289). Experiments by Rehren and Pusch (2007: 223-225) have shown that sedimentary material could also enter the glass through the silica preparation processes.

The REE constitute a difficult elemental group to quantify by most instrumental techniques without separation and pre-concentration, which makes ICP-MS in many ways an ideal analytical technique (Jarvis 1989). Each REE has at least one isotope free from isobaric overlap, and sensitivity is relatively uniform from 139La to 175Lu. The detection limits for the REE are considerably lower than for most other instrumental methods for comparable preparations. The chemical behaviour and normalised abundance change gradually over this series of elements (Gill 1997: 183-194).

The distribution of REE in nature corresponds to the Oddo-Harkins rule where a plot of concentration against Z shows that the abundances of individual REE vary in a sawtooth manner in all cosmic and terrestrial matter, even-Z elements being about ten times more abundant than odd-Z elements (Henderson 1984b: fig 1.2; Gill 1997: fig 13.2.1). The LREE are more abundant in the earth's crust than the HREE. To filter out the abundance variation between odd and even atomic number elements and to enable comparisons, it is conventional to normalise the concentrations in the sample to those in a chosen reference material. This means that the concentration of each individual REE in the sample is divided by the concentration of the same REE in the reference material. There are several reference materials used for normalising, including the 'average' chondrite meteorite, which is now known to be quite variable in composition, or 'average' sediment, such as the average **MU**d from **Q**ueensland, or **MUQ**, used as a proxy for the upper continental crust (UCC) (Kamber *et al.* 2005). No standardised value has been adopted in the literature for normalising the REE. MUQ is used in the present study, unless MUQ is being compared with the REE, in which case the average chondrites are used to normalise both data-sets. Values for MUQ and average chondrites are given in Table 2-7.

The plot of the REE is presented as the logarithm of the normalised abundance *versus* atomic number. The normalised concentrations give smooth rather than jagged patterns, thereby removing apparent differences to clarify the presentation of REE concentration data and to facilitate interpretation (Henderson 1984b: 6-9; Jarvis 1989: 563; Rollinson 1993: 135; Gill 1997: 213). For the origin of this normalising procedure see Henderson (1984b: 8). Concentrations at individual points on the plot of the REE are joined by straight lines. This is sometimes referred to as the Masuda-Coryell diagram. Trends on a REE diagram are usually referred to as REE 'patterns' and the shape of a REE pattern is of petrological interest (Rollinson 1993: 135) and can help us to assess whether or not glass raw materials are derived from similar geological origins.

Sometimes the plotted position of Eu does not fit the trend defined by the other REE. This is called an Eu anomaly, and is described as positive when the concentration of Eu is larger than that of Sm and Gd, or negative if it is less than the concentrations of Sm and Gd. Unlike the other REE, which only have a valence of +3, Eu can also have a valence of +2, so may be fractionated relative to the other REE. Similar anomalies may be recorded for other REE, especially Ce (see Henderson 1984b: 9; Rollinson 1993: 135; Pollard and Heron 1996: 131-132).

2.5.4 Presentation of Oxide and Elemental Data

SEM-EDS and EPMA data for the base glass constituents are reported in weight per cent (wt%) of the oxide of each element, while the ICP-MS trace element data are reported in ppb, with the exception of the colourant and related elements which are reported in ppm because they occur in comparatively large quantities which are more succinctly expressed as ppm. For further discussion of trace elements reporting see Rollinson (1993: 33, 37, 48). In general, averaged data are presented in the text of the present study, and the results for the analyses of the individual glasses are compiled in Appendices II to VI.

It is customary to present compositional data for ancient glasses in the main as bivariate plots (scatter plots) in which the variables are the elemental and oxide data. The principal aim of bivariate plots is to show variation between samples and to identify trends and correlations. The element or oxide plotted along the *x*-axis of the diagram should be selected to show the maximum variability between samples. Normally the oxide which shows the greatest range in the data-set would be selected (Rollinson 1993: 9, 66, 73). The reporting and discussion of oxide and elemental data form an important part of this dissertation. Traditionally geochemists have looked at the relationships between pairs of elements by plotting a large number of bivariate variation diagrams or by constructing a correlation matrix (Rollinson 1993: 22-28). All possible elemental and oxide pairs from the data obtained for the present study were inspected and the most informative were selected and reproduced in the text of the following chapters. The advice of Pollard and Heron (1996: 131-132), when discussing the reporting of ceramic analyses, has been applied in the present study: 'in general ... discussion of these correlations is usually restricted to simple observations of their occurrence' and what often distinguishes the different raw material sources 'is not the absolute concentrations of the various trace elements, but the degree of correlation between pairs of elements'.

In the present study, it has proved valuable and informative to present not only the REE data in line and dot plots (the Masuda-Coryell diagrams), but to develop a system using line and dot plots to present all of the grouped trace element data (SRE, alkalis and alkaline earths and so on) in a similar manner, and to observe the relative concentrations and patterns of the trace element groups belonging to the different glass types. This method of presenting the data also provides a convenient and useful way to compare the elemental concentrations in the glass with those found naturally in the UCC, as represented by MUQ. Previously, Freestone has employed line and dot plots to present and compare selections of elemental glass data (2000: 83; 2002: 264, 268; 2006).

2.5.5 Statistical Methods Used

Standard deviations (SD) are reported for the oxide data in wt% in keeping with established practices, and relative standard deviations (RSD) are reported for the elemental analyses, because they provide a meaningful and readily comparative value by which to assess the variation around the mean of the data. Averages are calculated for the elemental data with the omission of outliers of greater than 2 sigma variance from the mean.

The Analytical Methods Committee of the Royal Society of Chemistry technical brief (No.5 April 2001) states that 'Reporting the value found, accompanied by its uncertainty, is clearly the best method of reporting because it provides the most information.' Therefore, results that were below detection limit including negative results have been recorded and used for averaging, but not for normalising. If negative values are converted to zero, the mean result is biased

Table 2-7 MUQ and average chondrite elemental values (ppb) used by ACQUIRE for normalisation of ICP-MS data (Kamber et al. 2005; Sun and McDonough 1989)*

	Mud of Queensland (MUQ)	Average Chondrites		Mud of Queensland (MUQ)	Average Chondrites
Element	ppb	ppb	Element	ppb	ppb
Li	28,240	1570	Ce	71,090	612
Be	1920	no value	Pr	8460	95
Sc	16,490	no value	Nd	32,910	467
Ti	6,909,000	445,000	Sm	6880	153
V	120,000	no value	Eu	1570	58
Cr	64,500	no value	Gd	6360	206
Co	22,390	no value	Tb	990	37
Ni	31,570	no value	Dy	5890	254
Cu	32,360	no value	Ho	1220	57
Zn	73,470	no value	Er	3370	166
Ga	19,120	no value	Tm	510	26
Rb	79,510	2320	Yb	3250	170
Sr	142,000	7260	Lu	490	25
Y	31,850	1570	Hf	5320	106.6
Zr	199,000	3870	Ta	1120	14
Nb	15,330	246	W	1600	95
Sn	2770	1720	Tl	420	140
Cs	5380	188	Pb	20,440	2470
Ba	396,000	2410	Th	11,130	29
La	32,510	237	U	2830	8

***There are no values for Mo, Cd, As and Sb in MUQ.**

upwards and the variance is reduced, which would be misleading.

Where appropriate, the elemental and oxide data have been normalised using MUQ, which represents the mean elemental and oxide abundances of the UCC (section 2.5.3, and Kamber *et al.* 2005), and average chondritic values, an alternative data-set, are used to normalise the REE and MUQ when they are compared with each other. The values for each element for the MUQ and average chondrite normalisers are presented in Table 2-7 and the MUQ value for each oxide is presented in Table 2-8.

The process of dividing the glass data by values from another set of data, such as MUQ, is referred to in this study as normalising. This is distinct from the practice that is commonly termed normalisation of the base glass oxide data when those data are summed for each sample and are then re-summed to equal 100%. The re-summed data are referred to as reduced data and are denoted by the chemical abbreviation for the oxide followed by an asterisk, for example MgO*. This process of reducing the oxide data is here termed re-summing to avoid confusion with the process of normalisation that is applied to the elemental data. In the text, the reduced oxides are referred to by their oxide names, for example magnesia, and the chemical form of the oxide name is used in graphs and tables, for example MgO*. The normalisation or re-summing of compositional components of the glass is a way to compare the compositions of different glasses (Brill 1987: 1-25; Lilyquist *et al.* 1993: 40; Rehren 2000b: 14). Re-summing of base glass compositions facilitates the comparison of

Table 2-8 MUQ values for normalisation of SEM and EPMA oxide data (Kamber *et al.* 2005)*

Oxides	SiO_2	Al_2O_3	FeO	MgO	CaO	Na_2O	K_2O	P_2O_5
MUQ	67.53	17.31	8.64	1.57	1.1	0.8	1.54	0.18

*** The FeO value is calculated from the Fe_2O_3 value quoted in Kamber *et al.* 2005.**

different glasses without the complications of additives such as colourants and opacifiers (Lilyquist *et al.* 1993: 36; Brill 1970a: 122).

In the manipulation and interpretation of some of the data, it was found necessary to apply an unpaired, two-tailed t-test to this data. The t-test was deemed the most appropriate statistical analysis to assess whether the means of the oxides of two groups of glass are statistically significantly different from one another. Values of less than $p = 0.05$ were considered to be statistically significant. Analyses were performed by Dr Melanie Fillios (University of Sydney) using *Graphpad Prism Software* and were checked by Mr Dennis Dwarte (University of Sydney).

Chapter 3

Nuzi:
Late Bronze Age Glasses
Results of Compositional Analyses

3.1 Introduction to the Site

The site of Nuzi, modern Yorgan Tepe, is located in north-eastern Iraq, thirteen kilometres south-west of Kirkuk, or ancient Arrapha (see Figure 3-1). Nuzi is a small, square mound, measuring approximately 200 metres in length and in width, and an average of five metres in height. Nuzi lies on the edge of a broad, flat plain between the Tigris River, the River Zab and the foothills to the north-east, in a region with a long history of a large population and well-watered cultivation (Starr 1939: xxx-xxxi). It was occupied continuously from the late 3rd millennium BC to the third quarter of the 2nd millennium BC and is unique in providing a background for the artistic and architectural developments of the Mitannian period in Stratum II.

Apart from the discovery of several thousand cuneiform documents which give information on this provincial eastern city of the Mitannian/ Hurrian empire, Nuzi is known for the large number of very early glass objects unearthed mostly from its temple and palace. The fortified citadel encloses a palace and a sanctuary, which was transformed from a single temple (G29) to a double temple (G29, G53) complex during the Ur III to Isin-Larsa periods (Figures 3-2, 3-3). These iterations of the temple are arranged on the standard Mesopotamian bent-axis scheme. Neither version of the temple was altered significantly before its destruction in the mid 14th C BC. G29 contained glass and glazed terracotta, and is attributed to Ishtar (Astarte). The palace is also conventional in its layout of small rooms grouped around large courtyards (Wilhelm 1989: 81).

It is thought that the Hurrians did not infiltrate Syria until the 2nd millennium BC, although recent work at Urkesh in north-eastern Syria has revealed that they could have been present much earlier (Lawler 2008: 47) when they occupied the upper reaches of the Tigris and its eastern tributaries.

By the later 17th C BC, northern Syria had a strong Hurrian population with Hurrian rulers in the eastern part of the country. The Hurrian-Mitannian kingdom was well established by the 16th C BC (Klengel 1992: 86; Wilhelm 1989: 5-7, 18-21). Nuzi was inhabited in the 15th C BC and 14th C BC by a Hurrian speaking population, known from surviving texts. The Hurrian language was in use at least as far away as central Syria in the 14th C BC, when the Mitannian names of rulers were found in the south of Syria and into Palestine (Wilhelm 1989: 13-14, 18; section 1.2).

At the height of the Mitannian kingdom in the LBA, the Hurrians had spread across several cultural frontiers from the Zagros foothills to the Mediterranean coast. The kingdom of Mitanni stretched from the southern part of Kizzuwatna in the west to the ʻAmuq plain on the lower Orontes (Alalakh), and included the Aleppo region (Halab), the region around Hama and Homs on the upper Orontes (Qatna and Qadesh), the Euphrates Valley to the north of Meskene (Emar), the north-eastern Syrian arable plain (Mitanni/ Hanigalbat), Assyria, and the Kirkuk region (Arrapha), where Nuzi formed an eastern outpost, that was connected to Syria, south-western Anatolia and the Mediterranean region by virtue of being part of the Mitannian kingdom (Wilhelm 1989: 42; Kuhrt 1995: 283-300; Kühne 1999; Van De Mieroop 2004: 142-145).

During this period of the Mitannian hegemony over the northern Fertile Crescent, glass objects were produced for the first time in large numbers and, together with a range of vitreous products, amongst other goods, were distributed widely by trade and tribute through an inter-regional network of diplomatic and military connections which included Mitanni, Egypt and Babylonia (Wilhelm 1989: 80, 86, 89; Moorey 1989: 281-2; Moorey 1999: 196; Peltenburg 1987: 21-2). Akkermans and Schwarz (2003: 354) emphasise the technological and artistic innovation that saw the introduction of the core-formed glass vessel at this time, and indeed glass is well represented at the towns of Tell al-Rimah, Tell Brak, Alalakh and Nuzi in the Mitannian kingdom (Figure 1-1). For discussions on the movement or exchange of craftsmen, which falls outside the scope of the present study, see, for example, Zaccagnini (1983), Moorey (1999: 194), and Nicholson (2007: 3-4, 8) who discuss links to Mitanni, and Mitanni as a source of glass from the campaigns of Thutmosis III and into the Amarna period.

In the mid 14th C BC when the power of the Mitanni began to decline, Šuppiluliuma, the first great king of the new Hittite kingdom, took advantage of internal Mitannian conflict, and secured the support of Assyria

Figure 3-1 Satellite map showing location of Nuzi between the Tigris River and the River Zab, in ancient Mesopotamia (modern Iraq), adapted from Google Earth.

which had freed itself from Mitanni supremacy. The Babylonians, on the Mitannians' south-western front, marched into Arrapha as far as the Lower Zab, and the more northerly regions occupied by the Assyrians (see Van de Mieroop 2004:142-145). The simultaneous threats from more than one side led to the rapid downfall of the Mitannian kingdom which did not survive beyond the end of the 13th C BC, approximately a century after the destruction of Nuzi (Klengel 1992: 111; Wilhelm 1989: 34-35, 40, 87; Akkermans and Schwarz 2003: 329).

3.1.1 The Excavations

Edward Chiera led the first excavation of Nuzi in 1925-6 for the Iraq Museum and the American Schools of Oriental Research. The following four campaigns were directed by Chiera, 1927-8, Robert H. Pfeiffer, 1928-9, and Richard F.S. Starr, 1929-30 and 1930-31. These seasons were sponsored jointly by the Fogg Art Museum, the Harvard Semitic Museum and the American Schools of Oriental Research. The glass analysed in the present study comes from R.F.S Starr's excavations of 1930 and was kindly provided by Dr James Armstrong of the Harvard Semitic Museum, Boston.

The main features excavated include a temple, a palace, a city wall and related buildings, and suburban dwellings. Starr's excavations concentrated on the ridge on the north western side of the mound resulting in the clearing of the temple (Squares G to H) opposite the palace (Squares L, M and R), and of housing adjacent to these buildings. The temple is the only building having a 'direct and unbroken sequence' through to the uppermost levels of Nuzi. The chronological periods of Hurrian occupation at Nuzi were divided into Strata, labelled from IV to I, which are dated relatively, and are not given absolute dates (section 3.1.2 regarding dating of the glass). The period of occupation relevant to this study belongs to Stratum II with which the latest iteration of the temple, A, is contemporary. The temple phases are labelled A (latest) to G (earliest) (Starr 1939, xxxvii-xxxviii).

A site plan (Figure 3-2) is reproduced below. The temple and palace are the two largest and most important buildings excavated, and contained significant artefacts (Bjorkman 1999), including the glasses analysed in the present study.

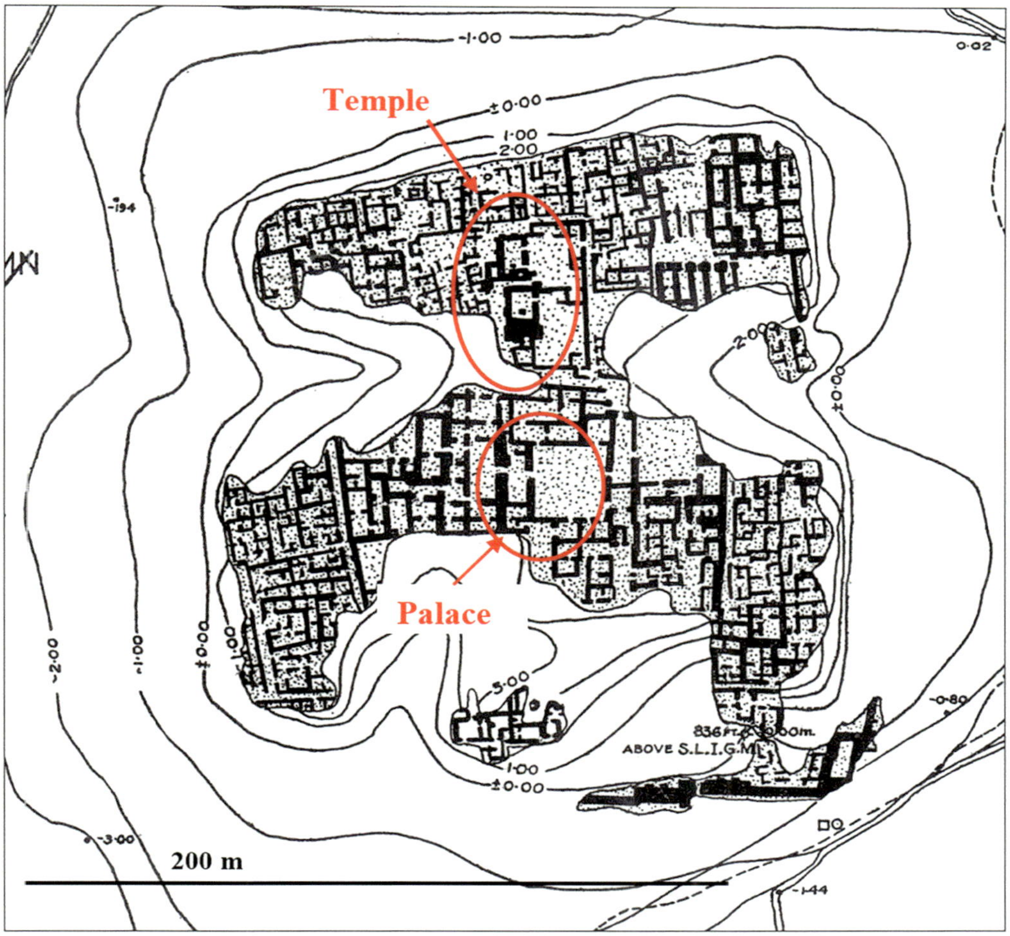

Figure 3-2 Plan of city of Nuzi, adapted from Starr (1939), Plan 2, highlighting the size and position of the Temple and Palace, where most of the glass was found.

3.1.2 The Glass

Approximately 11,000 glass objects were found in Stratum II, most likely dating from the late 15th or early 14th C BC through to the mid 14th C BC, when Nuzi was destroyed by the Assyrians. Dating has been discussed by Eliot (in Starr 1939: 507-522), Speiser (1929), Barag (1970: 136), Drower (1973: 417-420), Stein (1989: 36-60; 1997: 172-3) and Moorey (1999: 194, 196). Currently, 1350-1330 BCE is considered to be the latest date for the Nuzi tablets and the Stratum II destruction, with the occupation of the site not extending far beyond this date (Starr 1939: xxxv; Stein 1997: 172).

With a destruction date for Nuzi of *c.*1350/1330 BC, i.e. mid-14th C BC, the Nuzi glass is most likely to have been manufactured in an earlier period than the glass found at the great Egyptian glass-making site of Amarna, the construction of which began in *c.*1347 BC, in the 5th year of the reign of Akhenaten (1352-1336 BC) (Shaw 2000: 481). Amarna was probably abandoned during the reign of Tutankhamun (1336-1327 BC) (Kuhrt 1995: 195), around the lowest date for the destruction of Nuzi. The palace of Malkata where glass was discovered, immediately preceded the Amarna period, dating to the reign of Amenhotep III (1390-1352 BC). Even earlier glass from the reigns of Thutmosis III to Thutmosis IV (1479-1390 BC) is termed 'pre-Malkata' by Lilyquist *et al.* (1993: 33-35). Nicholson (2007: 1-9) presents the evidence for early Egyptian glass dating back to the early 18th Dynasty reign of Ahmose,[38] *c.*1500 BC, before which sporadic glass finds are not considered to represent 'deliberate and regular production of glass' (see also Roehrig 2005: 67; Nicholson and Henderson 2000: 195; section 1.2 to 1.4). It is possible, if not probable, that Nuzi glass was contemporary with Egyptian glasses from at least as early as the pre-Malkata period, to be addressed in the following discussion.

Vandiver (1983: 239) argued that the glass found at Nuzi must have been the product of an established technological tradition, and does not represent the

[38] For further information on the beads from the reigns of Ahmose to Hatshepsut see Roehrig (2005: 67).

early stages of a new technology. The consistent composition of Nuzi and other 2nd millennium BC Near Eastern glasses, and the existence of earlier and sporadic glass finds, all support the observation that this glass-making technology had its origins in earlier times, the products of which are not well preserved (see Moorey 1999: 190-192; section 1.3). Early glass of the mid 16th C BC is reported from Tell Atchana (Alalakh) but the contexts and dating are unreliable (Moorey 1999: 193, 195; Gates 1981; McClellan 1989).

It is estimated that over 16,000 beads of glass, frit and stone, were scattered over the floor of the temple, particularly Temple A (Figure 3-3). They came in a large range of sizes and forms, including spherical, flattened spherical (i.e. the diameter exceeds the height slightly but not enough to be described as ring or annular in shape), cylindrical, elliptical, fluted, ribbed, flat and rectangular amongst many other variations 'blue-green' was the most common colour, followed by white, yellow and black. Stone beads were relatively scarce. There were also round, flat mould-formed pendants or plaques of glass, and core-formed vessel fragments, from both the temple and the palace. The vessels came in various shapes, and were primarily blue with thread decoration in other colours. Only blue and green glasses have been sampled in the present study. Blue frit objects and mostly blue-green glazed pottery were common (Starr 1939: 92-4, 445-452, 457, plate 120; Barag 1970: 135-141, 191-192; McGovern 1980: 220).

The temple is ascribed to the goddess Ishtar (Astarte) based on finds of statuary in Room G29, including eyes with irises of green glass, and green-glazed terracotta. It is impossible to tell whether G29, the cella (the central room of the temple), and temple furnishings belonged to the latest period (Temple A of Stratum II), or whether they accumulated from the time of Temple C when the cella achieved its final state (Starr 1939: 110). Room G29 remained architecturally unchanged through several iterations of the temple, indicating that G29 was important and that the objects found here were probably accumulated throughout the time of Temples A, B and C, being consequently highly valued and originating earlier than Stratum II and the final destruction (Starr 1939: 88-90).

Starr (1939: 92-3) discussed the use of glass eye beads as architectural decoration found within G29, suggesting that the majority of beads found here were strung and hung on glazed wall nails that studded the interior to decorate the room, rather than being offerings for the goddess, Ishtar. Indeed, the mortar in which the paving of G29 was laid contained several hundred glass beads, and the chapel wall within the palace (L6-L10 wall) contained small glass objects. In this way these objects were deliberately incorporated into the fabric of the buildings (Starr 1939: 109-110) and presumably must have dated earlier than the destruction. Barag (1970: 136) believed from his study of the glasses from Nuzi that some of the glass may have been old when the city was destroyed.

The pavement of G50 in the preceding Temples B and C revealed many glass, frit and stone beads, as in Temple A. Temple C dates to the major period of Stratum III, with Temple B being present during the change from Stratum III to Stratum II (Starr 1939: 62). Considering this further, Starr formed the opinion that these objects are largely the products of the latest phase of Temple A. As beads, some of which are analysed in the present study, were found on the pavement of G50, unspecified as Temple A, B or C, it seems reasonable to acknowledge a potentially older date of manufacture for these. or the beads from G29, which might date to the latest phase, Temple A, there is the possibility that, as architectural decoration, they could date back to the time of Temple C when the last structural change was made to the cella. The beads that were incorporated in the paving must also date from earlier than the destruction. It is reasonable to consider that, with the evidence presented by Starr, the glass from the latest temple might have been handed down from an earlier time as heirlooms, perhaps of religious significance, contrary to Stein's (1989: 54) proposition that these are 'delicate objects with a restricted lifetime of no more than one generation'.

Mixed and strung together with the glass beads in G29, were large numbers of small seashells, three species of which were found only in the Mediterranean Sea. Others came from the Indo-Pacific, most likely the Persian Gulf. These finds indicate that there was trade between Nuzi and these regions, and are important because they are found only in the temple, highlighting their cultic significance. This forms part of the evidence for trade across the Hurrian empire with (Syrian) coastal regions to the west (Starr 1939: 94, 489; see also Akkermans and Schwarz 2003: 354).

In summary, most of the Nuzi glass was found in the temple, palace and environs. While the temple underwent several iterations, the incorporation of glass beads into the fabric of the building indicates the possibility that the glass beads date from some time earlier than the currently accepted destruction date of the mid 14th century BC, making them earlier than glass from Amarna and possibly contemporary with the Malkata and pre-Malkata glasses, and hence they are some of the earliest 'mass-produced' glasses known.

3.1.3 Summary of Samples

For the present study, samples of 16 opaque and translucent blue and translucent green glasses were

Figure 3-3 Plan of Temple A and immediate environs. Most of the glass was found in G29 and G50. Adapted from Starr (1939), Plan 13.

Table 3-1 Description and provenance of Nuzi glass samples.*

Sample Number	Analysis Number	Object	Colour	T or O	Provenance	Starr (1939) Ref.
1930.66.90  Figure 3–4 Nuzi 1930.66.90 half turquoise cylinder bead (diameter 12 mm)	16	bead cylinder	turquoise	O	area of Temple	
1930.67.45	17	pendant disk	dark blue	T	Room G50 in Temple	Plans 8,11,13
1930.67.94	18	pendant disk	green	T	area of Temple	
1930.68.1	19	pendant disk	turquoise	O	area of Temple	
1930.68.15	20	bead spacer	dark blue	T	Room G50 in Temple	Plans 8,11,13
1930.68.15	21	bead cylinder	dark blue	T	Room G50 in Temple	Plans 8,11,13
1930.68.26	22	bead cylinder	dark blue	T	Room G64 to SW of Temple	Plans 10,11,13
1930.68.27	23	vessel or pendant fragment	green	T	Room G29 in Temple	Plans 8,11,13
1930.68.27	26	bead cylinder	dark blue	T	Room G29 in Temple	Plans 8,11,13
1930.69.39	27	bead	green	T	Room F24 to W of Temple	Plan 13
1930.82.10  Figure 3-5 Nuzi 1930.82.10 blue translucent vessel fragment (thickness 3 mm)	28	vessel	dark blue	T	probably outside Temple	
1930.82.15	29	vessel	dark blue	T	Room M79 in Palace, SE of Temple	Plan 13
1930.82.17	30	vessel	dark blue	T	Room C22 central mound to N of Temple	Plan 13

Sample Number	Analysis Number	Object	Colour	T or O	Provenance	Starr (1939) Ref.
1930.82.17	31	pendant - disk	dark blue	T	Room C22 central mound to N of Temple	Plan 13
1930.82.48	32	pendant - disk	dark blue	T	Probably area C central mound to N of Temple	Plan 13
1930.82.50	33	fragments	dark blue	T	Room C62 central mound to N of Temple	Plan 13
1930.82.55  Figure 3–6 Nuzi 1930.82.55 blue translucent vessel? fragment (thickness ~4 mm)	34	vessel?	dark blue	T	Room C87/C100 central mound to N of Temple	Plan 13

*The description 'turquoise' is reserved for opaque blue glasses. Translucent glasses are described as blue or green. Sample numbers are Harvard Semitic Museum registration numbers.

collected from the Harvard Semitic Museum holdings of Nuzi glass, for chemical analysis by EPMA and solution ICP-MS (sections 2.3, 2.4). The identification, provenance and description of the glasses appear in Table 3-1. Seven cylindrical, ribbed spacer, and spherical beads, five or possibly six pendants of flat circular or disk shape with a pierced lug for suspension, four to five vessels and unidentified fragments were sampled.[39] All glasses were found in the temple or in buildings surrounding the temple. The blue and green glasses were generally in poor condition, but those sampled contained fresh glass beneath their weathered surfaces. The freshest glass was sampled.

3.2 Base Glass Composition

The Nuzi glasses were analysed for the reduced or base glass oxide composition using EPMA (section 2.3.4). This analysis defined a coherent compositional group of soda-lime-silica glass containing 64% to 74% (average 68%) silica, and 14.5% to nearly 18% (average 16%) soda. Soda is the second most abundant oxide in these and all soda-lime-silica glasses. The average reduced composition of the Nuzi glasses is presented in Table 3-2. Complete raw oxide and trace element data for each sample are tabulated in Appendix II.

The Nuzi glass contains high-magnesia, high-potash concentrations (HMHK) consistent with the use of a plant ash flux, typical of the LBA, or 2nd millennium BC, Near Eastern HMHK plant ash glasses as defined by Sayre and Smith (1961: 1824-1826; section 1.5).

The level of silica in the Nuzi glasses is comparable to the level found in the Upper Continental Crust (UCC), as represented by MUQ (section 2.5.5). As might be expected from a material based largely on silica, alumina and iron oxide concentrations are much lower than those found in MUQ. The plant ash-related oxides magnesia, potash, lime and especially soda are enriched relative to the natural content of the earth's crust, reflecting their deliberate addition in the form of plant ash.

The Nuzi glasses have levels of magnesia and potash elevated on average to 4.4% and 2.9% respectively, with ranges of approximately 3% to 6% and 2% to 5% each. The two oxides are positively correlated (r^2 = 0.69) suggesting that they were introduced into the glass batch together, probably as part of the plant ash (Figure 3-7). A positive correlation between potash and magnesia is often found in ancient plant-ash glasses (Barkoudah and Henderson 2006: 315, 319; section 1.3). The level of phosphorus at an average of 0.17% (range of 0.11% to 0.24%) is slightly lower than that commonly expected in a plant-ash glass. Four of the 16 glass samples contain in excess of 0.2% phosphorus, but this oxide does not appear to correlate with any others.

The average alumina concentration is 0.6% and the iron concentration is low at 0.3%,[40] suggesting that a

[39] The uncertainty about sample numbers is because of the uncertainty in identification of two pieces, one of which might be a vessel or a pendant (sample 23) and the other of which is likely to be a vessel (sample 34).

[40] In the calculation of this average, the value for sample 27 was

Table 3-2 Average reduced compositions of 16 Nuzi glasses in wt% oxide.*

Samples	Na2O*	MgO*	Al2O3*	SiO2*	P2O5*	K2O*	CaO*	FeO*	Total
Average n=16	16.30	4.38	0.62	68.00	0.17	2.91	7.32	0.28	100.00
SD	1.24	0.96	0.23	2.48	0.04	0.75	2.11	0.10	
Av. glass/ MUQ	20.38	2.79	0.04	1.01	0.94	1.89	6.65	0.03	

Note that FeO values are based on 15 samples excluding number 27 which has an exceptionally high iron oxide content.

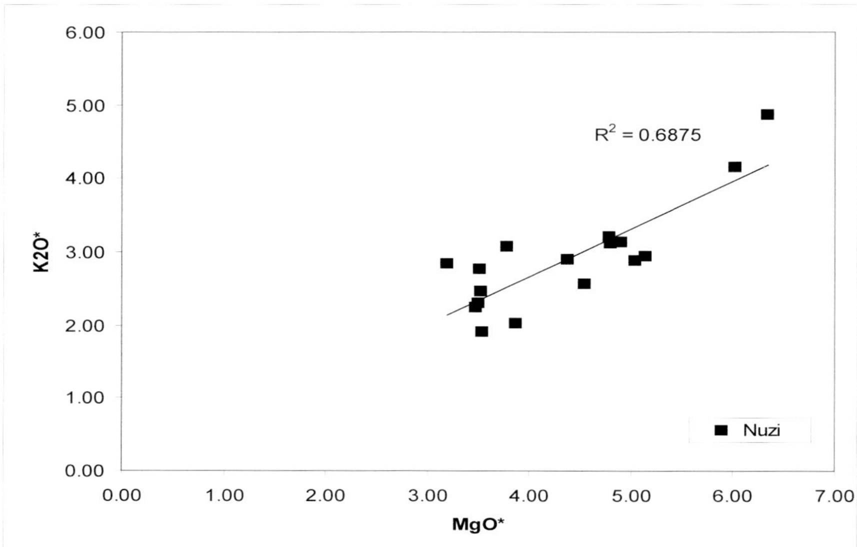

Figure 3-7 Scatter plot of magnesia vs potash in wt% oxide for Nuzi glasses.
* indicates reduced composition in this and all following plots.

relatively pure form of quartz, such as vein quartz or quartz pebbles, or a pure, or mature sand source has been used by the glass-makers. Quartz pebbles contribute essentially only silica to the glass. If a colourant-free glass is made using the simple two-component model of pure silica plus a plant ash soda flux, the remaining major and minor elements must be derived from the plant ash (Brill 1970a: 122; Lilyquist *et al.* 1993: 41-2; Freestone and Gorin-Rosen 1999: 112; Shortland 2005: 1-2; sections 1.5, 1.8), or be contaminants from some other source (Rehren and Pusch 2007; sections 1.6 to 1.9). The concentration of silica does not appear to correlate with that of other oxides and elements, which is likely to reflect its purity (section 1.6).

Iron oxide and alumina are well correlated (r^2 = 0.80) (Figure 3-8). Alumina and iron oxide are usually contaminants from soil minerals that are introduced with the silica or through other ingredients such as the plant ash (Barkoudah and Henderson, 2006: 313) (see REE sections 2.5.3, 3.3.1). The ratio of alumina to iron oxide concentration is 2.4:1 which closely approximates that found naturally in the UCC,[41] indicating that these oxides have probably entered the glass as soil impurities. Figure 3-8 shows iron oxide *versus* alumina; note that there is a group of glasses, including the green ones, with higher concentrations of alumina and iron oxide, which incorporated more of the contributing ingredient into the batch.

These glasses have high lime levels averaging a little over 7% (range of 4.5% to 11%), typical of plant ash glass, although the lime does not correlate with the other plant ash related oxides.

Examination of the bivariate plot of lime *versus* magnesia (Figure 3-9) shows that the glasses may be divided into two groups based mainly on lime concentration. The higher lime group has calcium oxide concentrations in excess of 8.5% (average 9.4%, range of 8.5% to 11.5%), while the other glasses form a separate group of lower

omitted as it was inexplicably high. This sample was slightly weathered and one measurement rather than four were taken so the precision was not as reliable.

41 Calculated from data presented in McLennan (1989).

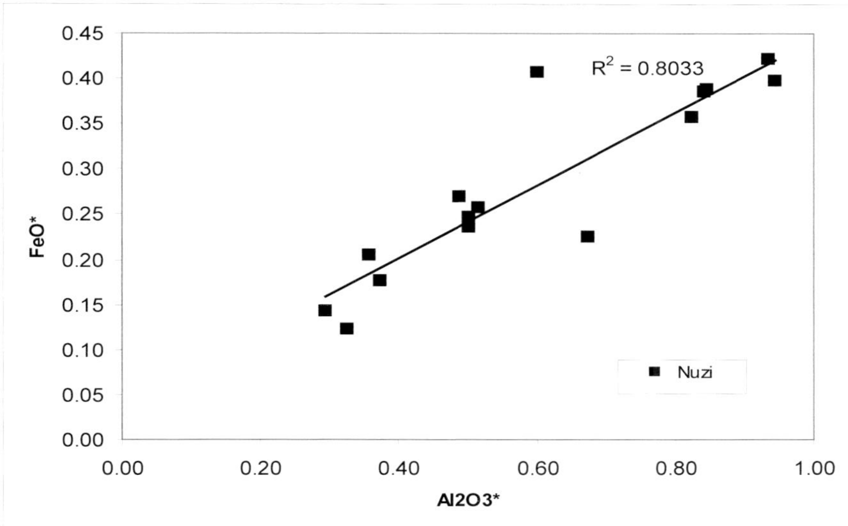

Figure 3-8 Scatter plot of alumina *vs* iron in wt% oxide for Nuzi glasses.

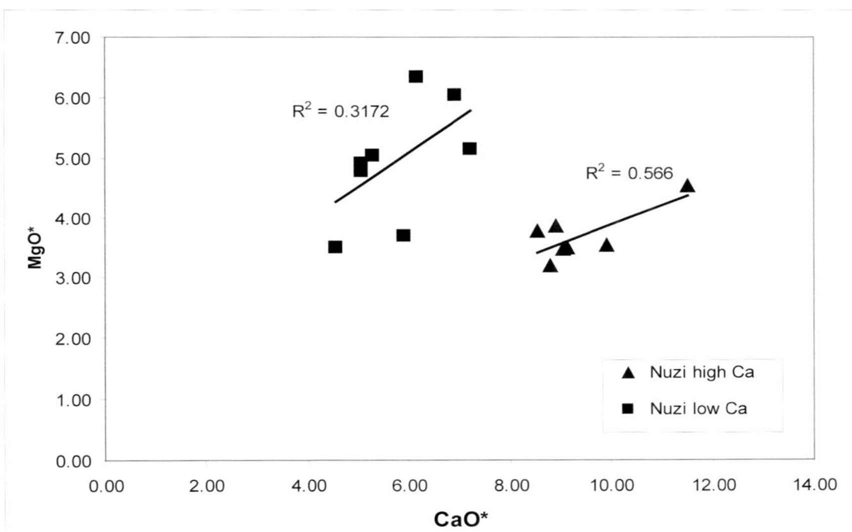

Figure 3-9 Scatter plot of lime *vs* magnesia in wt% oxide for Nuzi glasses.

lime concentration (average 5.7%, range of 4.6% to 7.2%). It is only when the glasses are examined as two lime-based groups that the lime concentrations appear to correlate positively with magnesia, although the correlation in the low-lime group is weak (low-lime r^2 = 0.32, high-lime r^2 = 0.57). As seen in Figure 3-9, however, the correlation in the high-lime group is dependent on a single sample and as such might be misleading. There is no correlation between these oxides when the glasses are considered as one group.

When the glasses are examined as two lime-based groups (Figure 3-10), there is no correlation between magnesia and potash in the high-lime group, but a strong positive correlation is found in the low lime-group (r^2 = 0.75), which is similar to that described as possible for plant ashes of the *Salsola* type grown in limestone environments (Barkoudah and Henderson 2006: 311-312, 319). This strong correlation in the low-lime group is responsible for the correlation of the concentrations of magnesia and potash observed in Figure 3-7 for all glasses when grouped together.

The question arises as to whether the high-lime glasses are low-lime glasses which had extra lime added to them, or whether a different plant ash or combination of ashes was used to make the glasses in this group. This issue will be discussed in section 8.4.

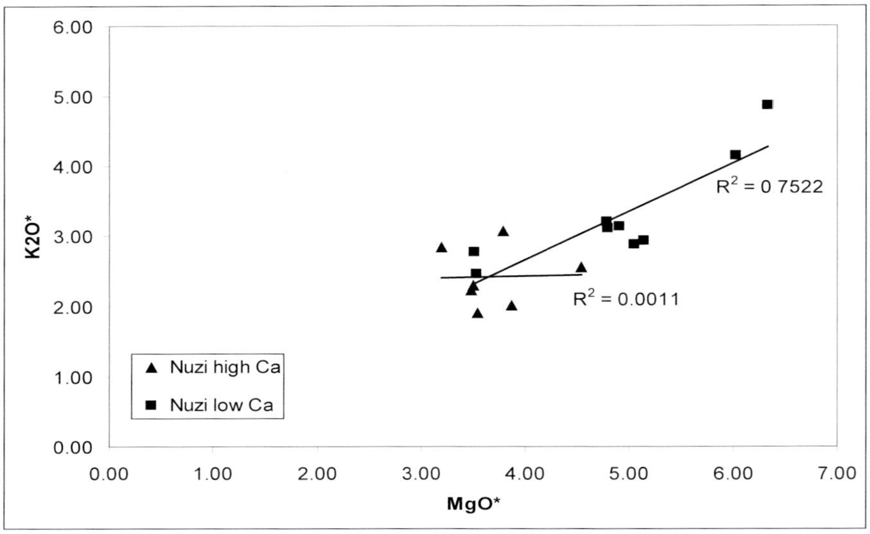

Figure 3-10 Scatter plot of magnesia vs potash in wt% oxide for high- and low-lime groups of Nuzi glass showing a powerful positive correlation in the low-lime group and no correlation in the high-lime group.

Table 3-3. Average reduced compositions of low- and high-lime Nuzi glasses in wt% oxide.*

	Na_2O*	MgO*	Al_2O_3*	SiO_2*	P_2O_5*	K_2O*	CaO*	FeO*	Total
Low CaO *n=9*	16.54	4.92	0.65	68.47	0.18	3.29	5.70	0.29	100.00
High CaO *n=7*	16.01	3.70	0.58	67.46	0.16	2.42	9.40	0.27	100.00

Note that FeO values for the low-lime group are based on eight samples excluding sample 27 which has an exceptionally high iron oxide content.

The only two opaque glasses analysed from Nuzi fall into the high-lime group. One explanation could be that a high lime content was needed to crystallise the opacifying compound calcium antimonate ($Ca_2Sb_2O_7$) (see Figures 2-1, 2-2, 2-3; section 1.10.2), although it is possible to opacify glasses with lower lime contents (section 8.5). The high-lime group also includes five translucent glasses where elevated lime concentrations are not associated with opacification or elevated Sb contents.

The average reduced data for the low and high lime groups are presented in Table 3-3. Examination of these data indicates that while there is little difference in the concentrations of most oxides between the low- and high-lime groups, both magnesia and potash are elevated on average in the low-lime glasses. Soda levels are slightly elevated while phosphorus, alumina and iron oxide levels remain relatively constant. The significance of these differences and similarities will be discussed in section 8.4.4.

3.2.1 Summary

The Nuzi glasses may be described as typical Near Eastern plant ash type or HMHK glasses, containing an average magnesia concentration of 4.4%, average potash concentration of 2.9% and average phosphorus of 0.2%. The Nuzi glasses are probably made with quartz pebbles as the data suggest a pure silica source, with an average alumina concentration of 0.6% and average iron concentration of 0.3%. There are two groups of glass based on lime concentration. These findings show that overall the Nuzi glasses are of relatively similar base glass composition, which suggests that they were made from similar raw materials to a similar recipe. The significance of the two sub-groups with low- and high-lime contents is explored in section 8.4.

3.3 Trace Element Composition

Trace elements were investigated through analysis with solution ICP-MS and the results are presented below following the protocol established in section 2.5.

3.3.1 Rare Earth Elements (REE)

Averaged absolute concentrations of REE in the Nuzi glasses are presented in Table 3-4. When normalised to the average continental crust (MUQ) as shown in Figure 3-11, REE patterns are reasonably homogeneous and, when averaged (Figure 3-12) are relatively flat. There

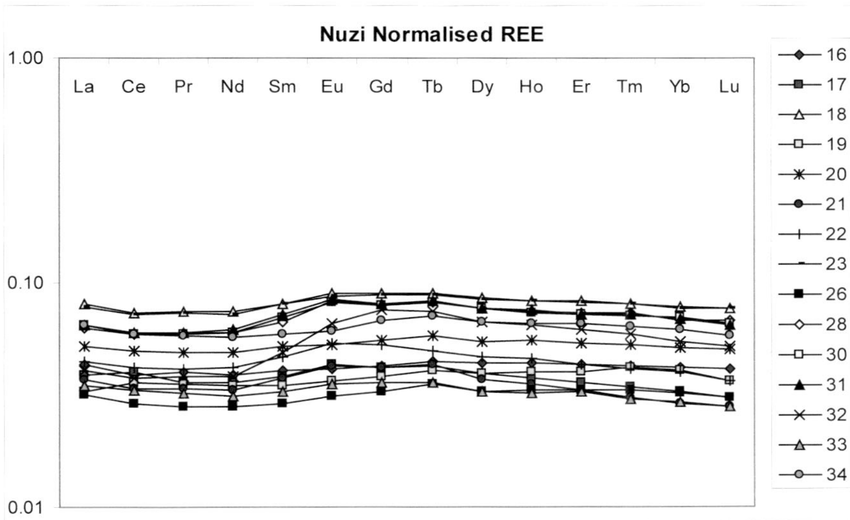

Figure 3-11 Line plot of the REE of individual Nuzi glasses (sample numbers given) showing similarity of pattern. A logarithmic scale was used. Data are normalised to MUQ.

are slight variations within the group (for example, particularly sample 32), but they do not appear significant enough to constitute different geochemical groups. Slight variations in REE pattern and variations in the absolute concentration might reflect natural variations in raw materials, or intrinsic variations in the glass batch, perhaps as a result of contamination introduced during the preparation and mixing of the raw ingredients (sections 1.6 to 1.9).[42]

The REE do not appear to be entering the glass mix with the lime, the Cu colourant or Sb opacifier, as there is a lack of correlation between the REE and these ingredients. The REE are strongly correlated ($r^2 \geq 0.7$) with the following SRE: Al, Fe, Be, Ti, V, Cr, Ga, Zr, Nb, Mo, Ba, Hf, Ta, W, Th and U. It is noted that there are no apparent correlations between the REE and the plant ash related elements K, Mg, Ca, Li, Rb, Sr and Cs.

The correlation of the REE with alumina and iron oxide and the other sediment-related elements (SRE: section 3.3.2) indicates that they entered the glass with the sediment or soil contaminant already inferred (see above). Despite the lack of correlation between the concentrations of the REE with plant ash related alkalis and alkali earths, there is a possibility that the REE entered with the plant ash, because the plants could have had soil attached to them when harvested. In such circumstances, there would be no correlation with the alkalis as the amount of soil in a batch of plant ash could be arbitrary. In addition, REE could have been added as contamination during the ashing process if it were carried out in contact with REE containing materials like soil (Shortland 2005: 2).[43] Different parts of the plants could have had different total REE contents,

Table 3-4 Average REE for Nuzi glasses in ppb.

Nuzi Rare Earth Elements (ppb)

n=15

	Averages	RSD	Average glass/ MUQ
La	1903	36	0.05
Ce	3419	30	0.05
Pr	404	31	0.05
Nd	1567	32	0.05
Sm	361	33	0.05
Eu	93	36	0.06
Gd	381	34	0.06
Tb	61	34	0.06
Dy	337	34	0.06
Ho	68	34	0.06
Er	185	34	0.06
Tm	27	35	0.06
Yb	167	34	0.05
Lu	24	36	0.05

[42] See also Freestone and Gorin-Rosen (1999: 111-112) regarding minor variations in raw material composition.

[43] Turner (1954: 441) proposed the contamination of glasses by the melting of the crucible during firing which Saleh *et al.* (1972: 143-172) discussed further in their study of glass-making at Wadi el-Natrun. Shortland (2005: 1-2, 4) and Henderson (Nicholson and Henderson 2000: 217) conclude that this phenomenon is unlikely.

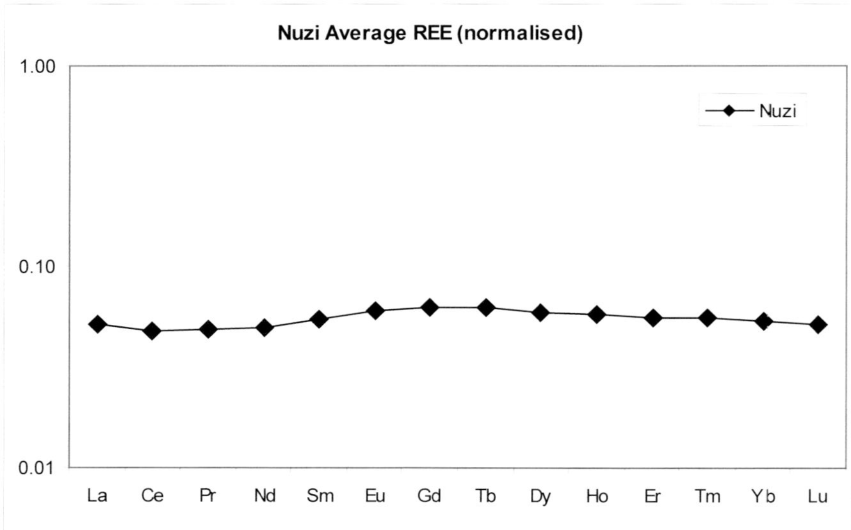

Figure 3-12 Line plot of averaged REE from Nuzi glass. Data are normalised to MUQ. A logarithmic scale is used.

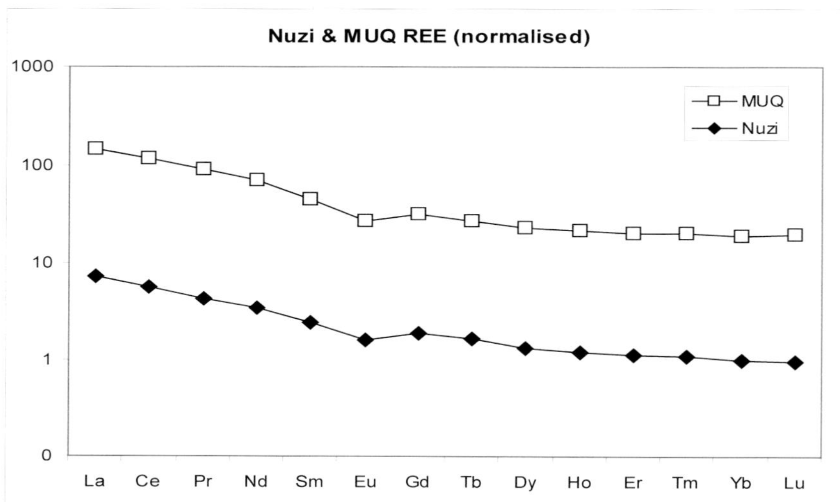

Figure 3-13 Line plot of averaged Nuzi REE data and MUQ REE data both normalised to average chondrites for comparison (chondrite data from Sun and McDonough (1989), reproduced in Table 27 of Normalisers)

though it is likely that the whole plant was ashed (see Brill 1979: *The Glass Makers of Herat* video; Barkoudah and Henderson 2006: 298).

Rehren and Pusch (2007: 223-225) have shown that the grinding of quartz pebbles to prepare them for mixing into the glass batch might add up to 2wt% of elements including Al, Fe, K, Ca, Ti, Sr, Zr, Ba and transition elements derived from grinding tools such as were found at Qantir in Egypt. The known association of the REE with alumina, iron oxide and the SRE suggest that the concentrations of the REE might be increased also during the quartz grinding process (section 1.6).

There is greater variation amongst the light REE with samples 28, 30, 31 and 32 having similar LREE patterns, but with lower LREE concentrations. There is generally a lower concentration of LREE than of the high REE (HREE), with either no, or a very small, positive Eu anomaly relative to MUQ (Figure 3-13). Samples 17, 19 and 21 have small positive Ce anomalies.

Figure 3-13 compares MUQ and the average REE for glasses from Nuzi. As might be anticipated from the almost horizontal trends of the MUQ-normalised plots of the individual samples (Figure 3-13) there is a close similarity between the patterns of the REE from Nuzi

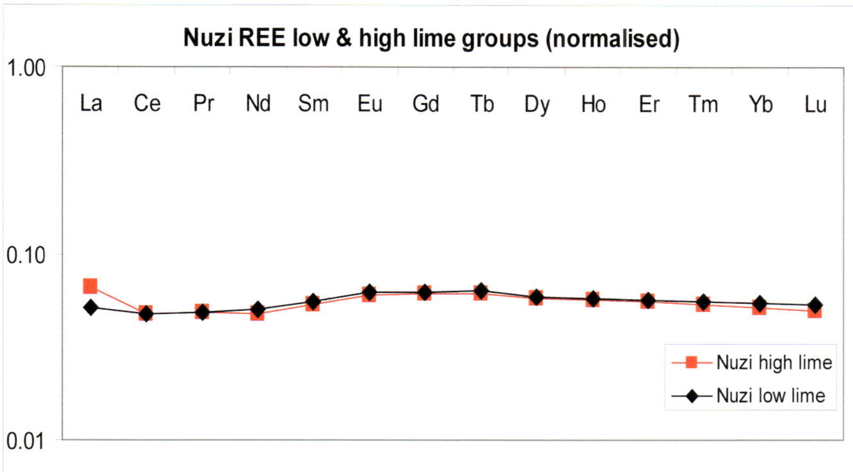

Figure 3-14 Line plot of the averaged REE normalised to MUQ for high- and low-lime groups.

glass and from the UCC, which is consistent with the derivation of the Nuzi REE from sediment, such as soil. Over all, the ratios of the REE in Nuzi glass to those in the UCC are within the range of 0.05 to 0.06 reflecting the pure nature of the silica source. It is assumed that this ratio is essentially a reflection of the amount of sedimentary material incorporated into the glass, and this is termed the *sedimentary background signal* (SBS).

In Figure 3-14 the REE of the high- and-low lime groups of the Nuzi glasses have been plotted for comparison. There is a slight difference in concentration of La between the two groups. The high-lime group, which contains the opacified glasses, has a higher La content. This may be due to an analytical problem in this particular run (section 2.4.2). As high concentrations of Sb may cause analytical error in the measurement of La, the elevated La concentrations in the high lime group may be due to interference from the high Sb in the opaque glasses. That La correlates with Sb, when the other REE do not, suggests that this is an artefact of the ICP-MS analysis.[44] Otherwise the patterns of the high and low lime groups are identical, showing that the REE are not associated with the lime or Sb sources as has been suggested by the lack of correlation between these elements.

In summary, the REE in Nuzi glasses exhibit a continental crustal pattern, though of lower concentration, due to the pure nature of the raw materials used to make the Nuzi glasses. The ratios of the concentrations of REE in the glasses to their concentration in the UCC are found to be from 0.05 to 0.06 establishing a constant base-line ratio or SBS with which to compare the ratios of concentrations of other trace elements found in the Nuzi glasses. The REE contents and patterns are the same for both the high and low lime groups of glass. The REE could have been introduced into the glasses as part of the plant ash, as contaminants of the plant ash or as contaminants of the otherwise pure silica source. The REE correlate with iron oxide and the SRE, supporting the proposition that they were all associated with sedimentary contaminants.

3.3.2 Sediment-Related Elements (SRE)

The concentrations of the SRE are reasonably homogeneous and are averaged and presented in Table 3-5. In the high-lime glasses, the concentrations of Y, Tl and Th are lower than in the low-lime glasses. Slight variations in SRE concentration are probably a result of the slightly varied proportions of raw ingredients used to make these glasses, or of the use of different tools for the preparation of the raw materials. Essentially they are the same kind of glass, exhibiting small internal differences that form subgroups such as the high- and low-lime groups.

Generally, the SRE are strongly positively correlated with each other and with alumina and iron oxide, Be, and the REE in the order of $r^2 = 0.8$ to 0.9. A weaker positive correlation is observed with manganese ($r^2 = 0.6$). The SRE are strongly negatively correlated with Cu, as are the REE, alumina, iron oxide and Ba in the order of $r^2 = -0.7$ to -0.8. The concentrations of manganese and soda are negatively correlated with Cu ($r^2 = -0.5$ to -0.6). These correlations confirm that the SRE and soda were not added with the Cu colourant. In their study of Nuzi glasses, Shortland and Eremin (2006: 590) found positive correlations between the SRE they analysed: Al, Ti, V, Fe, Y, Zr, Nb, and the REE (section 3.3.1). They noted that the presence of insoluble elements such as

[44] The opacified samples had very high Sb contents and the data had to be corrected for Sb oxides. The largest correction was for La and a slight error in the correction will have a significant impact on the La content.

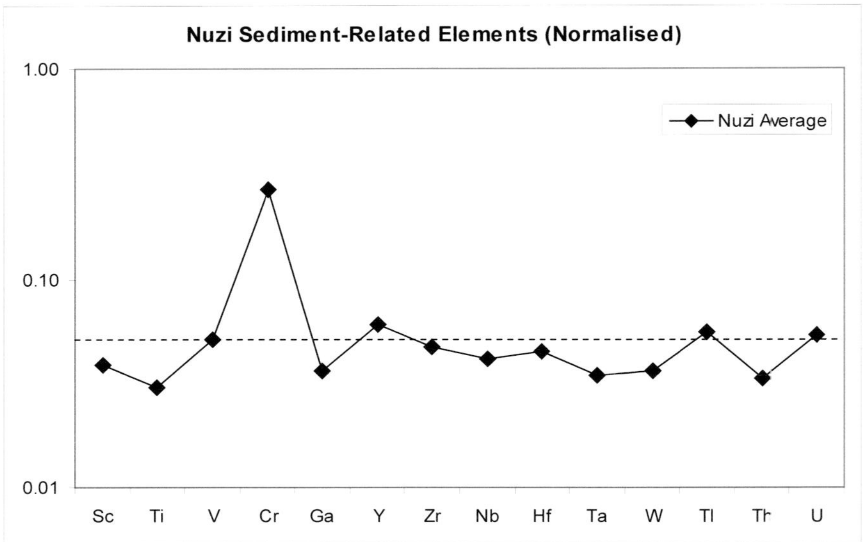

Figure 3-15 Line plot of the averaged concentrations of sediment-related trace elements in Nuzi glasses, normalised to MUQ. A logarithmic scale is used. Mo is not plotted as there is no MUQ value by which to normalise it. The dotted line crosses the Y-axis at 0.05 to represent the SBS of the Nuzi glasses as inferred from the REE concentrations.

Table 3-5. Averaged compositions in ppb of sediment-related elements from Nuzi glasses.

Nuzi Sediment-Related Elements (ppb)

n=15

	Average	RSD	Average glass/MUQ
Sc	634	83	0.04
Ti	208,763	38	0.03
V	6131	38	0.05
Cr	17,247	49	0.27
Ga	693	37	0.04
Y	1915	34	0.06
Zr	9404	37	0.05
Nb	634	42	0.04
Mo	531	48	
Hf	243	38	0.05
Ta	39	45	0.03
W	40	75	0.03
Tl	25	118	0.06
Th	381	37	0.03
U	154	33	0.05

Zr makes it unlikely that the SRE come from the plant ash, but could be introduced either accidentally or deliberately from other sources such as clay, sand or rock powder.

The normalised concentrations of the SRE are represented in Figure 3-15 which shows in particular that Cr is the only element significantly enriched above the SBS, with a ratio to its concentration in MUQ of 0.27.

The ratios of the concentrations in Nuzi glass to those in MUQ for the elements V, Y, Zr, Hf, Th and U are in the range of 0.05 to 0.06, at the SBS level defined by the REE. Those elements with values which fall below the SBS of 0.05 to 0.06 are Sc, Ti, Ga, Nb, Ta, W and Th, and alumina and of iron oxide, which range from 0.03 to 0.04. Even so, all of the SRE and REE, except Cr, show closely similar concentration ratios to MUQ, supporting the view that the glasses can be seen to have incorporated approximately 0.05 (5%) sedimentary material.

Figure 3-16 is a line plot illustrating the average concentrations of SRE in Nuzi glasses compared with those in MUQ. Here, again, the patterns are similar with slight variations being due to the relative enrichment of Cr and slight relative depletions of some of the other SRE.

The SRE concentrations in the glasses from both the high- and low-lime groups, identified from base glass oxide analysis, are plotted together in Figure 3-17 showing that as was found with the REE, the division into groups based on lime concentration is not manifested in the concentrations of the SRE.

3.3.3 Alkali and Alkaline Earth Elements

The Nuzi glasses are fluxed with plant ash. The alkali and alkaline earths magnesia, potash, soda, lime

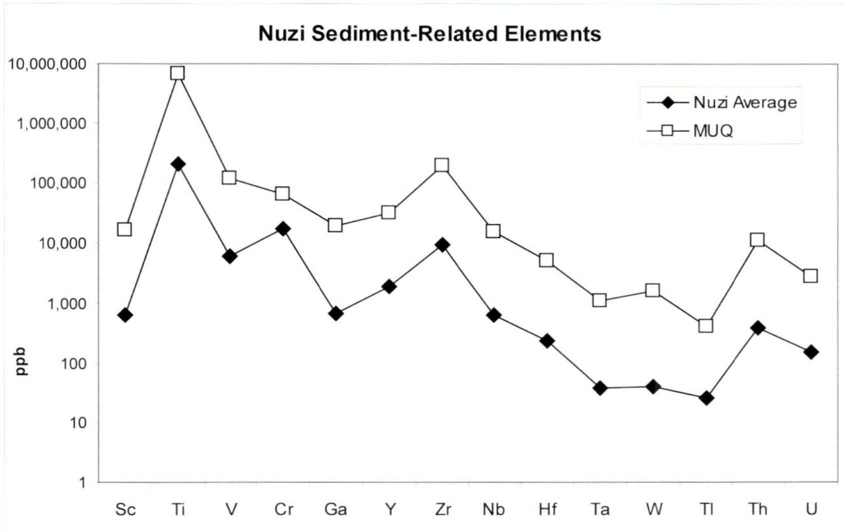

Figure 3-16 Line plot of raw Nuzi sediment-related trace element data in ppb compared with MUQ on a logarithmic scale.

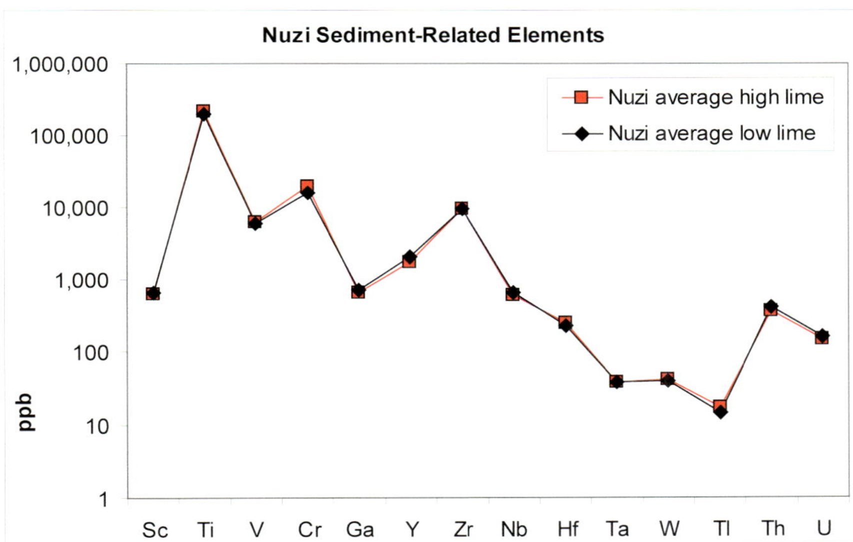

Figure 3-17 Line plot comparing averaged raw concentrations of sediment-related trace elements in the high and low lime groups. A logarithmic scale is used. Calculation of averages for W omits sample 28 and for Tl omits sample 26 because they are outliers.

and phosphorus are constituents of plant ashes, and have been reported in section 3.2. Other alkali and alkaline earth elements, present in trace element concentrations and assayed for by ICP-MS in the present study, are Li, Be, Rb, Sr, Cs and Ba. As the elements in the alkali and alkaline earth groups of the periodic table are chemically related, it is possible that they might be present together in plant ash as was shown by Barkoudah and Henderson (2006: 306-311). The averaged results of these analyses are presented in Table 3-6.

In Figure 3-18 the averaged alkali and alkaline earth elements of the Nuzi glasses are compared with MUQ. This comparison shows that, apart from elemental enrichment of Li and Sr in particular, the patterns are closely similar, with the concentrations of elements in the Nuzi glasses being depleted compared to MUQ. This phenomenon was observed in the REE and SRE patterns and concentrations.

When the averaged concentrations of alkali and alkaline earth elements are normalised to MUQ, the resultant

Table 3-6. Average compositions in ppb of alkali and alkaline earth elements from Nuzi glasses.

Nuzi Alkali and Alkaline Earth Elements (ppb)							
		Li	**Be**	**Rb**	**Sr**	**Cs**	**Ba**
n=15	**Average**	20,887	107	12,826	393,723	418	38,819
	RSD	39	37	34	18	55	32
Average glass/MUQ		0.74	0.06	0.16	2.78	0.08	0.10

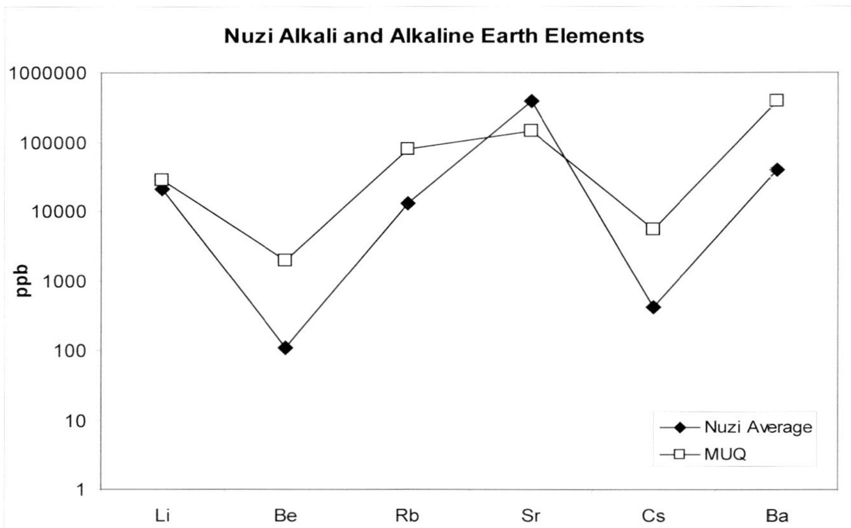

Figure 3-18 Line plot of averaged raw data for Nuzi alkali and alkaline earth elements compared with MUQ. Note logarithmic scale.

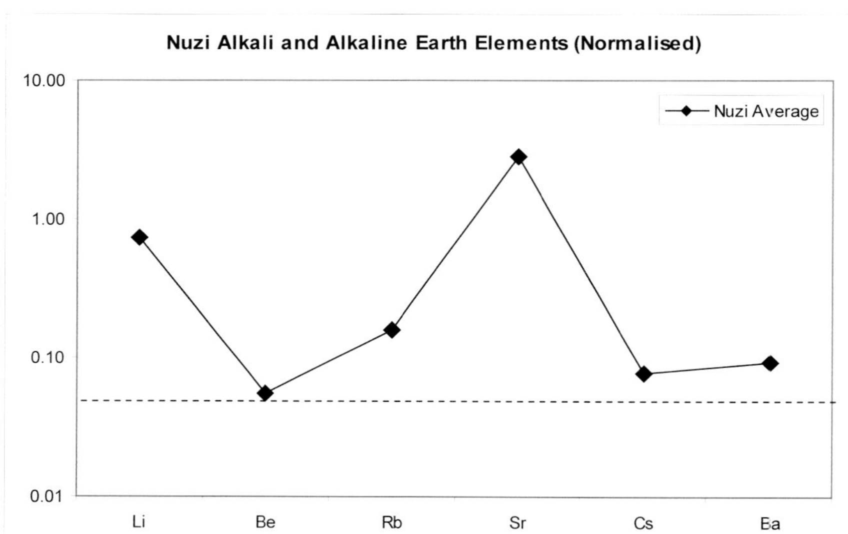

Figure 3-19 Line plot of averaged Nuzi alkali and alkaline earth trace elements normalised to MUQ, on a logarithmic scale. The dotted line represents the approximate sedimentary background signal as determined by the ratios of Nuzi REE to MUQ REE, at 0.05.

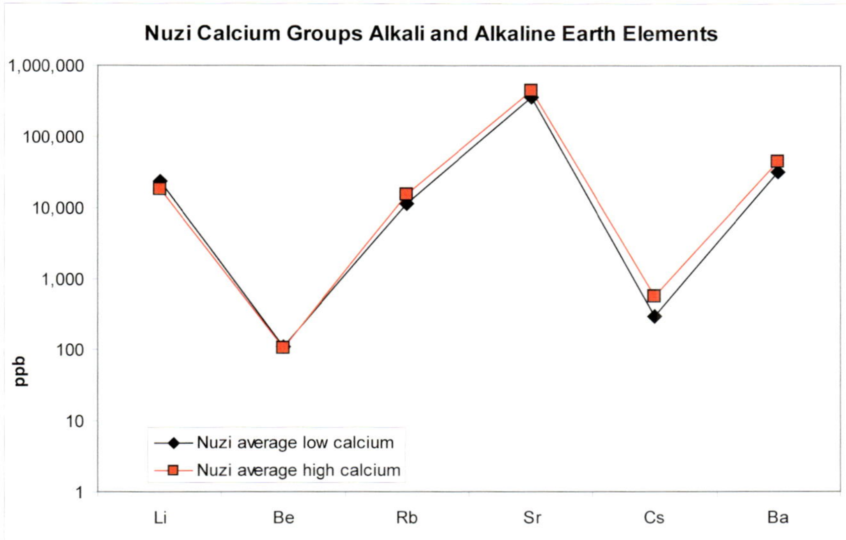

Figure 3-20 Line plot of averaged raw data for low- and high-lime groups of Nuzi glass. A logarithmic scale is used.

ratios indicate that all elements are enriched in the Nuzi glasses compared with average concentrations in the UCC, with the exception of Be (Figure 3-19). The ratio of Be to MUQ is 0.06, indicating that this element was present at the SBS level. Together with the strong correlation of Be with the SRE, it is likely that, for Nuzi glasses, this element was not incorporated in the plant ash flux.

The alkali and alkaline earth element data are consistent with the flux having been derived from plant ashes. Sr and Li are particularly enriched, having ratios to MUQ of 2.78 and 0.74 respectively. The concentrations of Rb, Cs and Ba are elevated to a lesser degree (between 0.08 and 0.16 relative to the SBS of 0.06). Ba correlates with the SRE and the REE ($r2 \approx 0.7$), rather than with the alkali and alkaline earth elements, and hence a substantial Ba contribution may come from the sedimentary contaminant, rather than the flux.

Figure 3-20 illustrates that the high- and low-lime groups have closely similar concentrations of the alkali and alkaline earth elements, with the low-lime group having slightly lower contents of most of these elements. This could be explained by the association of these elements with plant ashes. The relationship between the plant ash elements in these glasses is supported by the moderate positive correlations that magnesia and potash have with Li (r^2 = 0.5 to 0.6), and lime has with Sr and Cs (r^2 = 0.7), and weakly with Rb (r^2 = 0.4).

3.3.4 Colourants and Opacifiers

Most of the Nuzi glasses analysed are translucent blue, with two being light or turquoise blue opaque, and three green translucent. Results of ICP-MS analyses of colourant and opacifying elements, and those elements associated with them, are presented in Table 3-7. These elements include Mn (reported as MnO), Co, Ni, Cu, Zn, Cd, Sn, Sb and Pb.

Colourant-related trace element concentrations from opaque and translucent Nuzi glasses are compared in Figure 3-21 to show the differences in concentrations of Ni, Sn, Sb and Pb between the two groups and compared with MUQ.

The colour of blue glass is usually achieved by the addition of Cu or Co. In these glasses, CuO is present at between 0.2% and 1.5%, average 1.2%. The higher concentrations would impart the blue colour. The average concentration of elemental Cu in the blue samples is 9160 ppm. For the green glasses (18 and 23), the Cu concentration averages 1871 ppm (0.18% to 0.2% CuO), not enough to impart the green colour.

The Co contents, at just a few ppm, are well below the levels needed to colour glass. Both opaque and translucent glasses have low average Co levels of 3 ppm to 4 ppm, with a range in the translucent glasses of approximately 1.5 ppm to 6 ppm, reflecting natural variation in the raw material which carries it, and resulting in an RSD value of 41.

While Cu was added to the glass batch to produce the blue colour, the green is probably produced by iron. An average of 0.4% iron oxide in glasses can impart a green or blue tinge under certain oxidation conditions (section 1.10). Several of the Nuzi glasses, both green and blue translucent, have iron oxide concentrations at or approaching this level. All three green glasses have

Table 3-7 Averaged compositions of colourant and related elements from Nuzi glasses, presented in ppm.*

	Nuzi Colourant and Related Elements (ppm)									
Opaque		MnO wt%	Co	Ni	Cu	Zn	Cd	Sn	Sb	Pb
n=2	Average	0.04	4.06	35.7	8582	17.6	0.056	48.2	10,979	84.3
	RSD	20	18	30	3	13	29	116	23	25
	Translucent				*n=11*					
n=13	Average	0.04	3.06	13.3	9265	16.2	0.050	7.05	9.84	5.84
	RSD	36	41	19	25	12	48	66	121	47
Av. opaque/MUQ		0.25	0.18	1.13	265.19	0.24		17.38		4.13
Av. translucent/MUQ		0.25	0.14	0.42	251.15	0.22		2.55		0.29

*MnO is determined by EPMA and presented in wt%. Sample 27 is omitted as it was analysed once rather than multiple times, is slightly weathered, so its data may be less reliable.
Note there are no values for Sb or Cd in MUQ. Data for opacified and translucent glasses have been averaged separately due to significant differences in the concentrations of several key elements. The Cu average for translucent glasses omits the two green glasses with very low Cu levels.

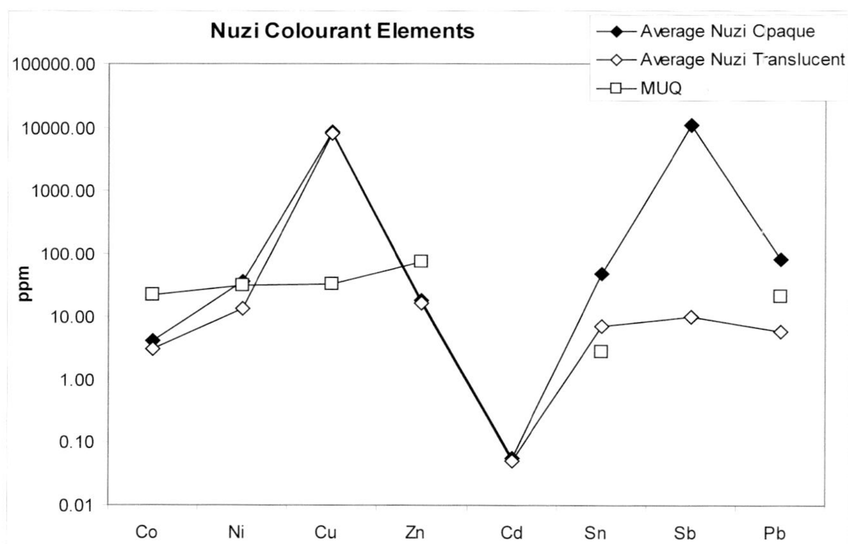

Figure 3-21 Line plot of averaged raw colourant and related elements in Nuzi opaque and translucent glasses, and MUQ, from ICP-MS data. A logarithmic scale is used to avoid compression. There is no MUQ value for Sb and Cd.

high iron oxide concentrations (at least 0.4%) which is likely to be contributing to the colour. The blue translucent glasses, 20, 28, 30 and 31, similarly have high iron oxide levels, without the relatively low Cu concentration observed in two of the green glasses.

Sb was added to translucent glass to combine with calcium, to form the opacifier calcium antimonate (sections 1.10.2, 2.3.2). Calcium antimonate was identified as the opacifier for opaque turquoise blue glasses (samples 16 and 19). The differences between the opaque and translucent glasses occur in the concentrations of Sb, Ni and Pb, all of which are found in greater quantity in the opacified glasses. The two opaque glasses have between 9200 ppm and 12,800 ppm Sb, while the translucent glasses have an average of 7 ppm with a range of 1 ppm to 38 ppm, giving a large RSD value. Zn is present at an average of 16 ppm to 17 ppm in all glasses.

The two glasses opacified with Sb have a large variation in Sn content (8 ppm and 88 ppm), while three of

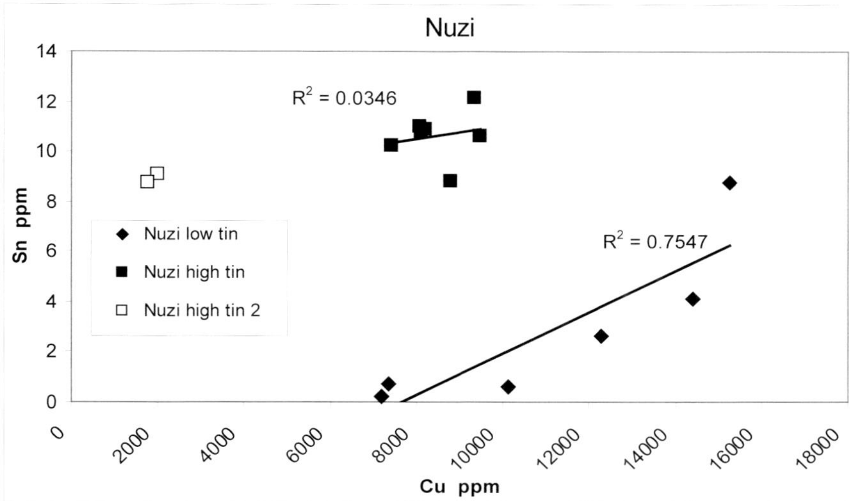

Figure 3-22 Scatter plot of Cu vs Sn in ppm for Nuzi glasses, excluding sample 16 which has an exceptionally high Sn content.

the translucent glasses (samples 20, 21 and 22) have Sn contents of 2 ppm to 7 ppm, with the remaining samples having Sn concentrations at levels an order of magnitude greater. The ratios of Sn to Cu concentration are much lower than would be expected if bronze scrap were used as the Cu source. Sample 16 has a Sn:Cu ratio of 1:100, while the rest of the glasses analysed have a ratio an order of magnitude smaller. This does not support the addition of Sn in the form of bronze, which would have a ratio closer to 1:10 of Sn:Cu (section 1.10).

Figure 3-22 is a bivariate plot of Cu *versus* Sn, showing a range of concentrations in both elements. There appear to be two groups of glasses, excluding the two low Cu, high Sn green glasses. These groups do not correspond with the high- and low-lime groups. The two opaque glasses are part of the high Sn group, along with six low-lime translucent glasses. The low Sn group consists of three high- and three low-lime glasses.

Sn and Cu are not correlated in the higher Sn group, whereas the low Sn group exhibits a strong positive correlation between these elements (r^2 = 0.76), indicating that they were added together, perhaps as part of an ore. These data suggest that there could be more than one source of Cu for the Nuzi glasses.

Table 3-7 presents the ratios of the averaged data for opacified and translucent colourant-related elements to the MUQ values for these elements (Co, Ni, Cu, Zn, Sn and Pb). The ratios of these concentrations are all elevated above the SBS of 0.05 to 0.06 (see section 3.3.1), indicating that substantial proportions of these elements have been introduced into the glass batch with the intended colourant or opacifier, even though their concentrations are very low.

The two opacified glasses, samples 16 and 19, have noticeably higher concentrations of Sn, Sb, Ni and Pb, and have slightly higher concentrations of Co and Zn, than the translucent low Sb glasses (Figure 3-21). MnO concentrations are low, averaging 0.04% in both groups. MnO is somewhat elevated above the SBS, with a ratio to MUQ of 0.25.

Figure 3-23 is a bivariate plot which illustrates the higher concentrations in opaque glasses of Pb and Ni relative to those found in the translucent glasses. Pb and Ni are very strongly correlated (r^2 = 0.93) if all the glasses are examined as one group. Without the two opaque glasses, which bias the correlation, these elements are only weakly correlated (r^2 = 0.36) in the translucent glasses.

The concentrations of the colourant elements in both groups are compared in Figure 3-24, which illustrates the colourant-related element differences between the low-lime glasses, the opaque, or high-lime, high Sb, glasses, and the high-lime, low Sb, translucent glasses. Co, Cu, Zn and Cd are of closely similar concentration between the three categories of glass plotted, suggesting that these elements are independent of the lime concentrations and the opacity of the Nuzi glasses. The concentrations of Sn, Pb, and to a lesser extent Ni, are highest in the opacified glasses, and lowest in the high-lime, low Sb glasses, as noted previously, suggesting that these elements are associated with Sb.

Figure 3-25 shows the elevated concentrations of lime and Sb in opaque glasses, compared with the translucent glasses. As discussed earlier (section 1.10.2), it is plausible that the association of lime and Sb is technically favourable to the manufacture

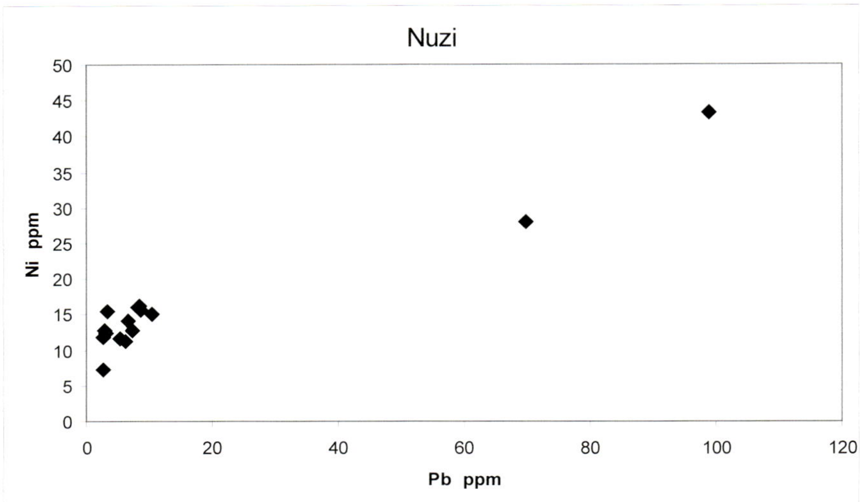

Figure 3-23 Scatter plot of Pb vs Ni for Nuzi glasses.

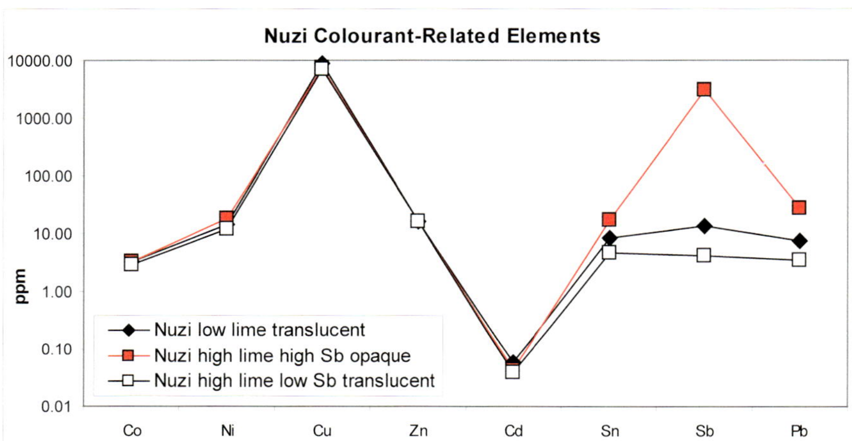

Figure 3-24 Line plot comparing averaged raw data for colourant elements in high- and low-lime glasses.

of opaque glass by promoting the precipitation of calcium antimonate. The origins of the higher lime concentrations are discussed in section 8.4.

In summary, the blue Nuzi glasses are coloured by the addition of Cu. Iron oxide probably enhances the colour of the blue glasses with low Cu concentrations and is in particular responsible for the colour of the green, low Cu glasses. The glasses opacified with calcium antimonate are distinguished by higher concentrations of Sn, Sb, Pb and to a lesser extent Ni, than are found in the translucent glasses, which suggests that these elements were associated in the raw materials. It appears that there may be two sources of Cu, a high and a low Sn source, and neither appears to be derived from bronze related ingredients, a phenomenon recorded to date only in Egyptian glasses. It is possible that the two Cu-based groups indicate the use of different Cu ores.

3.4 General Conclusions

Overall, the Nuzi glasses of the LBA have been identified by base glass and trace element analyses as a coherent group of typical 2nd millennium BC soda-lime-silica glasses of the HMHK type, fluxed with plant ash. These glasses are compositionally homogeneous with the exception of their lime contents. The glasses can be divided into two groups based on high- and low-lime concentration which will be discussed in section 8.4.

The REE and SRE data show correlations with alumina and iron oxide which indicate that the raw material which introduced them to the glass batch, most probably the plant ash, or the process of preparation of the ash and the silica, was derived from sediment of uniform geological origin, close in composition to the average continental crust. The sediments of a major

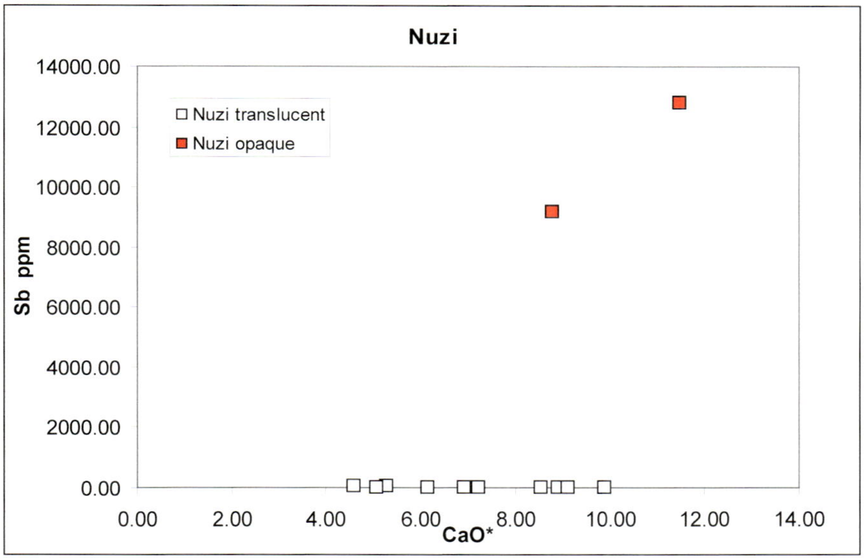

Figure 3-25 Scatter plot of CaO* vs Sb for Nuzi glasses.

river system, such as those of the Tigris-Euphrates alluvial plain, are a very likely source for such geochemically well-mixed material. Cr is considerably enriched relative to the SBS. The concentrations of the alkali and alkaline earth elements are mostly enriched relative to the SBS, and are thus likely to have been components of the plant ash. Be is an exception, which appears to be associated with sediments rather than plant ash, and Ba, Rb and Cs may have contributions from both sources.

The data for colourants describe a group of glasses predominantly coloured blue with Cu or green with iron oxide, when the Cu content was too low to colour the glasses. Two glasses were opacified with calcium antimonate. The glasses can be divided into two groups based on Sn content, suggesting that there were two sources of this element, and correlation with Cu in one case, and elevated concentrations relative to the SBS in both cases, indicates that these two groups reflect different sources of Cu colourant. The low ratios of Sn to Cu indicate that the Cu was not sourced from bronze scrap, consistent with the previous findings of Shortland (2005: 4) for Mesopotamian glasses.

The compositions of the Nuzi glasses of the 15th to 14th centuries BC represent the reasonably consistent production of some of the earliest glass objects known over a period of perhaps some 200 years. These glasses have closely similar chemical compositions which suggest controlled used of selected raw materials combined in particular recipes. A variation occurs in the lime contents, perhaps as a result of the selection of two different lime sources or of an, as yet, unexplained variation in the lime concentrations of different glasses, which is explored further in section 8.4. It is possible that a choice of two Cu sources was used when colouring the blue glasses. It is difficult to ascertain how many workshops were involved in the production of these glasses, but it seems reasonable to suggest that the number was probably limited. The evidence implies that the glass-makers operated within the parameters of a well-defined recipe, but used varying lime and Cu raw materials, perhaps depending on availability at the time, or perhaps on conditions of batch preparation and firing. Examination of the LBA glasses from Pella in the following chapter (4) will inform further our understanding of glass-making practices at this early time.

Chapter 4

Pella: Late Bronze Age Glasses Results of Compositional Analyses

4.1 Introduction to the Site and the Excavations

Pella is a multi-period site which lies in the foot hills of the eastern side of the north Jordan Valley, along the main trade route north-south from Egypt through the Levant to Syria and Anatolia in the north, and eastward along the Tigris and Euphrates into Mesopotamia (see map, Figure 4-1). Pella is situated at the south-east end of the Jezreel Valley which runs north-west from the Jordan valley to the plain of Akko/Acre on the Mediterranean coast. Thus Pella is placed at the cross-roads of the major north-south and east-west routes of the region, in a position that was central to the movement of goods between different and far-reaching lands (Smith 1987). While the lack of evidence for glass-making at Pella does not preclude the possibility that it could have occurred there, its location raises the question of whether the glasses found there were products of the glass-making technologies of either Egypt or the Near East, or were perhaps derived from both. Pella has been excavated initially by a joint College of Wooster, Ohio, and University of Sydney team, from 1979 to 1985, and thereafter solely by the University of Sydney (Hennessy 2001) (Figure 4-2).[45] Excavations on the tell of Khirbet Fahl have explored the massive Migdol temple and environs (Area XXXII) since 1997, providing much LBA and IA glass. The earliest construction of this temple dates from the MBA (*c.*1850 BC), and the latest of six reconstructions to the IA II (*c.*950-800 BC) (Figure 4-3). This gives close chronological parameters for the glass found here (Bourke *et al.* 2003; Bourke 2004).

From the LBA temple (phase II) and environs come six samples of opaque and translucent blue glasses from two ingots, beads, a disk pendant and a vessel (samples 79, 86, 93, 95, 136 and 144).

Two of more than 100 tombs at Pella also provided glass examined in this study. Tell Husn is a largely natural hill to the south across the Wadi Jirm from the main tell. From Tomb 62 (Area XI) (Figure 4-4), on its north-eastern side, come some of the earliest glasses known. This very rich tomb, excavated in 1984, dates to the MBA to LBA (1600–1500 BC) (Smith and Potts 1992: 69-81; Potts *et al.* 1985: 205-6). Tomb 62 was by far the largest and best equipped of the 20 MB/LB period tombs excavated at Pella over the years, and one of the largest tombs excavated in the southern Levant (Bourke 1997: 107; Richards 1992: 3; Eriksson 2001; Donnelly 2001; Bourke 2006).

Tomb 62 is entered through a short dromos which leads to three rock-cut chambers. More than 2000 objects were recovered from the tomb, including an assemblage of domestic ceramics of typical Jordanian and Palestinian types, as well as a number of imported Cypriot and Syrian pieces. In addition to pottery, small finds included scarab and cylinder seals, gold jewellery, arrow heads, bone inlay and spindle whorls, calcite flasks, and not least, 123 glass beads (Smith and Potts 1992: 77), of which 32 have been sampled and analysed in the present study.

Approximately 100 to 150 people were buried in this tomb, each successive burial causing some disturbance of the contents (Bourke 1992: 217). The date of the tomb is determined by the well-recognised MBIIC/LBI pottery styles and Egyptian Dynasty XV/XVII (Second Intermediate Period) scarabs (Richards 1992: 4). Most of the glass was found in the south-east corner of Chamber I, which was undisturbed and may be dated to the early LBA I, *c.*1550–1525 BC. This period is contemporary with the Egyptian 18th Dynasty king, Ahmose, first ruler of the New Kingdom, and with glass from the tomb of Ahhotep (Lilyquist *et al.* 1993: 43). Tomb 62 glass is thereforeearlier in date than the pre-Malkata glass of Thutmosis III and following kings examined by Lilyquist *et al.* (1993).

The Pella LBA glasses are dated to an earlier period than that commonly accepted for the LBA glass from Nuzi, which is considered to be one of the earliest collections of regularly produced Near Eastern glass (section 3.1.2). The LBA finds from Pella (Bourke 2006) indicate the existence of widespread contacts in the eastern Mediterranean and Egyptian regions, reinforced by textual references such as are found in the Amarna Letters (for those letters specifically referring to Pella, see Moran 1992: EA 255, EA256; Knapp 1993: 46-48).

[45] Thanks are due to Dr S. Bourke for our many discussions concerning the dating of the contexts from the Migdol temple and Tombs 62 and 89 which are included in this study.

Figure 4-1 Satellite map showing position of Pella in the north Jordan valley, reproduced from Google Earth.

The second intact tomb to provide glass for this study is Tomb 89 from Area II, excavated in 1987 (Figure 4-5). It is a single chamber tomb dating to the IA IB/IIA (*c*.1050–850 BC) and is situated to the north-east of the main tell, in a modest IA cemetery on the lower north-western slopes of Jebel Abu el-Khas. This area comes into use at the end of the LBA when burials appear to cease on the northern slopes of Tell Husn (Bourke 1997: 112-113). From this tomb four beads, from in excess of 200 objects in total, were sampled and analysed in the present study. These four beads are contemporary with 19 beads sampled and analysed from the temple service areas (Areas XXXIIA and XXXIIB) and from the temple courtyard (Area XXXIIW), dating to the IA IIA – IIB, *c*.1050–900 BC.

Occupation of Pella was continuous from the MBA throughout the LBA, ending in a fiery destruction around 1150–1100 BC. IA occupation was continuous from *c*.1050 BC through to the final destruction of *c*.800 BC. Most of the LBA glass examined in the present study comes from the LBIIA–IIB phase of the temple (*c*.1300–1200 BC), the period of greatest Egyptian influence. There is also a small but detectable Kassite Mesopotamian influence from 1350 BC onwards (Bourke 2004; Bourke *et al.* 2003). Tomb 62 is contemporary

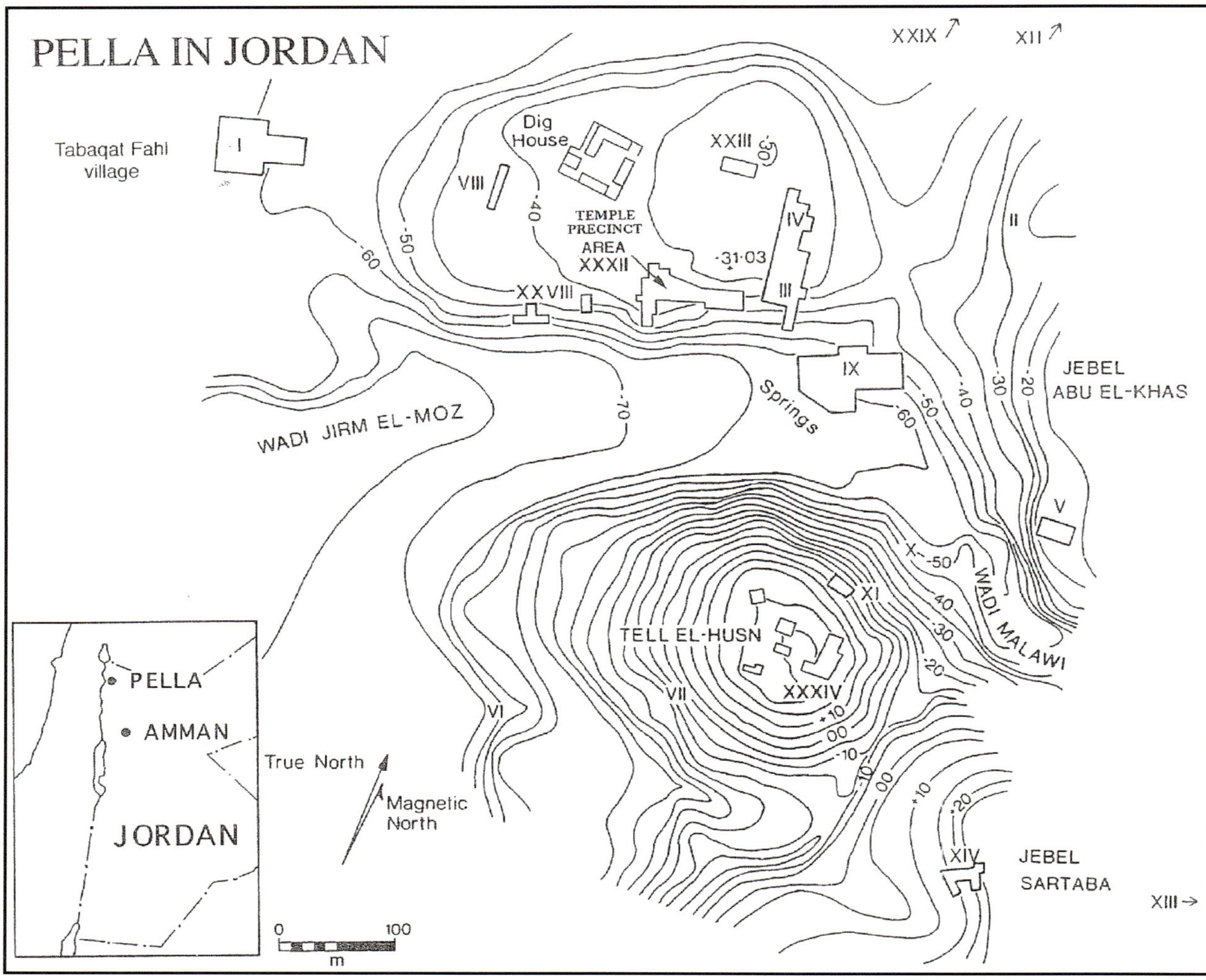

Figure 4-2 Contour map of site of Pella indicating excavation areas referred to in text. Reproduced courtesy of the Pella Excavation Project.

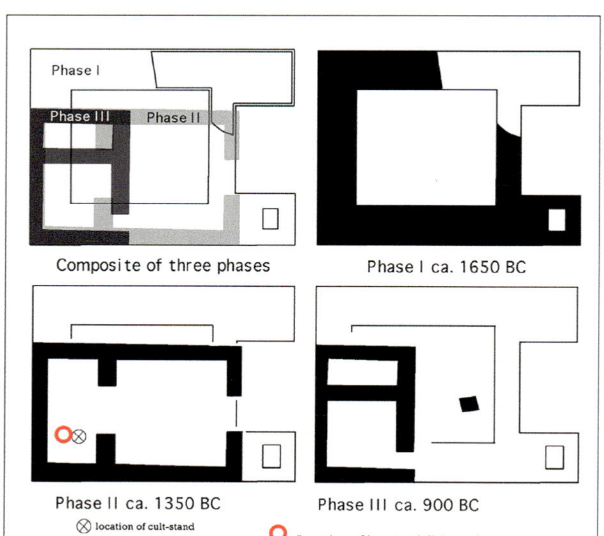

Figure 4-3 Three phases of the Migdol Temple, Area XXXII, showing the location in the Holy of Holies of two important LBA glass objects, an ingot (sample 86) and a disk pendant (not sampled). Plans adapted from Bourke 2004: Fig. 3.

with the massive stone Migdol MB/LB temple phase, while Tomb 89 corresponds to its final Iron IIA/B reconstruction.

From excavated evidence, it is believed that the period from the 10th C BC until the final destruction in *c.*800 BC was considered one of prosperity and represented a cultural high-point (McNicoll *et al.* 1980: 72; Smith *et al.* 1981: 14-17; McNicoll *et al.* 1982: 49-63; Smith 1982: 327-328; Sauer 1986: 9-13). After the end of the 9th C BC there was a gap in major occupation of the site until the middle Hellenistic period, *c.*200 BC. Glass from this period just precedes and possibly overlaps chronologically the earliest IA glass from Nimrud (Reade *et al.* 2005; section 6.2.2).

4.2 The LBA and IA Glasses Analysed

The LBA glasses analysed in the present study were found in the temple and environs, domestic area IVE, and Tomb 62. The IA glasses were found in the temple

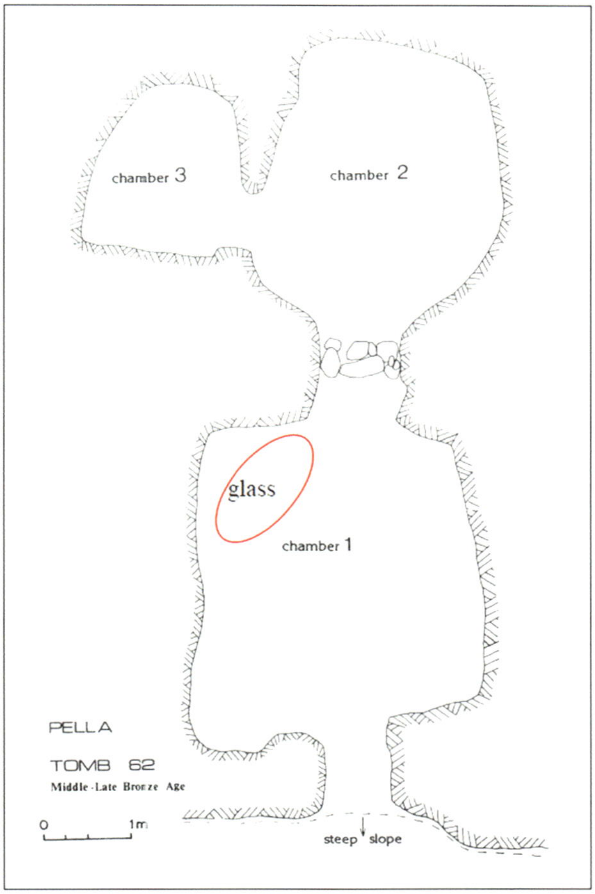

Figure 4-4 Plan of Tomb 62, in Area XI, adapted from Potts *et al.* 1985: 193.

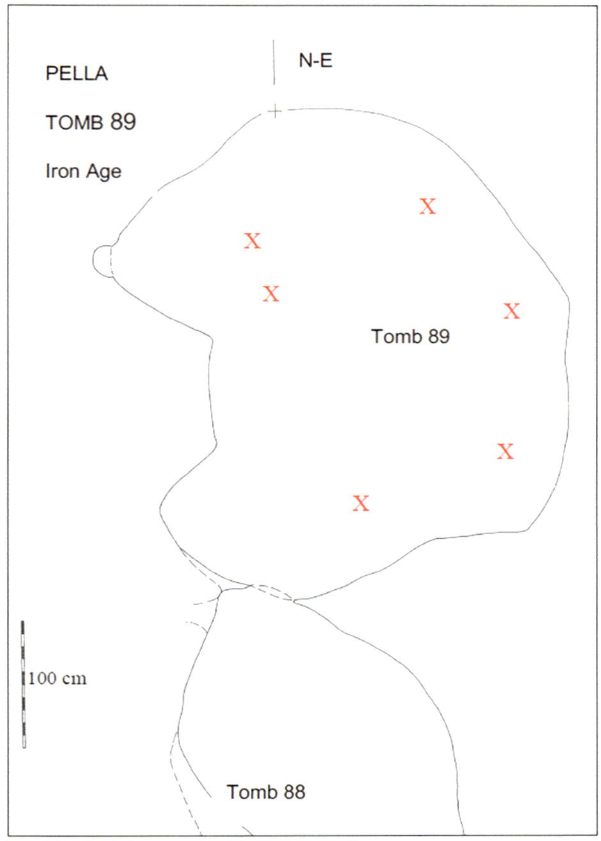

Figure 4-5 Plan of Tomb 89, in Area II. Adapted from a plan provided courtesy of the Pella Excavation Project. Red Xs mark the positions in which the glasses were found.

area and Tomb 89, giving a variety of functional contexts, predominantly ritual and burial for both periods (Table 4-1). Base glass analysis defined two coherent compositional groups of soda-lime-silica glass: HMHK and LMLK. The results for the LBA to IA HMHK glasses are presented first, followed by the results for the analyses of the IA LMLK glasses in Chapter 5 (there were no LBA LMLK glasses).

The LBA glass assemblage includes several object types, while the IA assemblage has produced only beads. The beads of the LBA are larger and more diverse in shape than those of the IA, and are all plant ash fluxed (HMHK). Except for one black (dark green) bead, the beads are coloured blue, as are more than 70% of all glass objects found at Pella. The black glass (sample 74), from a sondage in Area IVE, dates to the LBA, 13th C BC, but is unusual in being the only black glass from this period. Of the 27 IA beads, 19 from the temple and environs and four from Tomb 89 are small and black coloured, and have been identified by base glass analyses as being of the LMLK type, which might indicate the use of a mineral soda flux. The remaining four beads from IA contexts from the temple and environs include three HMHK, translucent and opaque blue, and one colourless bead, and will be reported with the HMHK LBA glasses.

4.2.1 HMHK Glasses

The number of plant ash glasses found in IA contexts at Pella is small (*n*=4). Sample 98, a turquoise bead, was found in the temple in an IA pit (*c.*1050–900 BC) which was dug into a LBA context. The trench was excavated carefully and was well recorded, strongly indicating that the bead was part of the IA material recovered from the pit. Sample 127, is from a blue annular bead found in the IA temple Area XXXIIR, cut into the preceding LBA temple, and dated to *c.*1050–900 BC. Sample 119, a colourless bead also found in the temple forecourt (Area XXXIIT), is thought to date to the IA I, probably 11th to 10th C BC, but this context is much disturbed by pits and might not provide a reliable date for this bead. Sample 77, a blue bead, is the latest in date, *c.*950–850 BC, or Late IA I- IA II, coming from Area XXXIIG of the temple exterior. On balance, it seems probable that these four beads could indeed have been manufactured

Table 4-1 Summary of glasses analysed from Pella in the LBA to IA.

	Tomb 62	Temple and environs	Area IVE	Temple and environs	Tomb 89
Period	*MBA-LBA*	*LBAII*	*LBAII*	*IA*	*IA*
Date	*c.1550-1525 BC*	*c.1300-1200/1150 BC*	*c.1300-1200 BC*	*c.1050-900 BC*	*c.1050-850 BC*
n =	*32*	*6*	*1*	*23*	*4*
Colours	*opaque and transl. blue*	*opaque and transl. blue*	*black*	*opaque and transl. blue (3), black (19), colourless (1)*	*black*
Type	*beads*	*ingots, beads, pendant, vessel*	*bead*	*beads*	*beads*

in the LBA, as their colours and shapes are consistent with the LBA assemblage and their contexts are not all securely dated. In view of these factors, the likelihood of an LBA origin for these beads is strong even when the context is reliable (Area XXXIIG, sample 77), and especially when the heirloom status of the Nuzi temple beads is recalled.

It is not the purpose here to provide a typological survey of the glass forms of the LBA or of the IA, but to concentrate on those glasses that have been analysed for composition, putting them into context as necessary. Glass object types are similar at many of the major LBA glass sites, including Pella, Nuzi and Tell Brak.[46] Published results of chemical analyses of Nuzi and Tell Brak glasses are compared in Chapter 8 (section 8.3.1).

Beads are the predominant glass objects of this period, forming over 75% of the glasses found at Pella. Bead forms are similar between Nuzi, Pella and Tell Brak, with a predominance of blue spherical, flattened spherical and annular (otherwise called 'ring' or 'disk') beads. These forms constitute in excess of 60% of beads found at Pella. Ribbed spacer beads are found at both Nuzi and Pella, and are of the same type as that illustrated in Spaer (1984: Figs 1-2) and James and McGovern (1993: Fig. 66). Thick-walled glass cylinder beads or blanks were found at the three sites (Vandiver, 1983: 246; Oates *et al.* 1997: 86; Bourke *et al.* 2003: 361-2). Disk-shaped pendants (sometimes called 'plaques'), with a pierced lug for suspension, are also known from Nuzi, Pella and Tell Brak (Starr 1939: pl. 120WW; Barag 1970: 191-2, figs. 15, 100; Oates *et al.* 1997: Fig. 222.83-4; Bourke *et al.* 2003: 361-2, Fig.7). Notably, one disk pendant from Beth Shan (McGovern 1980: 408, No. 305) is closely similar to those found at Nuzi and Pella in greater numbers. McGovern noted another disk pendant from Megiddo Stratum IX of the LBA (1980: 424, no. 306). For further comparisons and discussion the reader is referred to McGovern who classified these as Syro-Palestinian, and not Mesopotamian (Near Eastern), types of inherent value (1980: 220-222, 282, 301).

Two unusual tetrahedrally shaped, blue translucent ingots form part of the LBA corpus from Pella, and both objects (samples 86 and 95) were sampled for the present study. They were found on the floor (Area XXXIIG 107.9; see Bourke *et al.* 2003: 361, Fig. 5) and in the occupation deposit (Area XXXIIR 11.13) of the Holy of Holies in the second phase of the temple, presumably laid there as offerings, or possibly as part of the architectural decoration (as at Nuzi). They were found in association with luxury items, such as a lapis lazuli and gold bead, ivory inlays and imported pottery (Bourke forthcoming). One ingot (sample 86) was found together with a disk pendant which was not sampled. A sample was taken from a similar pendant (sample 93) from Area XXXIIS 4.42.

There have been few reports concerning LBA ingots of any shape. One fragmentary ingot is reported from Nuzi, probably of a circular, crucible shape (Barag 1970: 140, Fig. 16), and there are records of a further 13 ingots from the Mitannian palace at Tell Brak, which appear also to be crucible-shaped (Oates *et al.* 1997: 85-6, 89, Fig. 124). An irregular piece of raw glass was found at Beth Shan, Level VIII, a solid glass sphere was found below the altar steps in the Level VII temple (James and McGovern 1993: 151-2), and an ingot fragment was found at Ashdod Stratum XII (Dothan and Porath 1993: Fig. 38.8). It is interesting to note that a collection of crucible-shaped ingots was salvaged from the Ulu Burun shipwreck and have been linked to Egyptian

[46] While nearby Beth Shan in the Jezreel Valley has a large corpus of LBA glass objects, many are of Egyptian style (McGovern 1993a; 1993b; McGovern 1980).

glass-making at Amarna (Bass 1986; Nicholson *et al.* 1997:145; section 1.2).

From this brief survey, it is evident that glass of the LBA from the Syro-Levantine region and Mesopotamia was predominantly coloured blue, and was largely formed as beads and pendants in forms found at Nuzi, Pella, Tell Brak and Beth Shan, amongst other sites.

4.3 Summary of HMHK Samples

Samples of one black, one colourless, 16 opaque turquoise blue and 25 translucent blue glasses were collected from the Pella Excavation Project, University of Sydney, for chemical analysis by SEM-EDS and solution ICP-MS (Chapter 2). The identification, provenance and descriptions of the glasses appear in Table 4-2. There are 39 beads of various shapes including spherical, flattened spherical, cylindrical, ribbed spacer, annular, biconical and barrel beads. Annular and spacer beads form the majority of bead shapes. There are two ingots of unusual elongated tetrahedral shape, one disk pendant, one sherd that was probably part of a vessel, and one that is most likely to have been a pinhead. The glasses were excavated from the temple (Area XXXII), Area IVE and Tomb 62 (Area XI) (see site plan Figure 4-2). The LBA samples are blue translucent or blue opaque, with the exception of one black bead. One colourless, one opaque blue and two translucent blue beads from IA temple contexts were sampled. The glasses were all weathered to some degree, but those sampled contained unweathered glass beneath their weathered surfaces. The glass which appeared unweathered was sampled.

4.4 Base Glass Composition

The Pella glasses were analysed for base glass oxide composition using SEM-EDS (section 2.3). This analysis, consisting of 39 samples from the LBA and four possibly from the ensuing IA, defined a coherent compositional group of soda-lime-silica glass fluxed with plant ash (HMHK type).

The average reduced compositions of the Pella HMHK glasses expressed as weight % oxide are presented in Table 4-3. Complete raw oxide and trace element data for each sample are tabulated in Appendix III.

Examination of the glasses and the results of chemical analysis reveal that neither temple, domestic, nor tomb provenance has any bearing on composition or physical form of the objects from Pella, and neither are composition and physical form correlated.

All of the LBA (and IA) glasses from Pella were typical Near Eastern HMHK LBA, or 2nd millennium BC, plant ash glasses, similar to those analysed in the present study from Nuzi (section 3.2). The bead and pendant shapes found at both Pella and Nuzi are the same, as are the colours of the glasses. The four IA HMHK glasses are of closely similar composition to those from the LBA and, in the present study, have been grouped together with the LBA glasses.

A comparison of the oxide concentrations in the Pella HMHK glasses with those found in the UCC, represented by MUQ, shows that the concentrations of lime, magnesia, potash and particularly soda are enriched in the glasses as a result of the addition of plant ashes.

The concentration of silica in the HMHK glasses is comparable to that found in the UCC, but the concentrations of alumina and iron oxide are depleted relative to the UCC, indicating that the silica is from a relatively uncontaminated source. The concentrations of magnesia and potash are approximately two to two and a half times that found in the earth's crust, while the concentration of lime is over six times, and soda over 20 times the concentration normally found in the earth's crust. The concentrations of the oxides assayed are closely similar for the glasses from Nuzi and Pella.

In the Pella HMHK glasses, the concentrations of magnesia and potash are elevated on average to 4% and 3% respectively, with ranges of approximately 2.1% to 6.3%, and 2.1% to 4.6% respectively (Figure 4-20). The concentrations of magnesia and potash are weakly correlated (r^2 = 0.25) in the Pella HMHK glasses by contrast with the concentrations of magnesia and potash in the Nuzi glasses (r^2 = 0.69). The concentration of phosphorus in the Pella glasses, at an average of 0.14%, is low for that conventionally expected of a plant ash glass, and is similar to that measured in the Nuzi glasses. The Pella glasses have high lime concentrations ranging widely from 2.9 % to 10.4%, and averaging approximately 7%, which is typical of plant ash glass (Figure 4-22). The similarities and differences between the Nuzi and Pella HMHK glass compositions are described briefly in the following sections and are discussed in Chapter 8.

The Pella HMHK glasses contain 61% to 72% (average 68%) silica, and 12% to 20% (average 17%) soda. The alumina concentration averages 0.75% (range of 0.2% to 2.2%) and the iron oxide level is low at approximately 0.4% (range of 0.11% to 0.62%), suggesting that, as was found with the glass from Nuzi, a relatively pure form of quartz was used as the silica source to make the glasses. The Pella glasses have slightly higher contents on average of both alumina and iron oxide compared with the Nuzi glasses (Figure 4-21). While the higher iron oxide content of some Nuzi glasses contributes to their green colour, there are no green glasses from this

Table 4-2 Description and provenance of Pella HMHK glass samples*.

Sample Number	Analysis Number	Object	Colour	T or O	Provenance (Figure 42)	Date and Comments
940304	74	bead – flattened spherical	green- black	T	sondage IVE 111.9	LBII c. 13th C BC
960870	79	bead – cylindrical	turquoise	O	Temple XXXIIG 107.7	LB 1300-1200/1150 BC
961074 RN190083 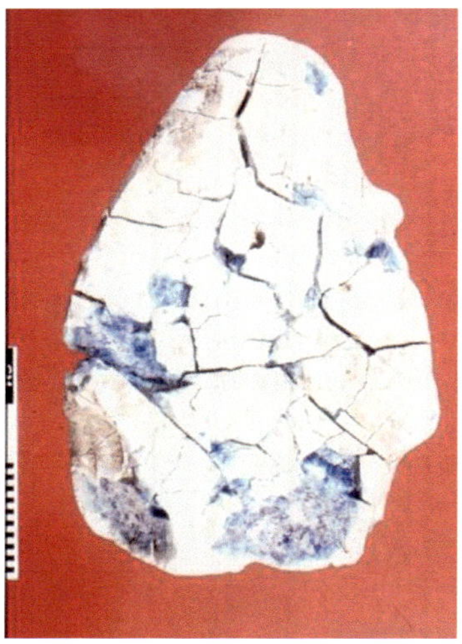 Figure 4–6 Pella 190083 blue translucent tetrahedral ingot (Image courtesy of the Pella Excavation Project)	86	ingot - tetrahedral	blue	T	Temple XXXIIG 107.9	LB 1300-1200 BC
990598  Figure 4–7 Pella 990598 turquoise opaque pendant disc with lug (Image courtesy of the Pella Excavation Project)	93	pendant – disc with lug	turquoise	O	Temple XXXIIS 4.42	probably LB 1300-1200 BC
990641	95	ingot – tetrahedral	blue	T	Temple XXXIIR 11.13	LB 1300-1200 BC
991149	136	bead – flattened spherical	blue	T	Temple XXXIIR 22.7	LB 1300–1200 BC
991212	144	vessel (?) fragment	pale blue	T	Temple XXXIIS 12.1	LB 1300-1200 BC
841969 RN70783	187	bead - spherical	blue	T	Tomb 62 Area XI	Early LBI 1550-1525 BC
841957 RN70773	188	bead – spacer	turquoise	O	Tomb 62 Area XI	Early LBI 1550-1525 BC
841930 RN70748	189	bead – annular	blue	T	Tomb 62 Area XI	Early LBI 1550-1525 BC

Sample Number	Analysis Number	Object	Colour	T or O	Provenance (Figure 42)	Date and Comments
841928 RN70746  Figure 4–8 Pella 70746 blue translucent annular bead	190	bead – annular	blue	T	Tomb 62 Area XI	Early LBI 1550-1525 BC
841714 RN70579D  Figure 4–9 Pella 70579D turquoise opaque annular bead	192	bead – annular	dark turquoise	O	Tomb 62 Area XI	Early LBI 1550-1525 BC
841626 RN70505  Figure 4–10 Pella 70505 turquoise opaque biconical bead	193	bead – biconical	pale turquoise	O	Tomb 62 Area XI	Early LBI 1550-1525 BC
842035 RN70845  Figure 4–11 Pella 70845 blue translucent pinhead showing fresh glass under weathered surface.	194	knob or pinhead	dark blue	T	Tomb 62 Area XI	Early LBI 1550-1525 BC
842168 RN70950	195	bead – annular	turquoise	O	Tomb 62 Area XI	Early LBI 1550-1525 BC

Sample Number	Analysis Number	Object	Colour	T or O	Provenance (Figure 42)	Date and Comments
842166 RN70948  Figure 4-12 Pella 70948 turquoise opaque annular bead	196	bead – annular	turquoise	O	Tomb 62 Area XI	Early LBI 1550-1525 BC
842161 RN70943	197	bead – cylinder (?)	blue	T	Tomb 62 Area XI	Early LBI 1550-1525 BC
842024 RN70834	198	bead – annular	blue	T	Tomb 62 Area XI	Early LBI 1550-1525 BC
842023 RN70833  Figure 4-13 Pella 70833 pale turquoise opaque annular bead	199	bead – annular	pale turquoise	O	Tomb 62 Area XI	Early LBI 1550-1525 BC
842033 RN70843  Figure 4-14 Pella 70843 turquoise opaque flattened spherical bead	200	bead – flattened spherical	turquoise	O	Tomb 62 Area XI	Early LBI 1550-1525 BC
841414 RN70342  Figure 4-15 Pella 70342 blue translucent barrel bead	201	bead – barrel	blue	T	Tomb 62 Area XI	Early LBI 1550-1525 BC

Sample Number	Analysis Number	Object	Colour	T or O	Provenance (Figure 42)	Date and Comments
841595 RN70478	202	bead – spacer	blue	T	Tomb 62	Early LBI 1550-1525 BC
841612 RN70491  Figure 4–16 Pella 70491 blue translucent biconical bead	203	bead – biconical	blue	T	Tomb 62	Early LBI 1550-1525 BC
841600 RN70483	204	bead – spacer	blue	T	Tomb 62	Early LBI 1550-1525 BC
841600 RN70483  Figure 4–17 Pella 70483 blue translucent spacer bead	205	bead – spacer	blue	T	Tomb 62	Early LBI 1550-1525 BC
841600 RN70483	206	bead – spacer	blue	T	Tomb 62	Early LBI 1550-1525 BC
841599 RN70482	207	bead – spacer	blue	T	Tomb 62	Early LBI 1550-1525 BC
841585 RN70470	208	bead – spacer	pale blue	T	Tomb 62	Early LBI 1550-1525 BC
RN70391	209	bead – annular	turquoise	O	Tomb 62	Early LBI 1550-1525 BC
841429 RN 70354D	210	bead – spacer	blue	T	Tomb 62	Early LBI 1550-1525 BC
841472 RN70391	211	bead – annular	turquoise	O	Tomb 62	Early LBI 1550-1525 BC
841472 RN70391	212	bead – annular	blue	T	Tomb 62	Early LBI 1550-1525 BC

Sample Number	Analysis Number	Object	Colour	T or O	Provenance (Figure 42)	Date and Comments
841472 RN70391  Figure 4–18 Pella 70391 turquoise opaque annular bead	213	bead – annular	turquoise	O	Tomb 62	Early LBI 1550-1525 BC
841472 RN70391	214	bead – annular	blue	T	Tomb 62	Early LBI 1550-1525 BC
841472 RN70391	215	bead – annular	turquoise	O	Tomb 62	Early LBI 1550-1525 BC
841430 RN70355A	216	bead – spacer	blue	T	Tomb 62	Early LBI 1550-1525 BC
841430 RN70355D 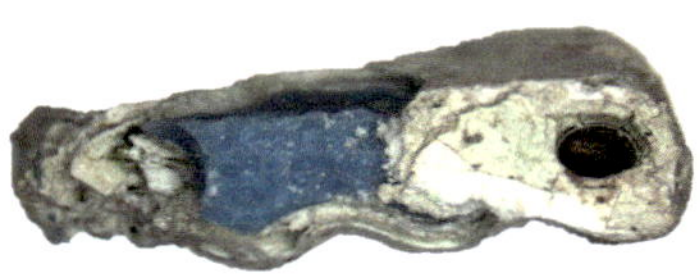 Figure 4–19 Pella 70355D turquoise opaque spacer bead (end view showing unweathered core)	218	bead – spacer	turquoise	O	Tomb 62	Early LBI 1550-1525 BC
841430 RN70355E	219	bead – spacer	blue	T	Tomb 62	Early LBI 1550-1525 BC
841988 RN70799	222	bead – spacer	pale turquoise	O	Tomb 62	Early LBI 1550-1525 BC
950534	77	bead– spherical	blue	T	Temple XXXIIG 6.14	late IAI - IAII 950-850 BC
990804	98	bead– flattened spherical (?)	turquoise	O	Temple XXXIIQ 11.15	IA pit in LB context 1050-900 BC
990982	119	bead – spherical	colourless-greenish tinge	T	Temple XXXIIT 11.36	late IAI c.11th – 10th C BC
990993	127	bead – annular	blue	T	Temple XXXIIR 14.5	IA 1050-900 BC

*'Turquoise' is used to describe only opaque light blue glasses, while 'blue' is reserved for blue translucent glasses. Under 'Object Number', RN = registration number. The other numbers are book numbers, i.e. running excavation catalogue numbers assigned when objects are processed on-site, before they are given a registration number. The first two digits of the book number record the year of excavation. Under 'Provenance', the Roman numerals refer to the area, the letter to the trench, and the following numbers separated by a point are locus and level.

Table 4-3 Average reduced compositions of 43 Pella HMHK glasses in wt% oxide. Included here are 39 LBA and four IA HMHK glasses.

Samples	Na_2O*	MgO*	Al_2O_3*	SiO_2*	P_2O_5*	K_2O*	CaO*	FeO*	Total
Average *n=43*	16.77	3.96	0.75	67.83	0.14	3.04	7.12	0.39	100.00
SD	1.92	0.93	0.42	3.36	0.07	0.59	2.17	0.16	
Average Glass / MUQ	20.96	2.52	0.04	1.00	0.77	1.98	6.47	0.05	

* = reduced composition of oxides in all tables and charts.
The ratio of average glass composition to MUQ is calculated from raw data.

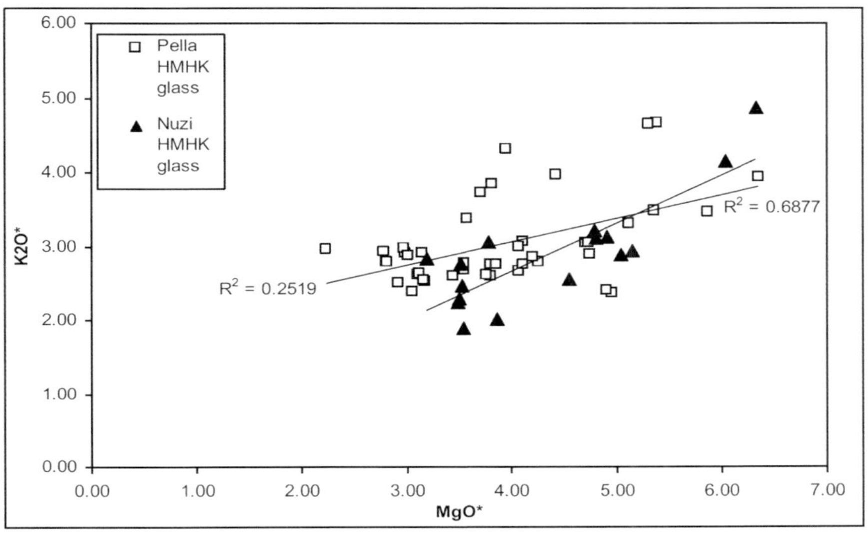

Figure 4-20 Scatter plot of magnesia vs potash (wt% oxide) for the Pella HMHK glasses compared with the Nuzi HMHK glasses showing similar oxide concentrations from both sites. Note that the correlation between the concentrations of the oxides is strong for the Nuzi glasses but not for the Pella glasses.

period at Pella, despite their higher average iron oxide content.

The ratio of alumina to iron oxide is 1.9:1, very close to that found naturally in the UCC, and is likely to reflect the natural ratio suggesting that these elements have entered the glass as impurities with one of the glass constituents.

With three high alumina outliers removed (1.6% to 2.3% alumina), iron oxide and alumina are well correlated (r^2 = 0.6) (see Figure 4-21). Two of the outliers (samples 77 and 119) are IA in date, and the third (sample 136) is LBA in date. The other two IA glasses plot with the LBA group, supporting the suggestion that these glasses originate from the LBA period, the outliers being representatives of product variability of the time, and not indicators of changes in production in the IA.

Whereas these glasses have lime contents typical of plant ash fluxed glass, lime does not correlate with other major elements when the glasses are plotted as one group. Examination of lime *versus* magnesia for both LBA and IA glasses, plotted in Figure 4-22, shows that the glasses divide into two groups based upon lime concentration, similar to that observed for the glasses from Nuzi. The group with the greater lime concentration has 8.2% to 10.4% lime (average 9.4%), while the glasses with lower lime concentration form a separate group of 5.6% average lime concentration (range of 2.9% to 7.8%).

Figure 4-23 shows that the high- and low-lime concentrations of the Pella HMHK glasses correspond largely to their opacity or translucency, with the following three exceptions. The single IA opaque glass (sample 222) assayed in the present study falls

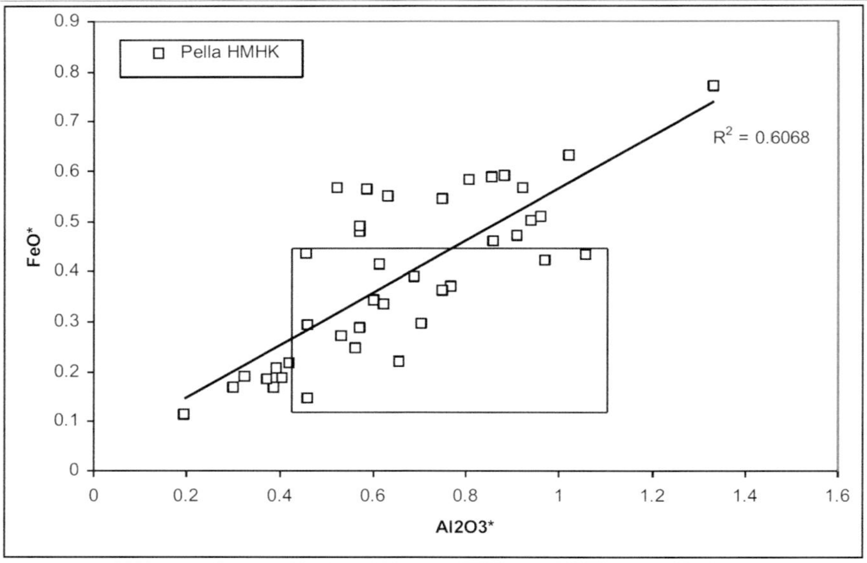

Figure 4-21 Scatter plot of alumina vs iron (wt % oxide) with three outliers removed revealing a strong correlation between these oxides, though it is not as strong as is found in the Nuzi glasses. The rectangular outline represents the range in which the Nuzi glasses plot for comparison, revealing the greater range in concentration of both oxides found in the Pella HMHK glasses.

into the high-lime group, whereas two translucent LBA glasses (samples 194 and 174) assayed have high-lime concentrations. In the present study, high-lime translucent glasses were also found at Nuzi. Although sample 74 from Pella has a high-lime concentration (8.2%), in this case its classification as a high-lime glass might be arbitrary. Figures 4-22, and 4-23 show that it plots with the low-lime glasses because of its high magnesia content (5.9%) which separates it from the tight group formed by the rest of the high-lime glasses.

The Pella group of glasses with a low-lime concentration shows a correlation of lime to magnesia of r^2 = 0.45, similar to that found in the corresponding group of glasses from Nuzi. There is, however, no correlation of lime to magnesia concentration in the glasses from Pella in the high-lime group, while for the corresponding group of high-lime glasses from Nuzi the apparent positive correlation between lime and magnesia is due to one glass, which when removed shows no correlation, suggesting that the same pattern occurs at both sites.

When the glasses from Pella Tomb 62 are examined as a sub-group with the advantage of their discrete context and reliable dating, the division of the glasses into low- and high-lime groups is sustained. One opaque glass has a low-lime concentration and one translucent glass with low Sb fits into the high-lime group, which is otherwise dominated by opaque glasses with high Sb concentrations. The average lime concentration in 19 low-lime glasses from Tomb 62 is 6%, while the average concentration in 13 high-lime glasses from this tomb is 9.5%, which is similar to the averages of lime content calculated from all Pella HMHK glasses. This indicates that both groups of glass were in use at the same time, and that the observed difference based on lime content does not indicate a chronological disparity. At Pella, the lime-based groups, with the exception of three glasses, correspond to groupings which are related to either the opacity or the translucency of each glass. The significance of these lime-based groups will be explored further in section 8.4.

When the magnesia and potash concentrations of the Nuzi glasses were examined as a single group the correlation of the concentrations of these oxides was moderately strong (r^2= 0.7) (section 3.2). The weak correlation (r^2 = 0.25) of the concentrations of magnesia and potash in the Pella HMHK glasses examined as a single group, shown in Figure 4-20, is not improved when the glasses are examined as two groups based on lime content (Figure 4-24); the group of low-lime glasses from Nuzi exhibits a very good positive correlation (r^2 = 0.75) between the concentrations of potash and magnesia, which is not evident in either of the lime-based groups of glasses from Pella.

The averaged, reduced data for the low and high lime groups of Pella glasses are presented in

Examination of these data indicates that there is little difference in the concentrations of silica between

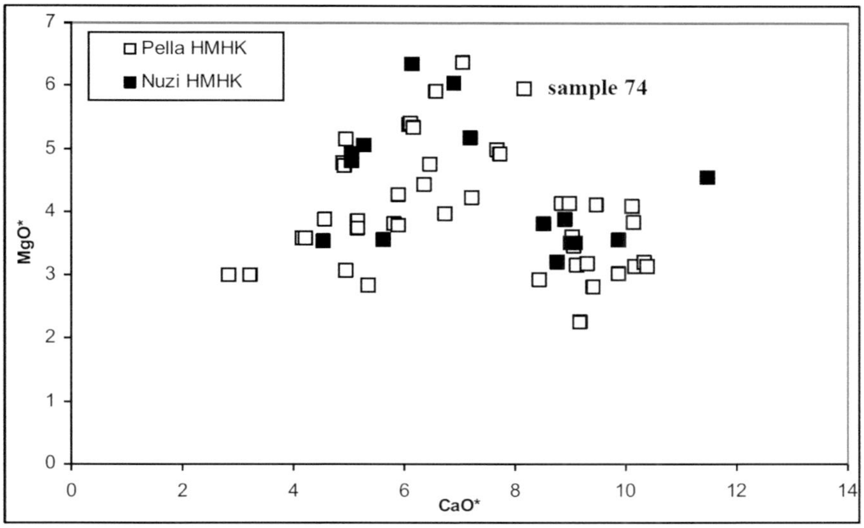

Figure 4-22 Scatter plot of lime vs magnesia (wt % oxide) for Nuzi and Pella HMHK glasses showing the existence at both sites of the same two lime-related groups.

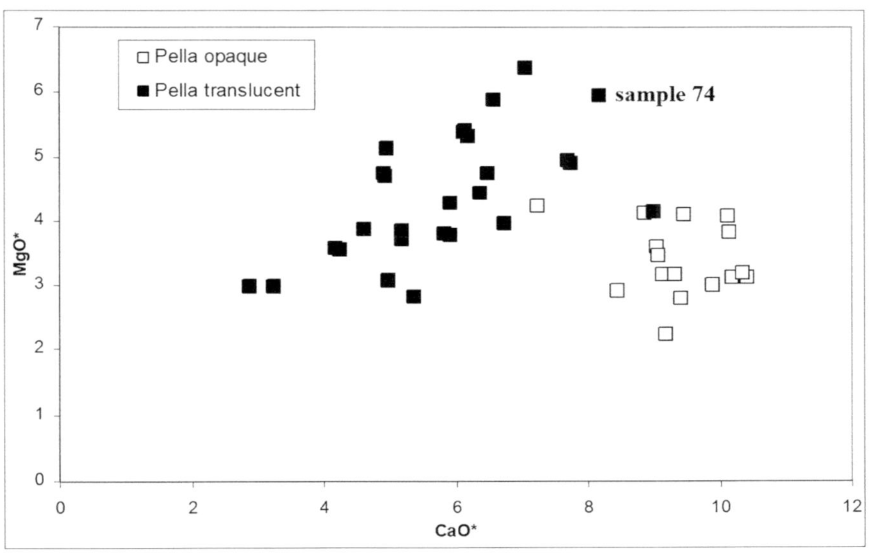

Figure 4-23 Scatter plot of lime vs magnesia (wt % oxide) for opaque and translucent Pella HMHK glasses.

the groups, as seen for the glasses from Nuzi, and a difference of 0.8% for magnesia, and 0.45% for potash between the groups, which is less than that measured in the glasses from Nuzi. There is, however, a greater difference (1.73%) in the soda contents, compared to 0.5% in the Nuzi glasses. The average lime contents for high- and low-lime groups are the same for the glasses from both Pella and Nuzi.

In summary, the SEM-EDS analyses of base glass oxides from the LBA Pella glasses identified them as typical Near Eastern HMHK glasses with potash and magnesia concentrations greater than 1.5% each, and phosphorus concentrations averaged at 0.14%. The lime concentrations of these glasses were found to fall into two sub-groups: high-lime glasses containing from approximately 8.2% to 10.5% lime, and low-lime glasses containing between approximately 2.9% and 7.8% lime. Similar groups have been identified in the HMHK glasses from LBA Nuzi, and are discussed in Chapter 8.

The Pella HMHK glasses have a poor correlation between the concentrations of magnesia and potash, in contrast to the strong positive correlation between

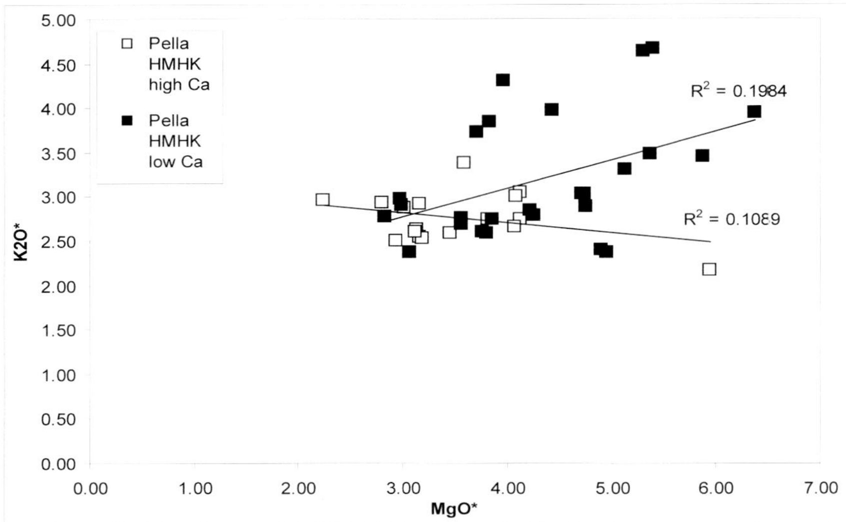

Figure 4-24 Scatter plot of magnesia vs potash (wt % oxide) for high and low lime groups of the Pella HMHK glasses showing a lack of correlation in both, contrary to the good correlation found in the low-lime group of glasses from Nuzi (section 3.2).

Table 4-4 Averaged, reduced concentrations of base glass oxides in low- and high-lime Pella HMHK glasses presented in wt%.

Samples	Na2O*	MgO*	Al2O3*	SiO2*	P2O5*	K2O*	CaO*	FeO*	Total
Low-lime *n=26*	17.52	4.32	0.83	67.93	0.13	3.20	5.65	0.42	100
High-lime *n=17*	15.79	3.52	0.63	67.37	0.15	2.75	9.44	0.35	100

the concentrations of these oxides for the glasses from Nuzi. There is a strong correlation between the concentrations of iron oxide and alumina for the glasses from Pella, which have a greater range in concentration of both these oxides than is found in the HMHK glasses from Nuzi. The concentration of silica does not correlate with those of other oxides for the Pella glasses, indicating that a relatively pure form of silica, such as quartz pebbles, was used to make these glasses. The comparatively low concentrations in the Pella glasses of iron oxide and alumina support the proposition that the silica source was quartz pebbles.

4.5 Trace Elements

Trace elements were investigated through analysis with solution ICP-MS (section 2.4).

4.5.1 Rare Earth Elements (REE)

The line patterns formed when the normalised REE concentrations of the Pella HMHK glasses are plotted in a line and dot plot are relatively homogeneous as illustrated in Figure 4-25.

It is apparent that the Pella HMHK glasses form one geochemical group, which has some slight variations in pattern and a range of concentrations that reflect small differences in the raw materials or glass-making processes (section 2.5.3). A similar pattern and the same range of concentrations of the REE are found in the glasses from Pella and from Nuzi. Samples 77 and 136, both translucent blue beads from the temple, one IA and one LBA in date, have small negative Eu anomalies, and are higher in REE concentration than the other glasses. Several glasses have small positive Eu anomalies and, as with the Nuzi REE, the Pella REE have a generally flat pattern with slight depletion of the HREE. While concentrations vary, the patterns are all closely similar, suggesting the same geological source of REE-containing raw materials was used to make the glasses from both sites.

Averaged raw concentrations of REE are presented in Table 4-5. The average concentrations of the REE in the

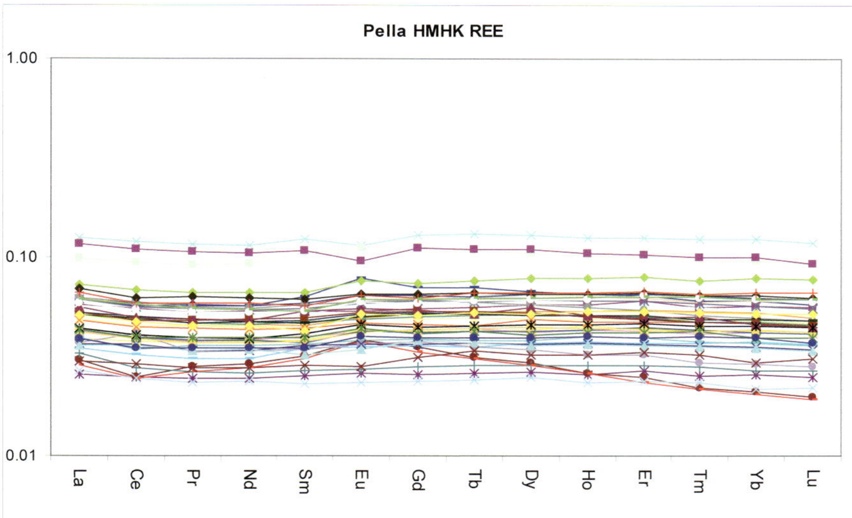

Figure 4-25 Line plot of the REE of individual Pella HMHK glasses showing general similarity of pattern. A logarithmic scale is used. Data are normalised to MUQ.

Pella HMHK glasses are closely similar to the average concentrations of the REE in the glasses from Nuzi. Reasons for variations in REE concentrations between individual samples are discussed in section 2.5.3.

The REE are correlated with iron oxide ($r^2 = 0.6$), alumina ($r^2 \geq 0.9$) and the other SRE (r^2 = 0.6 to 1.0), excepting Cr, Mo, W and Tl, as they are in the glasses from Nuzi. The REE have weak anti-correlations in the order of $r^2 \approx -0.4$ with Sb and no correlation with Cu, suggesting that the REE did not enter the glass batch with the blue colourant or the opacifier, and that the silica source is relatively uncontaminated. The REE were most likely derived from sedimentary contamination (sections 1.6 to 1.9, 3.3.1).

Figure 4-26 reveals the close similarity between the patterns of the REE from the UCC and from Pella HMHK glasses, divided into translucent and opaque sub-groups, revealing that they have the same pattern at slightly different concentrations. When REE data from the Nuzi glasses are compared with that of the REE from the Pella glasses, it is evident that the source of REE-bearing raw materials at these sites was derived from similar, if not the same, geological sediments. Like the Nuzi REE, the concentration of the Pella HMHK REE is depleted relative to the UCC, indicating relatively pure raw materials. Overall, the ratios of REE in Pella HMHK glass to those in the UCC are within the range of 0.05-0.06 (see Table 4-5), and are not enriched relative to it, establishing the same SBS as observed in the Nuzi glasses.

Figure 4-27 is a plot of the averaged Nuzi and averaged Pella HMHK REE data for opaque and translucent glasses, normalised to MUQ. Normalisation results in relatively flat REE patterns closely resembling the pattern of sediments found in the UCC. The REE patterns and concentrations for translucent and opaque Nuzi glasses are closely similar to those of the corresponding glasses from Pella; the Pella pattern is slightly flatter.

Table 4-5 Averaged raw values for Rare Earth Elements from Pella HMHK glasses in ppb.

Pella HMHK Rare Earth Elements (ppb)

	n=43 Averages All	RSD	Av. glass/ MUQ	*n=25* Average Low Ca	*n=18* Average High Ca
La	1709	40	0.05	1915	1423
Ce	3482	40	0.05	3918	2877
Pr	404	40	0.05	456	330
Nd	1566	40	0.05	1771	1281
Sm	336	42	0.05	382	272
Eu	82	37	0.05	93	68
Gd	338	41	0.05	382	276
Tb	52	41	0.05	59	43
Dy	312	40	0.05	350	259
Ho	64	39	0.05	71	54
Er	176	40	0.05	196	148
Tm	26	40	0.05	29	22
Yb	163	41	0.05	180	138
Lu	24	40	0.05	26	20

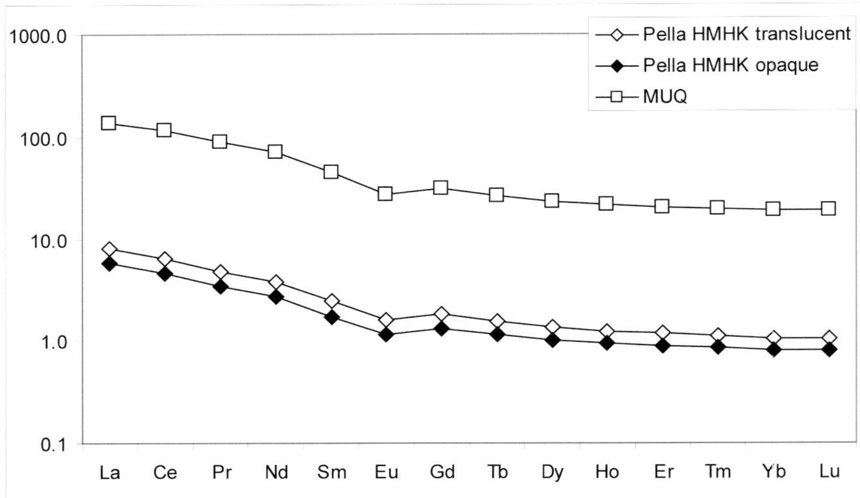

Figure 4-26 Line plot comparing averaged Pella HMHK REE data and MUQ REE data both normalised to average chondrites for comparison (chondrite data from Sun and McDonough 1989, reproduced in Table 27)

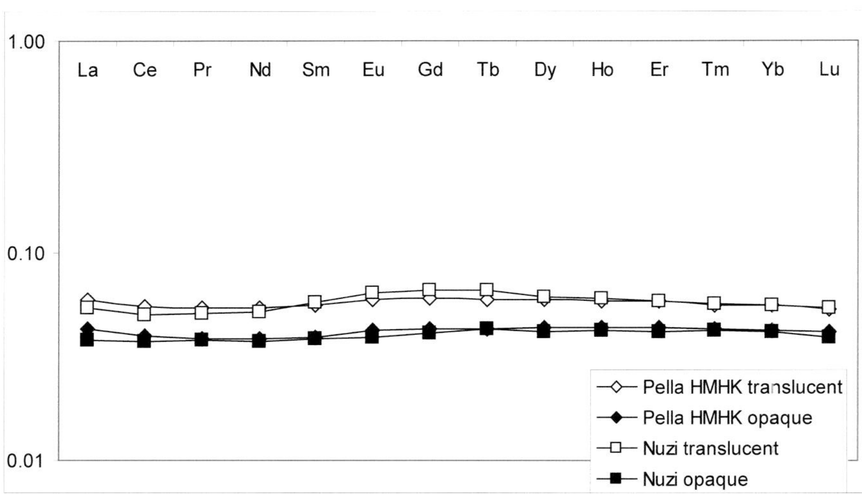

Figure 4-27 Line plot of the averaged translucent and opaque glass REE from Pella HMHK glasses compared with those from Nuzi glasses. Data are normalised to MUQ. A logarithmic scale is used.

The REE concentrations are uniformly slightly higher in the low-lime translucent glasses. This indicates that the opacifier did not add REE to the glasses. As the colourant is the same in both translucent and opaque glasses, and the silica source is relatively pure, it suggests that either the plant ashes and associated sediment incorporated in the opaque glasses contain lower levels of the REE, or the elemental concentrations in the opaque glasses are relatively lower due to the addition of lime and Sb.

It can be concluded that the REE in Pella HMHK glasses are close in pattern and concentration to those in the glasses from Nuzi, both exhibiting a continental crustal pattern, although of lower concentration than is found in the UCC, due to the purity of the raw materials used to make the glasses. The ratios of the concentrations of both Nuzi and Pella HMHK REE to the concentrations of these elements in the UCC are between 0.05 to 0.06, forming a constant base-line ratio (SBS) with which to compare the ratios of other elements. The most probable source of the REE is in association with the plant ash, possibly from sediment attached to the plants or incorporated during preparation of the ashes, while the silica, lime, colourant and opacifier sources appear to be unrelated to the REE (see section 3.3.1).

The REE patterns are the same for the opaque and translucent glasses, although the concentrations of REE in the translucent glasses are slightly higher.

4.5.2 Sediment-Related Elements (SRE)

The next group of trace elements to be considered is that related to sediments which were unintentionally added by the glass-makers and are present as contamination of the batch ingredients. Their concentrations are averaged and presented in Table 4-6.

Larger RSD values (e.g. for W and Tl) are due to the larger range of values for those elements. Several inconsistently high values have been excluded from the calculation of overall averages (see notes below Table 4-6 for a list of the number of samples averaged for each element). The samples 77, 119 and 136, in particular, have several elements present at exceptionally high concentrations, the reasons for which are unclear. While glasses 77 and 119 are purportedly IA in date, sample 136 is LBA, as are the other five samples (86, 188, 195, 198, 202), each with one inconsistently high value, which indicates that these compositional anomalies are not a result of the glasses having been produced in different chronological periods.

The SRE are generally well-correlated with each other (range of r^2 = 0.5 to 0.9) with the exceptions of Cr, Mo, W and Tl as noted previously, and with the REE, as found with the glasses from Nuzi. The normalised concentrations of the SRE are represented in Figure 4-28 with the SBS level of 0.05 marked by a dotted line. The average content of Cr is enriched as it is in the glasses from Nuzi, and W, Tl and U are enriched in the glasses from Pella compared to the SBS and to the levels of these elements in the Nuzi glasses. The remainder of the elements have contents similar to or below the level of the SBS and the pattern is otherwise closely similar in both sets of glasses.

Figure 4-29 is a line plot comparing the average concentrations of SRE in high- and low-lime Pella HMHK glasses with those in MUQ. The glasses have a similar pattern of SRE to that found in the UCC. Their concentration in the high-lime glasses is slightly higher,

Table 4-6 Averaged raw concentrations in ppb of sediment-related elements from Pella HMHK glasses.

Pella HMHK Sediment-Related Elements (ppb)

Elements	*n=43* Average All Glasses	RSD	Avge glass/ MUQ	*n=25* Average Low Lime	*n=18* Average High Lime
Sc	879	35	0.05	896	854
Ti	244,137	35	0.04	262,588	215,919
V	6920	34	0.06	7233	6441
Cr	20,648	60	0.32	22,524	17,779
Ga	759	41	0.04	774	671
Y	1760	40	0.06	1946	1476
Zr	9839	37	0.05	9949	8774
Nb	662	41	0.04	691	560
Mo	862	61		861	862
Hf	236	36	0.04	241	209
Ta	40	64	0.04	43	29
W	194	132	0.12	147	264
Tl	85	133	0.20	104	53
Th	340	30	0.03	357	316
U	221	64	0.08	239	197

After removal of outliers for calculation of averages for Average All Glasses n=42 for Tl, n=41 for Ga, Zr, Nb, Hf, and Ta, n=40 for W, Th, and U. For Average Low Lime n= 23 for Ga, Nb, Hf, Ta, and W, n=22 for Zr, Th and U. For Average High Lime n=17 for W and Tl.

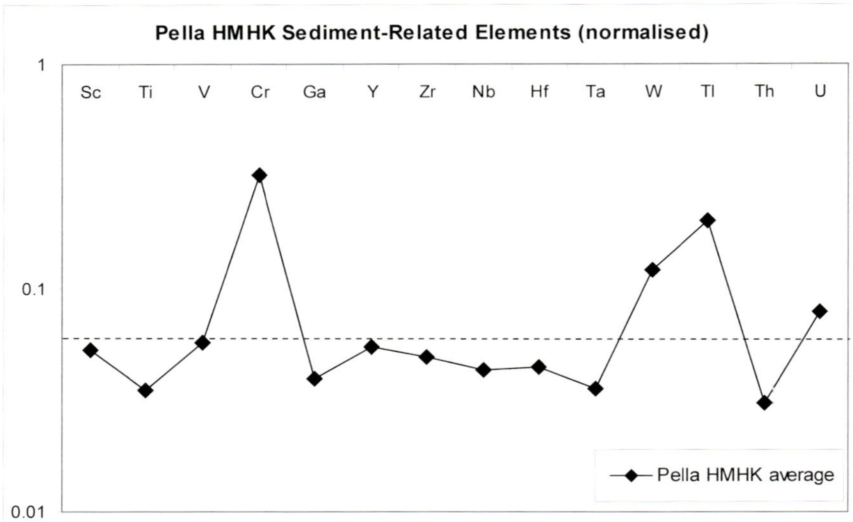

Figure 4-28 Line plot of the averaged concentrations of sediment-related trace elements in Pella HMHK glasses, normalised to MUQ. A logarithmic scale is used. Mo is not plotted as there is no MUQ value by which to normalise it. The dotted line crosses the Y-axis at 0.05 to represent the approximate SBS of the Pella HMHK glasses.

as is found with the associated REE, but on the whole the patterns and concentrations are closely similar to each other and to those found in the Nuzi glasses.

It can be concluded that the SRE are mostly at concentrations close to the SBS defined by the REE, with the notable exception of Cr, as seen in the glasses from Nuzi. It is most likely that the REE and SRE, including iron and alumina, which are also at the SBS level, entered the glass together, with the plant ash. It is evident also that there is a similarity in the pattern of the SRE in the UCC to the pattern of the SRE in both Nuzi and Pella HMHK glasses. There is little difference in the patterns and concentrations of the SRE between the opaque and translucent glasses from Pella, indicating that the raw materials were derived from similar geological sources.

4.5.3 Alkali and Alkaline Earth Elements

The alkali and alkaline earths, magnesia, potash, soda, lime and phosphorus are constituents of plant ashes used to flux HMHK glasses, and have been analysed by

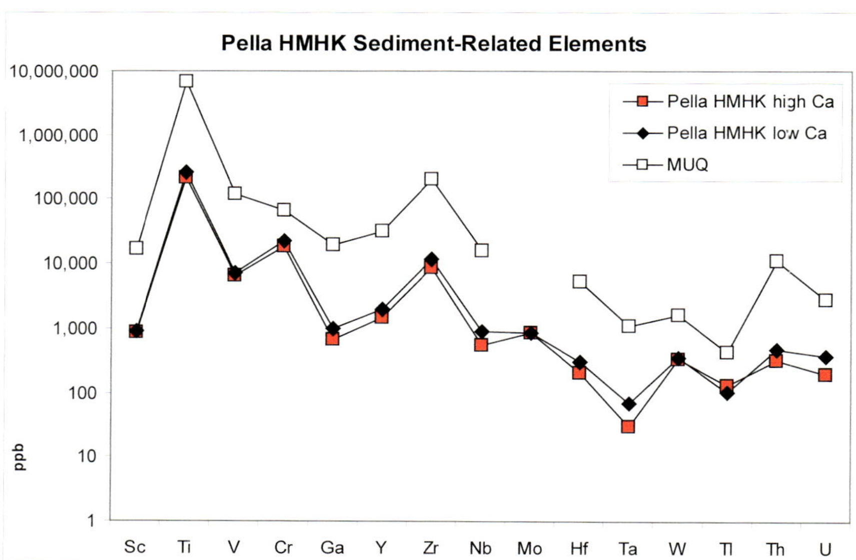

Figure 4-29 Line plot comparing averaged raw concentrations of sediment-related trace elements in opaque and translucent Pella HMHK glasses with MUQ. A logarithmic scale is used.

Table 4-7 Averaged raw concentrations in ppb of alkali and alkaline earth elements from Pella HMHK glasses*.

	Pella HMHK Alkali and Alkaline Earth Elements (ppb)	Li	Be	Rb	Sr	Cs	Ba
n=43	Avge All Glasses	16,351	99	16,460	381,501	443	46,318
	RSD	38	29	26	27	31	31
	Average glass/ MUQ	0.58	0.05	0.21	2.69	0.08	0.12
n=25	Average Low Lime	18,455	104	16,783	379,550	413	46,706
n=18	Average High Lime	13,134	93	15,966	384,484	488	45,724

*For Average All Glasses ***n***= 40 for Be, and for Average Low Lime ***n***= 22 for Be after outliers are removed.

SEM-EDS and reported in section 4.4. Other alkali and alkaline earth elements assayed for by ICP-MS in the present study are Li, Be, Rb, Sr, Cs and Ba. The averaged raw data are presented in Table 4-7. Three unusually high values for Be, from low lime samples 77, 119 and 136, were excluded in the calculation of averages and RSD values.

Figure 4-30 compares the averaged alkali and alkaline earth elements in the glasses from Pella and Nuzi with those in MUQ, and shows that they are near identical in elemental concentrations. Pella glasses exhibit a similar enrichment of Li and Sr to that found in glasses from Nuzi. Be, Ba and Cs are at low, almost SBS levels and Rb is slightly enriched (Figure 4-31). The patterns are similar to those found when MUQ is plotted, but at depleted concentrations, as observed in other groups of elements from the Nuzi and Pella HMHK glasses. These data are consistent with the use of a plant ash flux.

From Table 4-7 above and Figure 4-32 below, it is evident that there is essentially no difference in the contents of the alkali and alkaline earth elements between the high- and low-lime groups. The concentrations of the plant ash-related elements, soda, magnesia and potash, change slightly between the high- and low-lime groups, but the concentrations of the alkali and alkaline earth trace elements and SRE do not.

As was found in the glasses from Nuzi, it is likely that Be is related to contamination from sediments rather than to the composition of the plant ash because it is found at close to the SBS level.

The concentrations of the alkali and alkaline earth elements are closely similar to those of their counterparts in the glasses from Nuzi. Li, Rb and Sr are particularly enriched relative to their content in the UCC. Be remains at approximately SBS level indicating, as at Nuzi, that this element could be more closely related to the SRE in these glasses. Interestingly, the difference in the concentrations of alkali and alkaline earth oxides observed in the base glass compositions between the high- and low-lime groups are not observed in the concentrations of the alkali and alkaline earth elements in the trace element analyses

4.5.4 Colourants and Opacifiers

The Pella HMHK glasses analysed in the present study include glasses of the following colours: 25 blue translucent, 16 turquoise blue opaque, one unopacified black, and one colourless. Results of ICP-MS analyses of colourant and opacifying elements, and those elements associated with them, are presented in Table 4-8 in ppm. Manganese is omitted as it fell below detection limits of the SEM-EDS used to analyse it. Variability in the concentrations of these elements in the glasses from Pella is reflected in large RSD values for the data. Outliers omitted from the calculations of averages and RSD values are listed beneath the table.

All but two glasses are coloured blue. SEM-EDS results indicate that copper oxide is present in the blue glasses at concentrations ranging from 0.5% to 2.2%, and average 1.9%, as for the glasses from Nuzi (section 3.3.4). The higher concentrations of Cu would impart a

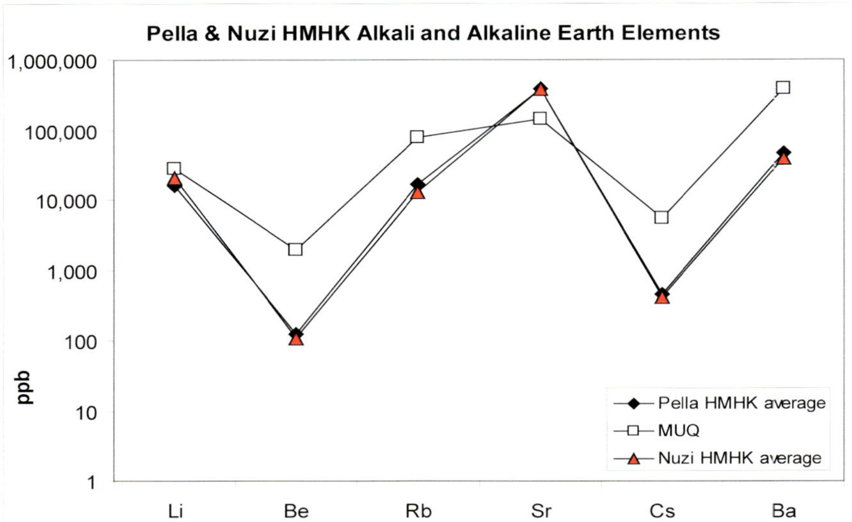

Figure 4-30 Line plot of averaged raw data for Pella HMHK alkali and alkaline earth elements compared with MUQ. A logarithmic scale is used.

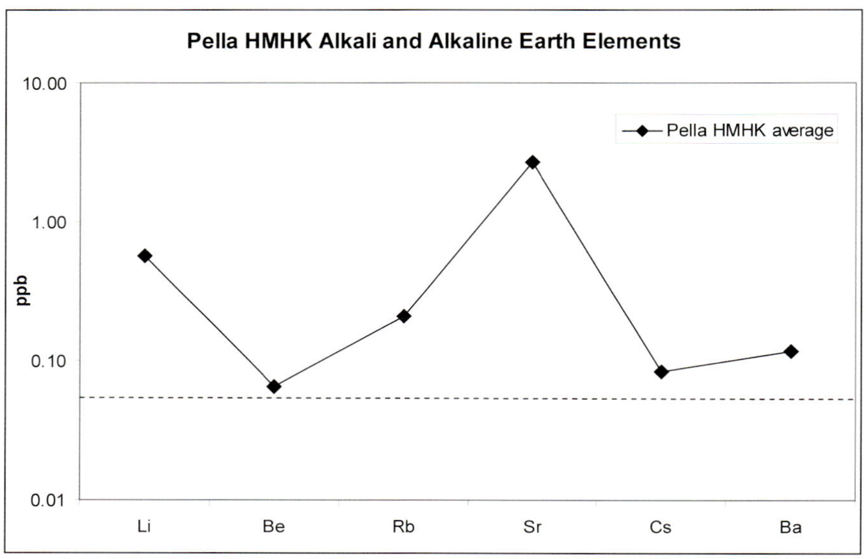

Figure 4-31 Line plot of averaged Pella HMHK alkali and alkaline earth trace elements normalised to MUQ, on a logarithmic scale. The dotted line represents the approximate SBS of 0.05, as determined by the ratios of Pella HMHK REE to MUQ REE.

blue colour to the glasses. From ICP-MS analyses, the concentration of elemental Cu averages 8800 ppm for all blue glasses (7613 ppm for opaque and 9525 ppm for translucent glasses, compared with an average in the blue Nuzi glasses, all but two of which are translucent, of 9160 ppm). Cu is particularly enriched compared with the average concentration of this element found in the UCC (Table 4-8).

While there are no green glasses in the assemblage from Pella, 20 of the blue glasses have iron oxide levels in excess of 0.4% and up to 0.8% (section 4.4). On average, the concentrations of iron oxide in the Pella HMHK glasses (0.39%) are marginally greater than those found in the glasses from Nuzi (0.26%). It is possible that iron contributes to the colour of the blue glasses, especially when their Cu concentrations are low, as is observed for both the green and some of the blue glasses from Nuzi.

In the glasses from Pella, the elemental Co concentrations have a wider range of values (1 ppm to 31 ppm, average 7 ppm) than is found in the glasses from Nuzi, and this element is considered unlikely to

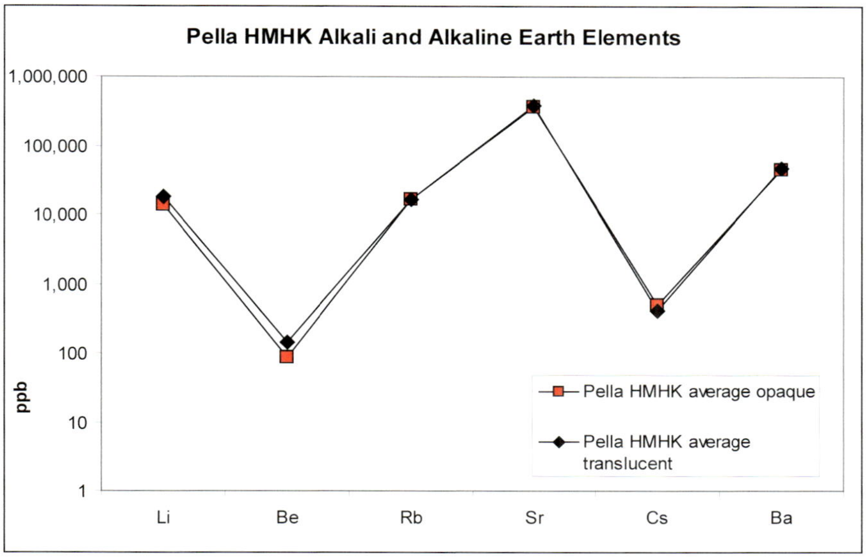

Figure 4-32 Line plot of averaged raw data for opaque (high lime) and translucent (low lime) Pella HMHK glasses. A logarithmic scale is used.

Table 4-8 Concentrations in ppm of colourant and related elements from Pella HMHK glasses.*

	Pella HMHK Colourant and Related Elements (ppm)	Co	Ni	Cu	Zn	As	Cd	Sn	Sb	Pb
n=16	***Opaque blue***									
	Average	4.71	20.5	7613	20.9	35.9	0.121	5.93	12,573	13.3
	RSD	54	32	38	43	78	53	84	32	61
n=25	***Translucent blue***									
	Average	7.82	20.3	9525	22.6	33.0	0.144	5.20	16.2	10.6
	RSD	59	63	30	30	86	26	136	90	110
n=1	***Green-black***	3.17	23.6	245	26.9	2.03	0.041	5.09	1456	3.16
n=1	***Colourless***	1.76	13.4	12.7	26.4	1.45	0.134	0.381	0.640	6.63
Av. opaque/MUQ		0.21	0.65	235	0.28			2.1		0.65
Av. transl. blue/MUQ		0.35	0.64	294	0.31			1.88		0.52

*Note that there are no values for Sb, Cd or As in MUQ. Data for opaque and translucent blue, green-black and colourless glasses have been presented separately because of significant differences in the concentrations of several key elements. Averages after removal of outliers for opaque blues n=15 for Co, Ni, Cd, Sb, and n=14 for Sn and Pb; for translucent blues n=24 for Co and Sn, n=23 for Pb, and n=22 for Sb.

act as a colourant at these low concentrations (section 1.10).

The concentrations of colourant-related elements are similar for the translucent and opaque blue glasses, except for the concentrations of Sb. Translucent glass samples 210 and 216 have higher than average Sb concentrations (approximately 1800 ppm compared with the average concentration in translucent glasses of 16.2 ppm). These values are low compared with the average Sb concentration in the opaque glasses (12, 573 ppm). One opaque sample (209) has a very low Sb concentration of 4 ppm and the reason for its opacity is unclear.

The differences in the Pb and Ni concentrations of opaque and translucent glasses from Nuzi were not observed in the glasses from Pella. The concentrations of Pb and Ni are well correlated (r^2 = 0.65) in the Pella translucent blue glasses.

It appears that the content of Sn is insufficient for it to have entered the glass associated with Cu in bronze, as ratios of Sn:Cu are 0.04:1 or less. For Sn and Cu to have been derived from bronze, the ratio of Sn to Cu would need to be at least 1:10 to reflect the proportions of these elements found in ancient bronzes (section 1.10; Shortland 2000a; Shortland 2005: 4). When Cu is plotted against Sn (Figure 4-33), there appear to be two groups of glass based on Sn concentration, similar to those observed in the glasses from Nuzi. Sn concentrations in the Pella high Sn group range from 11.34 ppm to 24.70 ppm (8.79 ppm to 12.13 ppm in the Nuzi high Sn group) and from 0.69 ppm to 8.14 ppm in the low Sn group (0.21 ppm to 8.74 ppm in the low Sn Nuzi glasses). In the glasses from Pella, the concentrations of Cu and Sn are well correlated in the high Sn group (r^2 = 0.79), as they are in the low Sn group of the Nuzi glasses (r^2 = 0.76). Glasses from Nuzi appear to be grouped as low Sn-high Cu and high Sn-low Cu, but Pella glasses have similar ranges of Cu concentration for both the low Sn (4098 ppm to 15,627 ppm) and high Sn groups (4672 ppm to 13,302 ppm).

Table 4-8 presents the ratios of the averaged data for opacified and translucent colourant-related elements to the MUQ values for these elements. The ratios are all elevated above the SBS, determined by the ratios of Pella HMHK REE to MUQ REE. The concentration of Cu is particularly elevated and is thus responsible for colouring most of the glasses blue. Although the concentration of iron oxide is not elevated above the SBS, it might be contributing to the blue colour particularly where the Cu concentrations are very low, as can be seen in the glasses from Nuzi (section 3.2). Calcium antimonate acts as the opacifying agent for the turquoise blue opaque glasses.

In Figure 4-34, the average colourant-related trace element concentrations of opaque and translucent Pella HMHK glasses of all colours are compared with the concentrations of the colourant-related elements found in MUQ, reported in Table 4-8. This comparison indicates that the concentrations of all elements are enriched compared with the average values found in the UCC.

On this line and dot plot, the opacified and translucent blue glasses exhibit similar patterns at similar concentrations to each other, except for Sb, which illustrates that they have closely similar compositions. The compositions of the blue Pella glasses are similar to their counterparts in the Nuzi glass collection (section 3.3.4). The black and colourless glasses have comparatively low concentrations of Cu and As, and similar concentrations of Ni, Zn and Cd compared with the blue glasses, suggesting that these last three elements are inconsequential to the colouring of the glass. The blue glasses have slightly higher concentrations of Pb and Co than the non-blue glasses, which could be a result of their association with Cu and As in the raw material used to colour the glasses blue.

The proposition that Cu and As might have been added together as components of the same raw material is supported because their concentrations are elevated in the blue glasses, and are comparatively low in the non-blue glasses. There is a positive correlation (r^2 = 0.54) between the concentrations of Cu and As in the translucent blue glasses (Figure 4-35), which is not seen in the opaque blue glasses. Glasses with As contents in the range of approximately 1.5 ppm to 25 ppm form a more compact group than glasses with higher As contents in the range of approximately 33 ppm to 92 ppm, suggesting that there might be two groups based on As content.

In Figure 4-36 the division of the glasses into a high and a low As group is clearly depicted when As is plotted against Co. The proposed As groups do not correspond with other identified compositional groupings, and both contain translucent and opaque blue glasses. The concentrations of As and Co are well correlated (r^2 = 0.66) in the low As group, with low average Cu content, but are not correlated in the high As group with wider ranging Cu content. There is no apparent correlation between Cu and Co in the Pella glasses. It appears that two different Cu sources were employed by the makers of the Pella HMHK blue glasses. As was not sought in the analyses of the glasses from Nuzi.

The black glass has very low concentrations of Co (3.17 ppm) and Cu (244.95 ppm) by comparison with the blue glasses, and has an iron oxide concentration of 0.4 % which is probably responsible for the black colour. This

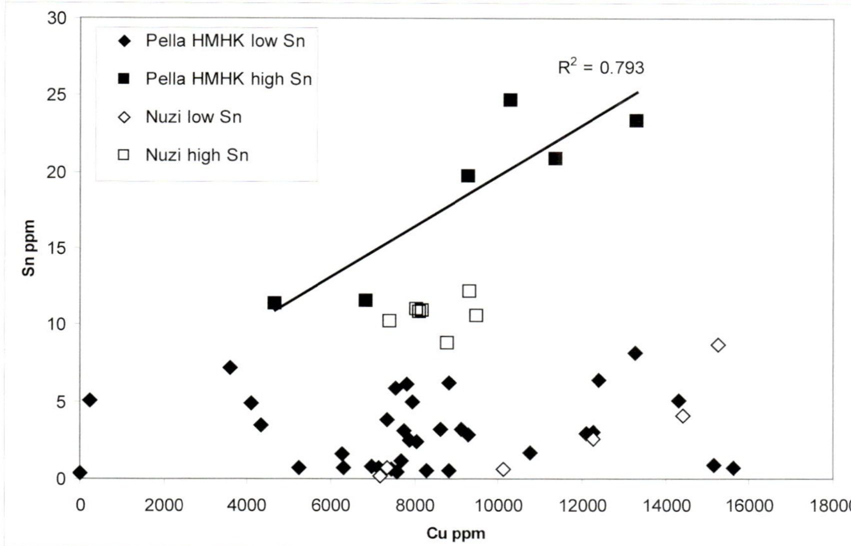

Figure 4-33 Scatter plot of Cu vs Sn in ppm for Pella HMHK and Nuzi glasses. Three glasses with exceptionally high Sn concentrations were removed to avoid compression of the graph.

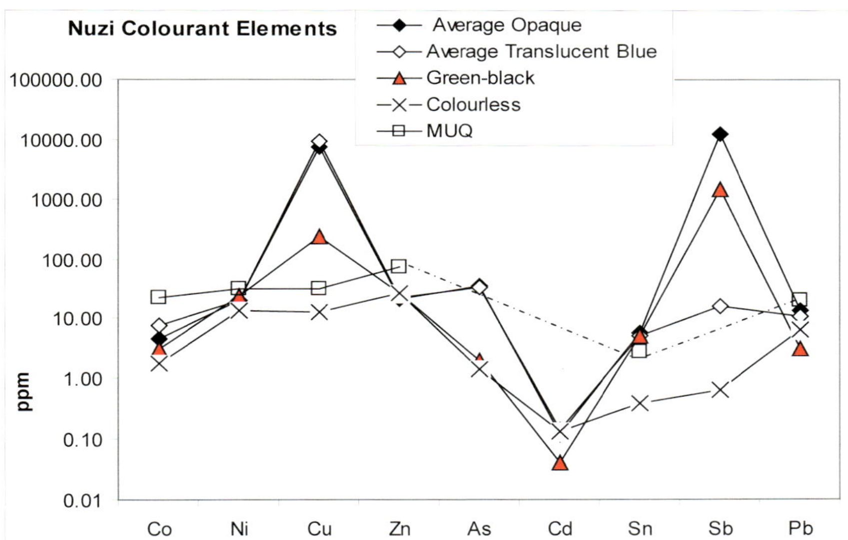

Figure 4-34 Line plot of colourant and related elements in Pella opaque and translucent glasses, and MUQ, from ICP-MS data. A logarithmic scale is used to avoid compression. There is no MUQ value by which to normalise Sb, Cd or As.

glass has a high lime concentration (8.2%), but has less Sb (1456 ppm) than the opaque glasses and more than the translucent blue and colourless glasses.

The colourless glass has the lowest concentrations of most elements and belongs in the low lime group (6.1%). The colourless glass has a relatively low Sb concentration (0.64 ppm) and contains 0.5% iron oxide, which contributes a greenish tinge.

In summary, the colour of the blue glasses is achieved by the addition of Cu to the glasses, with the colour being enhanced to some extent by iron oxide concentrations in excess of 0.4%. Based on a 'high' and a 'low' Sn content in the glasses from Pella, and 'high' and 'low' As groups, it is interesting to consider that the glass-makers might have used Cu from different sources, as has been suggested for the glasses from Nuzi in relation to Sn. The data for As from the Pella glasses adds strength to the proposition that Cu was sourced from

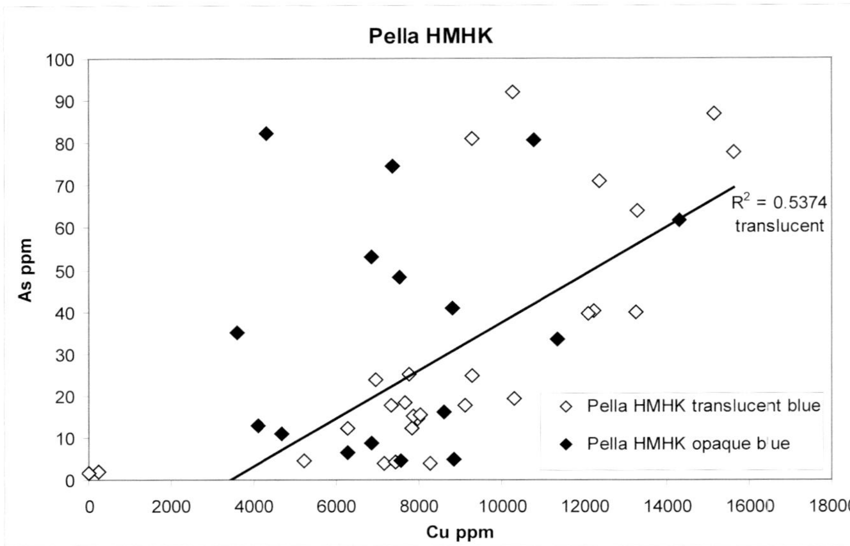

Figure 4-35 Scatter plot of Cu *vs* As for Pella HMHK glasses in ppm.

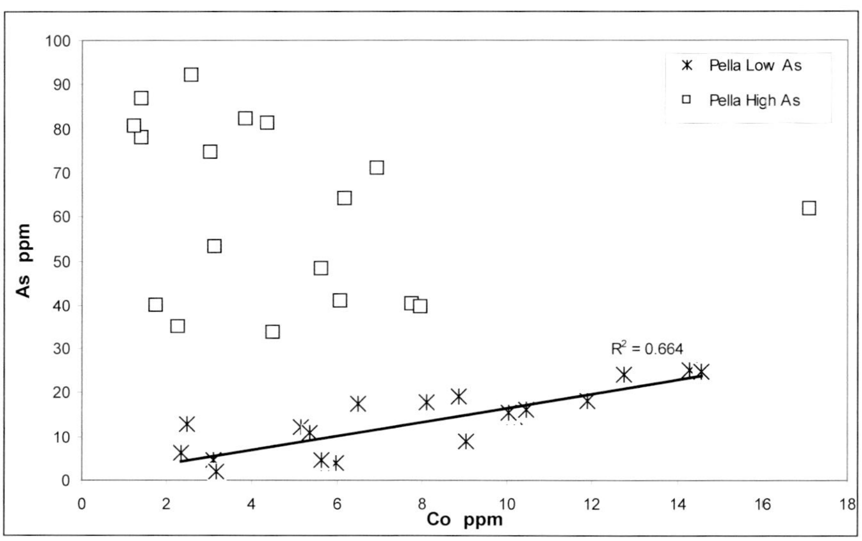

Figure 4-36 Scatter plot of Co vs As for Pella HMHK glasses divided into low and high As groups. Two high Co outliers (samples 189 and 207) are omitted from the low As group.

more than one place. The low ratios of Sn to Cu do not support the use of bronze scrap as a source of Cu or Sn, consistent with previous findings by Shortland (2005: 4) for Mesopotamian glasses.

4.6 General Conclusions

The base glass composition of the Pella HMHK glasses is nearly identical to that of the Nuzi glasses as reported in Chapter 3. The Pella glasses can be divided into similar high- and low-lime groups, which were identified in the glasses from Nuzi, although in the glasses from Pella it is clear that the lime groups are related to the opacity of the glasses: the high-lime group consists of glasses opacified with calcium antimonate, and the low-lime glasses are translucent. There are three exceptions: the low-lime opaque glass, sample 222, and the two high-lime translucent glasses, samples 74 and 194.

REE, SRE and alkali and alkaline earth and colourant data are closely similar to those for glasses from Nuzi, indicating that these two groups of HMHK glass are representatives of a constant and regulated glass-making tradition that spanned some four and a half centuries from *c.*1550 to *c.*1200 BC, from the Syro-Levantine region to Mesopotamia.

The four glasses attributed uncertainly to the IA are compositionally nearly identical to their HMHK counterparts from the LBA. The weight of chemical and physical evidence strongly suggests that they were made in the LBA and were used into the IA before being discarded in the contexts from which they were eventually excavated. It is doubtful, based on the present study, that the HMHK glass manufacturing tradition survived into the IA, even with what would be undeniably dwindling output.

Chapter 5

Pella: Iron Age Glasses Results of Compositional Analyses

5.1 Introduction

Glass-making traditions of the early 1st millennium BC (IA) Near East have been little investigated at the time of writing and are not well understood, although this is arguably one of the key periods in the development of glass-making. What little remains of this glass represents an important chronological and geographical link between the LBA glass-making technologies of Egypt and the Near East and the beginning of the long Greco-Roman tradition towards the end of the 1st millennium BC. In the following three chapters (Chapters 5, 6 and 7) the results of compositional analyses of the IA glasses from Pella and the Assyrian capital, Nimrud, are reported. This chapter will cover the Pella LMLK glasses from the IA (see Reade *et al.* 2009).

5.2 Summary of LMLK Samples from Pella

The HMHK glasses from Pella have been described in Chapter 4. The second compositional group found at Pella consists of IA LMLK glasses from the temple region and Tomb 89. Both contexts are dated from the early IAI to the IAII, 11th to 9th C BC. The LBA and IA glasses and their contexts have been described in the Chapter 4 with the general introduction to the site (4.1, 4.2).

The LMLK compositional group of glasses from IA Pella consists of 23 small, very dark green beads which appear black (sample 74 is an LBA black HMHK glass). Nineteen of these beads were unearthed from Area XXXII, trenches A, B and W, of the IA temple, dating between *c.*1050 – 850 BC. Sixteen of these beads were found together in the same locus and level of Area XXXIIB and could have belonged to a necklace or in some other collective form. They date from the IAI (late 11th C BC) to as late as the 9th C BC, according to the date of the latest pottery found in this pre-Aramaean IAI to IAII context. Comparable contexts in the IA temple area have been dated by Carbon-14 to the *c.*10th C BC (Bourke forthcoming). One bead (sample 46) was found in Area XXXIIA, and is dated to the late IAI (late 11th C BC). Of two beads found in Area XXXIIW, one probably dates to the 10th C BC (sample 132B), and the other is dated to the 9th C BC (sample 159).

From the contemporary Tomb 89, *c.*1050 – 850 BC, come four beads, barrel or spherical in shape, and which are of similar appearance to those described from the temple area. They are thought to date to the 10th C BC. All of the black LMLK glasses are dated to some 200 years from the 11th C to the 9th C BC. These beads were found in both burial and temple contexts, and their appearance and excellent preservation are consistent. The beads are all small (less than 10 mm maximum dimension) and vary slightly in shape to include spherical, slightly flattened spherical (i.e. the diameter exceeds the height slightly but not enough to describe them as ring or annular in shape), fluted spherical, barrel, and one unusual disc with a horizontal hole, that is to say the hole goes from side to side of the disc and not through the centre as in a ring bead (sample 159). The identification, provenance and descriptions of glasses appear in Table 5-1. These IA black glasses are visually quite different to those from the LBA at Pella, and provide a striking contrast, both physically and chemically, to the preceding LBA glass-making tradition.

5.3 LMLK Base Glass Composition

The IA LMLK glass samples were prepared and analysed for base glass oxides by SEM-EDS (section 2.3). This analysis defined a coherent compositional group of soda-lime-silica glass. Neither temple nor tomb context is related to the composition or the physical forms of the LMLK beads, as was found with the LBA HMHK glass from Pella. The averaged reduced compositions of the Pella LMLK glasses are presented in Table 5-2. Complete raw oxide and trace element data for each sample are tabulated in Appendix IV.

The Pella LMLK glasses are, by contrast with the HMHK glasses, low in magnesia (average of 1.5%, range of 0.7% to 2.8%) and potash (average 1.3%, range of 1% to 2%). When plotted as magnesia *versus* potash (Figure 5-5), the LMLK glasses form a relatively compact group distinct from the LBA HMHK glasses. Where magnesia and potash contents are correlated in HMHK glasses, this is not the case in LMLK glasses. Conventional reasoning would suggest that these are glasses fluxed with mineral soda, such as Egyptian natron. Reduced levels of lime, averaging 2.3%, and falling to less than 1% in some cases (range of 0.7% to 4%), support the view that these are indeed glasses made from natron which, by contrast with most plant ashes, contain little lime (Figure 5-6) (section 1.9).

Table 5-1 Description and provenance of Pella LMLK glass samples.

Object Number	Analysis Number	Object	Provenance	Date and Comments
900508	46	bead – spherical	Temple XXXIIA 19.26	probably Late IAI *c*.1050-1000BC
920047	52	bead – spherical	Temple XXXIIB 32.14	Following dates and comments apply to samples 52-67
920047	53	bead – spherical	Temple XXXIIB 32.14	The latest date of the pottery found in this context is 9th C BC
920047	54	bead – spherical	Temple XXXIIB 32.14	Carbon 14 dates for the IA temple area are *c*.10th C BC
920047	55	bead – flattened spherical	Temple XXXIIB 32.14	
920047	56	bead – flattened spherical	Temple XXXIIB 32.14	
920047	57	bead – flattened spherical	Temple XXXIIB 32.14	
920047	58	bead – flattened spherical	Temple XXXIIB 32.14	
920047	59	bead – flattened spherical	Temple XXXIIB 32.14	
920047	60	bead – flattened spherical	Temple XXXIIB 32.14	
920047	61	bead – flattened spherical	Temple XXXIIB 32.14	
920047	62	bead – flattened spherical	Temple XXXIIB 32.14	
920047	63	bead – spherical	Temple XXXIIB 32.14	
920047	64	bead – fluted spherical	Temple XXXIIB 32.14	
920047	65	bead – barrel/ spherical?	Temple XXXIIB 32.14	
920047	66	bead	Temple XXXIIB 32.14	
920047	67	bead – fluted flat spherical	Temple XXXIIB 32.14	
991059	132B	bead – spherical	Temple XXXIIW 1.13	IAII *c*.11th – 10th C BCE, probably 10th C BC white trailed decoration on end
010597	159	bead- disk-shaped, horizontal hole	Temple XXXIIW 2.60	IAII *c*.10th-9th C BC probably 9th C BC
871855 RN100242	223	bead – elongated barrel	Tomb 89 1.4	IAI-II probably 10th C BC
871706 RN100231  Figure 5-1 Pella 100231 green-black spherical bead, 7 mm L	224	bead – spherical	Tomb 89 1.4	IAI-II probably 10th C BC

871706 **RN100231** 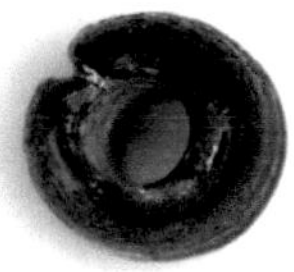 Figure 5–2 Pella 100231 green-black small barrel bead 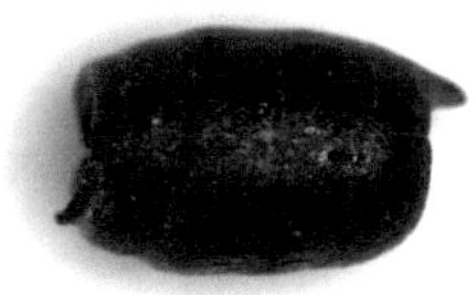 Figure 5–3 Pella 100231 green-black small barrel bead side view, 10 mm L	225	bead – small barrel	Tomb 89 1.4	IAI-II probably 10th C BC
870988 **RN100147**  Figure 5–4 Pella 100147 green-black small barrel bead, 8 mm L	226	bead – small barrel	Tomb 89 1.4	IAI-II probably 10th C BC

Table 5-2 Average reduced compositions of 23 Pella LMLK glasses in wt % oxide.*

Samples	Na_2O*	MgO*	Al_2O_3*	SiO_2*	P_2O_5*	K_2O*	CaO*	FeO*	Total
Average *n*=23	17.07	1.49	1.62	66.32	0.17	1.34	2.26	9.73	100.00
SD	1.38	0.48	0.46	2.04	0.06	0.23	0.76	2.87	
Average Glass /MUQ	21.34	0.95	0.09	0.98	0.94	0.87	2.26	1.13	

*The ratio of average glass composition to MUQ is calculated with raw data.

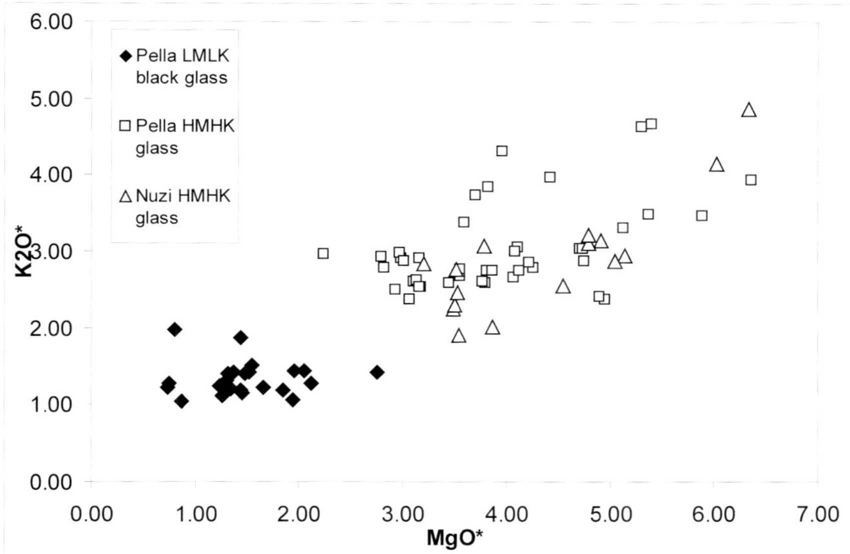

Figure 5-5 Scatter plot of magnesia vs potash in wt% oxide for Pella LMLK and HMHK groups, with Nuzi HMHK glass added for comparison.

Examination of the plot of lime *versus* magnesia (Figure 5-6) for LMLK and HMHK glasses shows that the LMLK glasses have lower concentrations of both magnesia and lime, and that the concentrations of these oxides are well correlated (r^2 = 0.8), indicating that they enter the glass in the same raw material, in this case presumably the sand source of silica. The phenomenon of high- and low-lime groups, identified in the HMHK glasses from both Nuzi and Pella, is not observed in the LMLK glasses.

The main differences between the Pella IA LMLK and the LBA HMHK base glass compositions are the highly enriched iron oxide (average of 9.7%) and alumina (average of 1.6%) contents and the much less enriched magnesia, potash and lime contents of the LMLK glasses. The high iron oxide and alumina contents of the LMLK glasses are suggestive of the use of a sand source of silica (section 1.6) rather than quartz pebbles, and this in combination with their low contents of potash, magnesia and lime are indicative of the use of

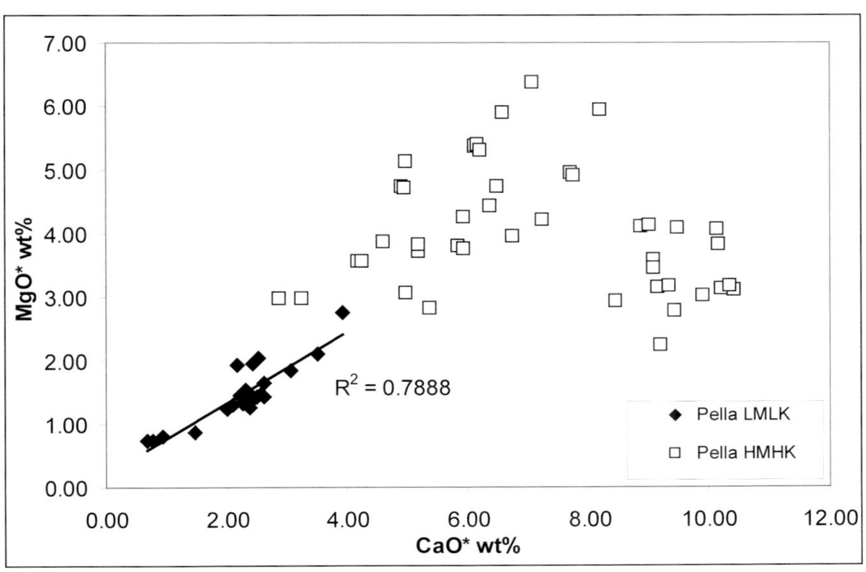

Figure 5-6 Scatter plot of lime vs magnesia for Pella LMLK glasses compared with Pella HMHK glasses.

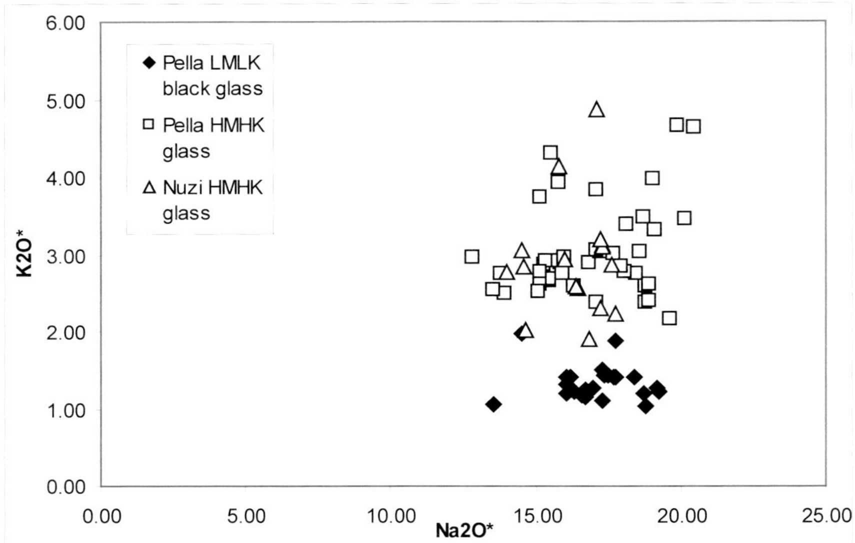

Figure 5-7 Scatter plot of soda vs potash in wt% oxides for Pella HMHK and LMLK glasses, with Nuzi HMHK glass added for comparison.

a mineral soda flux. Although the possibility that LMLK plant ashes were used must be considered in the light of current research into plant ash compositions (for example Barkoudah and Henderson 2006; Tite *et al.* 2006; section 1.8), the evidence from the data convincingly suggests a natron and sand composition.

The potash and magnesia concentrations of the LMLK glasses are close to those found in the earth's crust, while lime is present at twice the concentration (see Table 5-2). Phosphorus, which is associated also with plant ash, is found at concentrations in the LMLK glasses which average 0.17%, as might be expected in a plant ash glass, but the difficulty of measurement of this oxide at low concentrations with SEM-EDS makes it unwise to draw conclusions about its concentration here. The soda and silica concentrations of the LMLK glasses are comparable to those found in the HMHK glasses. In the HMHK glasses, iron oxide and alumina are present at SBS levels, but in the LMLK glass the iron oxide content is disproportionately higher than that of alumina, the reason for which will be addressed in sections 5.4.4 and 9.6.

The chemical data from the Pella glasses provide the essential characterisation of the typical soda-lime-silica base glass compositions for both HMHK and LMLK glasses: silica and soda concentrations remain relatively constant, while the choice of flux influences the concentrations of lime, magnesia and potash. Figure 5-7 provides a comparison of soda and potash levels from HMHK and LMLK glasses analysed in the present study. It has been shown that lime, potash and magnesia are all found at lower concentrations in the LMLK glasses. The soda concentration averages 17%, with a range of 14% to 19% in the LMLK glasses, and is closely similar to that of the HMHK glasses, indicating that even though the raw materials are different in these glasses, the total soda content remains the same.

The HMHK glasses contain an average of 0.8% alumina (with four samples of over 1%), while the IA LMLK group has relatively high concentrations of alumina averaging 1.6% (range of 1% to 2.6%), suggesting that the silica source was a relatively pure form of quartz for the HMHK glasses, such as vein quartz pebbles, but was an alumina-bearing sand for the LMLK glass (e.g. Brill 1999b: 474-477; Brill 1995: section 1.6). Grains of unmelted silica (up to approximately 250μm across), associated with cracking of the glass presumably due to differential thermal expansion during manufacture, can be seen in some of the LMLK beads (Figure 5-8), and their sizes and shapes are consistent with having been derived from a sand.

The average content of iron oxide is unusually high at 9.7%, with a range of 5.6% to 16.3%. At this level, iron oxide is assumed to be responsible for the black colour of these glasses (section 1.10). Figure 5-9 illustrates the concentrations of iron oxide *versus* alumina for LMLK and HMHK glasses. There is, however, a strong correlation between iron oxide and alumina in the HMHK glasses, because these components are derived from contaminant sediments and preserve the natural ratio found in the earth's crust. This is not the case in the iron-rich LMLK glasses where the ratio of alumina to iron oxide is only 0.17:1, and there is no correlation between them.

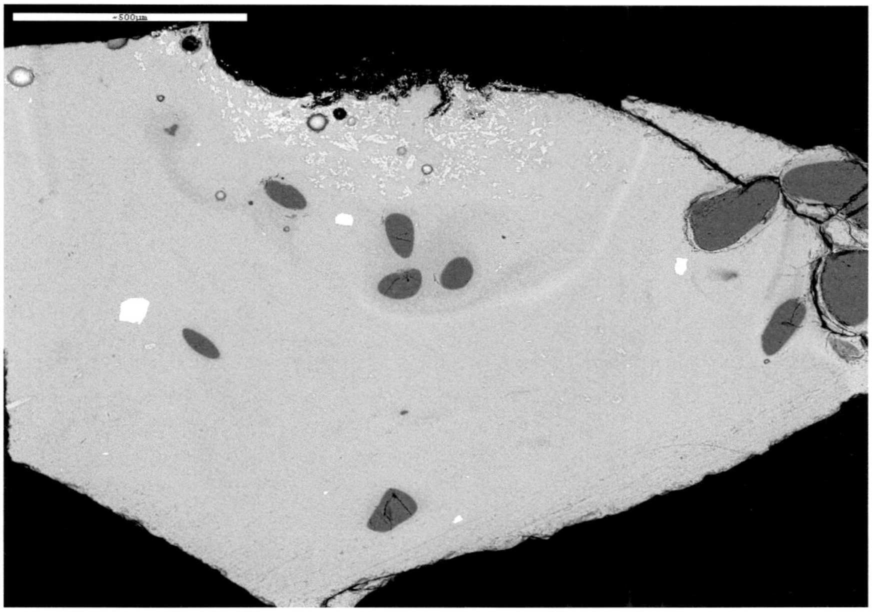

Figure 5-8 Backscattered SEM image of IA black glass, PEL 132B. White inclusions are chromite (the largest are indicated by arrows). Large gray inclusions are silica and a zone of small grains of a pyroxenoid devitrification product may be seen at the top of the micrograph.

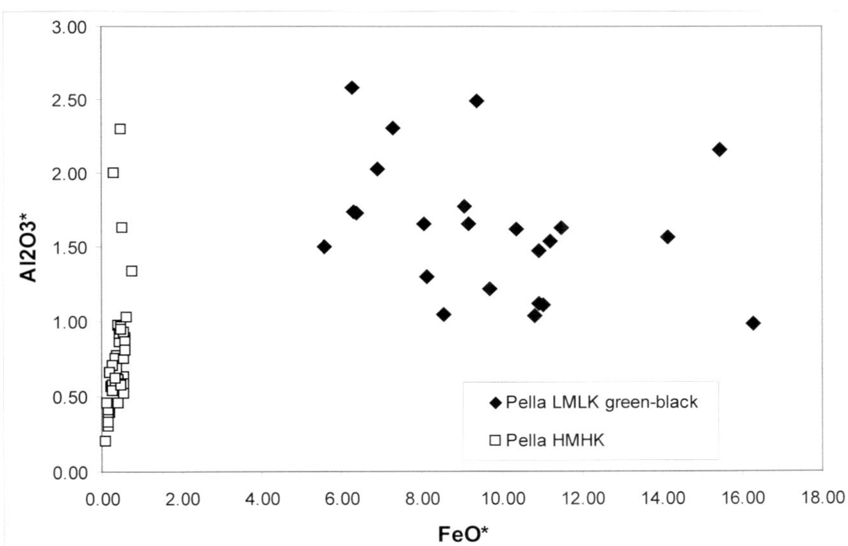

Figure 5-9 Scatter plot of iron vs alumina in wt% oxides for Pella LMLK and HMHK groups.

Using the SEM, bright grains of chromite, an iron-bearing chromium-aluminium oxide, were observed in a number of the IA black glasses (Figure 5-8). Chromite is a refractory mineral which is difficult to dissolve in glass. Its association with silica grains and its smaller size suggest that it may have been a heavy-mineral impurity in the glass-making sand. It is not clear at this stage if a separate iron oxide-rich colourant was added to a pre-existing glass or if the glass was made from a heavy mineral-bearing sand, which included a small amount of chromite as well as a large amount of iron oxide (section 9.6.1).

5.3.1 Summary

The Pella LMLK glasses are soda-lime-silica glasses believed to have been made with silica and lime from sand, and a mineral soda flux. The soda and silica levels

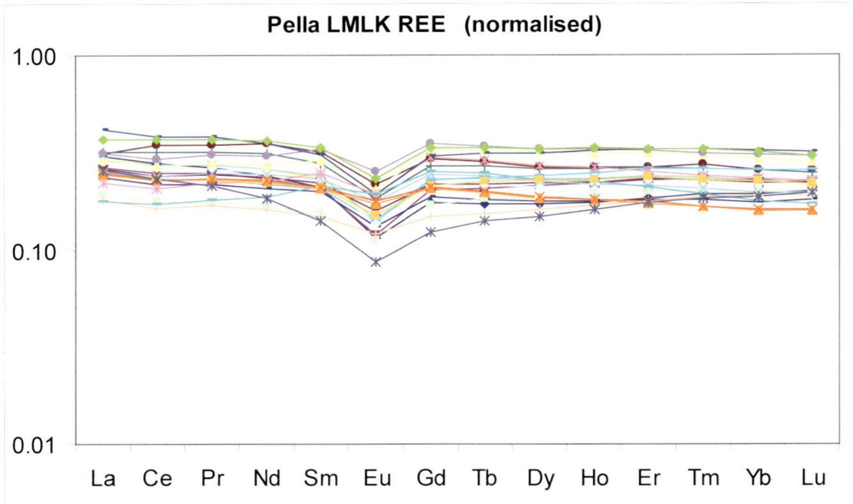

Figure 5-10 Line plot of the REE of individual Pella LMLK glasses showing similarity of pattern and concentration. A logarithmic scale is used. Data are normalised to MUQ.

are consistent with those of the HMHK glasses, whereas the use of a different flux alters the concentrations of lime, magnesia and potash. Alumina, from sand, is more concentrated in the LMLK glasses. They have unusually high iron oxide concentrations, which are assumed to provide their black colour and are responsible for their excellent preservation, where their low levels of lime might otherwise have allowed them to perish.

5.4 Trace Elements

Trace elements were investigated through analysis with solution ICP-MS, and the results are presented below following the protocol established in Chapter 2.

5.4.1 Rare Earth Elements (REE)

The REE patterns of the Pella LMLK glasses are relatively homogeneous as illustrated in Figure 5-10.

The consistency of the REE patterns indicates that the Pella LMLK glasses form a coherent geochemical group, with slight variations. The pattern and concentrations are different from those found in Pella and Nuzi HMHK glasses. The REE are more concentrated in the LMLK glasses and the relatively flat pattern is distinguished by a prominent negative Eu anomaly. HMHK glasses have no notable Eu anomaly when their concentrations are normalised with MUQ values. The negative Eu anomaly is indicative of REE derivation from a more fractionated crustal source than the REE of the HMHK glasses. It is probable that the REE are being introduced in greater quantities to the LMLK glasses, as impurities in the sand, than were present in either the ash or quartz pebble ingredients of the HMHK glasses. If the flux is indeed mineral soda from natron, then this material is known to be relatively pure and is therefore unlikely to be the source of the additional REE (section 1.9). Averaged raw concentrations of REE are presented in Table 5-3.

Figure 5-11 reveals the close similarity between the patterns of the REE from Pella LMLK and HMHK glasses, and from the UCC, when normalised with average chondrite values (Table 2-7), indicating the derivation of the Pella LMLK REE from sediments. The REE in the Pella LMLK glasses are less concentrated than in the earth's crust, and more concentrated than in the HMHK glasses, with a more pronounced negative Eu anomaly.

The ratios of the concentrations of the glass REE to those in the UCC are in the range of 0.23-0.27, excepting Eu which is 0.16, reflecting the negative Eu anomaly (Table 5-3). The ratios of the concentrations of the LMLK glass REE to those of the MUQ REE, excepting Eu, place the SBS of the LMLK glasses at a higher level than that determined for the HMHK glasses, due to their higher concentration of background sedimentary elements from a silica source less pure than that of the HMHK glasses. From the SBS it is clear that alumina (0.09) is depleted, while all other base glass oxides are enriched to some degree (Table 5-3).

Figure 5-12 illustrates the averaged Pella REE data normalised to MUQ. The different concentrations of the HMHK REE and the LMLK REE are highlighted here, as is the negative Eu anomaly of the LMLK glasses.

In summary, the REE in the Pella LMLK glasses are similar in pattern to those of the HMHK glasses, but they have a pronounced negative Eu anomaly. They are assumed to be present in the LMLK glasses at a higher

Table 5-3 Averaged raw values for Rare Earth Elements from Pella LMLK glasses in ppb.

Pella LMLK Rare Earth Elements (ppb)

n=23

	Averages	RSD	Average glass/ MUQ
La	8825	21	0.27
Ce	18,275	23	0.26
Pr	2184	22	0.26
Nd	8186	23	0.25
Sm	1608	23	0.23
Eu	257	26	0.16
Gd	1502	26	0.24
Tb	234	24	0.24
Dy	1366	23	0.23
Ho	285	23	0.23
Er	798	22	0.24
Tm	120	22	0.23
Yb	740	21	0.23
Lu	111	20	0.23

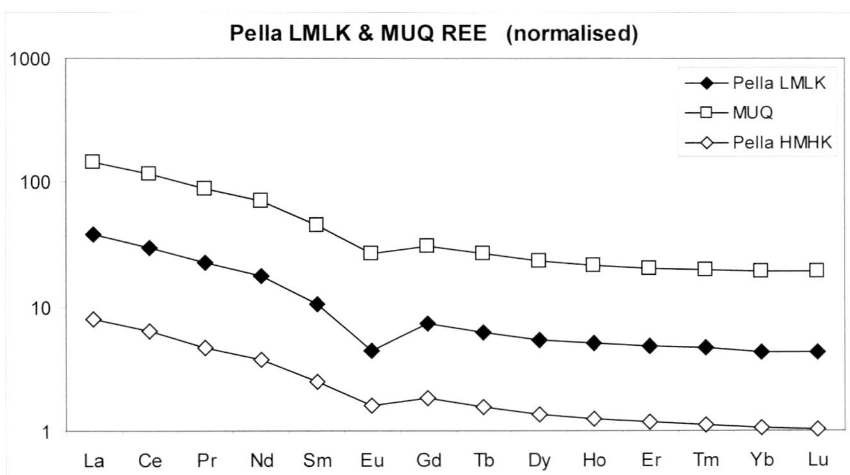

Figure 5-11 Line plot comparing averaged Pella LMLK and HMHK REE data with MUQ REE data, all normalised to average chondrites for comparison (chondrite data are reproduced in Table 27).

concentration due to the impurities in the sand source of silica and/or the iron-rich colourant material with which they are probably associated. The ratios of LMLK REE content with MUQ REE content are between 0.23 and 0.27, forming the SBS with which to compare the concentrations of other elements.

5.4.2 Sediment-Related Elements (SRE)

The concentrations of the SRE are averaged and presented in Table 5-4 below. The large RSD value for Cr may be explained by the highly variable quantities of this element in the glasses, ranging from 17,868 ppb

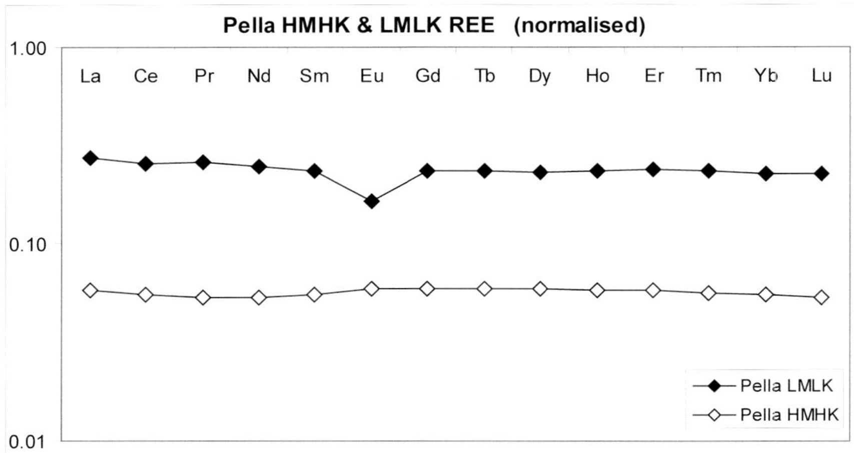

Figure 5-12 Line plot of averaged REE from Pella HMHK and LMLK glasses. Data are normalised to MUQ. A logarithmic scale is used.

Table 5-4 Averaged compositions in ppb of sediment-related elements from Pella LMLK glasses.

Pella LMLK Sediment-Related Elements (ppb)

*n=23**

	Average	RSD	Average glass/ MUQ
Sc	1950	49	0.12
Ti	1,069,615	33	0.15
V	64,187	44	0.48
Cr	98,725	92	1.30
Ga	2148	20	0.11
Y	7840	22	0.25
Zr	70,559	27	0.35
Nb	4254	20	0.28
Mo	7918	84	
Hf	1716	29	0.32
Ta	289	21	0.26
W	894	44	0.41
Tl	28	67	0.05
Th	3136	30	0.28
U	1942	48	0.69

*n = 22 for V, Cr and Tl. n = 21 for W.

to 432,285 ppb. This variation in the concentrations could be due to the presence of chromite grains in the glass (as seen in Figure 5-8) which could have caused individual samples to have been unrepresentative of the bulk glass in terms of Cr content. As outliers, the largest values for each of Cr, V and Tl were omitted from the calculation of the averages and RSD values for these elements. Large variations occur in the data for Sc, V, Mo, W, Tl and U. It is not possible to remove any values for average and RSD calculations for Mo and U as there is variation throughout the data-set for both elements and there are no clear outliers.

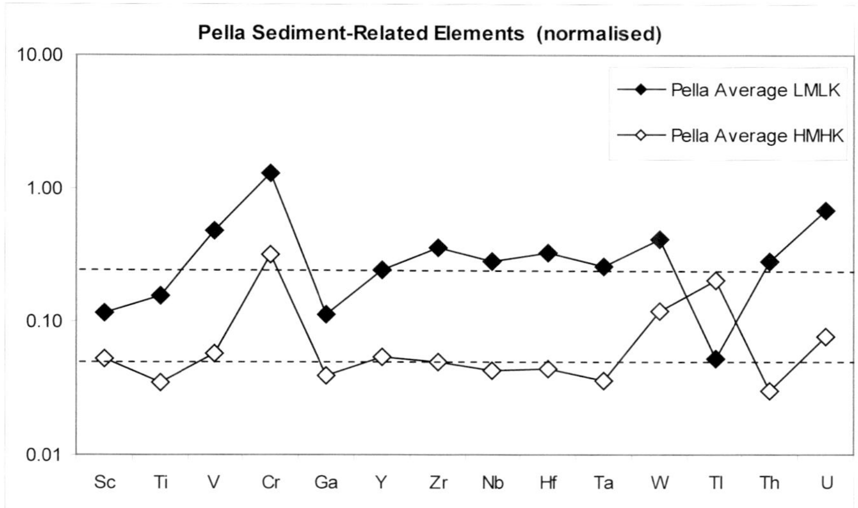

Figure 5-13 Line plot of the averaged concentrations of sediment-related trace elements in Pella HMHK and LMLK glasses, normalised to MUQ. A logarithmic scale is used. Mo is not plotted as there is no MUQ value by which to normalise it. Dotted lines cross the Y-axis at 0.25 to represent the SBS of the Pella LMLK glasses, and at 0.05 for the SBS of the HMHK glasses.

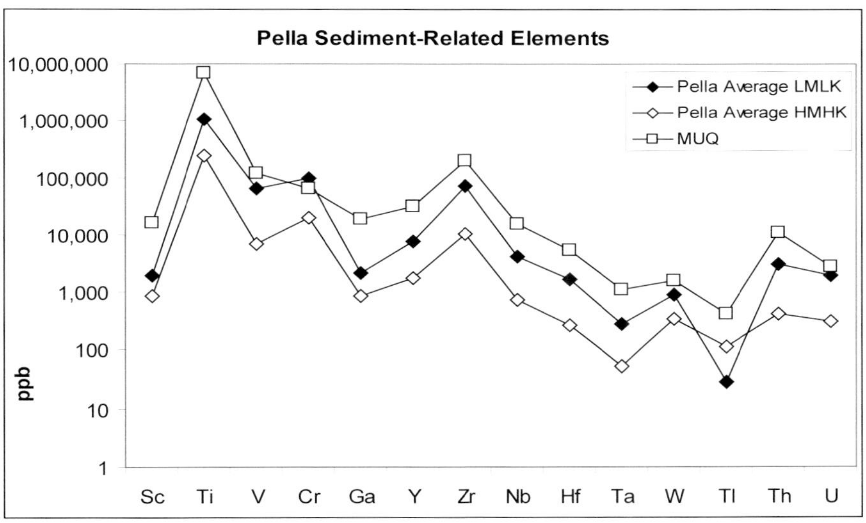

Figure 5-14 Line plot of raw Pella sediment-related trace element data in ppb for HMHK and LMLK glasses, compared with MUQ, on a logarithmic scale.

The ratios of the elemental concentrations of Y, Nb, Ta and Th from Pella LMLK glasses to those from MUQ are in the range of the SBS defined by the ratios of the concentrations of the glass REE to MUQ REE. Those elements with contents which are depleted relative to the SBS are Sc, Ti, Ga and Tl with ratios which range from 0.05 to 0.15. The remainder of the elemental concentrations are enriched, with Cr especially so at a ratio to MUQ of 1.3.

Figure 5-13 presents the normalised concentrations of the SRE for LMLK and HMHK glasses. Cr is highly enriched in the LMLK glasses due to the presence of chromite. Even so, the pattern of both groups is somewhat similar, disregarding the variations in W and Tl, and the slight relative increases in the contents of V, Zr and Hf in the LMLK glasses.

Figure 5-14 is a line plot illustrating the average raw concentrations of SRE in Pella LMLK glasses compared with those in MUQ and HMHK glasses. A general similarity of pattern is exhibited between the three groups, except for the particular variations noted previously. Figure 5-14 provides an alternative way to show the higher concentration of the SRE elements in LMLK glasses compared with those in HMHK glasses.

In summary, the SRE of the Pella LMLK glasses follow a similar pattern to that found in the earth's crust, but at depleted concentrations with the particular exception of Cr, which is enriched. It is suggested that a chromite rich raw material, such as sand, was used to make this glass. The SBS is higher than in the HMHK glasses, due to a larger concentration of REE and SRE impurities in the batch material.

5.4.3 Alkali and Alkaline Earth Elements

The averaged raw data for alkali and alkaline earth element concentrations are presented in Table 5-5. Data for Ba are variable, resulting in a large RSD value.

When the averaged concentrations of alkali and alkaline earth elements are normalised to MUQ, the resulting ratios indicate that compared with the SBS of approximately 0.25, established by the REE concentrations for LMLK glasses, the elements Li and Sr are enriched with ratios to MUQ of 1.62 and 0.77 respectively. Li and Sr are found also at elevated levels in the HMHK glasses (section 4.5.3), but Sr is considerably less enriched in the LMLK glasses than in the HMHK glasses. These two elements are, however, correlated weakly in both the Pella HMHK glasses ($r^2 = 0.2$) and the Pella LMLK glasses ($r^2 = 0.4$).

Sr is correlated strongly with magnesia ($r^2 = 0.7$) and with lime ($r^2 = 0.9$) in the LMLK glasses, which were presumably combined in a sand source of silica, or even in an LMLK plant ash, although this is difficult to establish. Lime and magnesia are present at relatively higher concentrations in the HMHK glasses than they are in the LMLK glasses. Sr and magnesia are correlated weakly ($r^2 = 0.2$) in the HMHK glasses and Sr and lime are correlated moderately well ($r^2 = 0.5$), as are soda and magnesia ($r^2 = 0.5$). In the HMHK glasses, the combination of these alkali components is assumed to have been found in the plant ash flux, if a pure silica source was used.

There is no correlation between soda and Li in either group of glasses, and as a consequence a link between the sources of these elements cannot be shown. Rb and Ba are found at concentrations near to the SBS level (0.22 and 0.24) in the LMLK glasses (see Figure 5-15), while Be (0.18) and Cs (0.08) are depleted.

Figure 5-15 illustrates the greater enrichment of alkali and alkaline earth elements, with the exception of Be, in HMHK glasses relative to their SBS levels. By comparison, only the concentrations of Li and Sr are enriched relative to their SBS in the LMLK glasses. The greater enrichment of the alkali and alkaline earth elements in the HMHK glasses is consistent with their higher potash, magnesia and lime concentrations derived from plant ash. The LMLK glasses by comparison are more likely to be mineral soda than plant ash fluxed. Although the possibility that plant ashes which exhibit compositional similarity to natron in the final glass were used was proposed by Brill (1970a: 111; section 1.8), it would be difficult to prove that these glasses had been made with a low magnesia, low potash and low lime ash with associated elevated Sr and Li.

Figure 5-16 provides an alternative view of the alkali and alkaline earth data by comparing the averaged raw alkali and alkaline earth element concentrations of the LMLK glasses with those of the HMHK glasses and MUQ. The LMLK glasses are more enriched than the HMHK glasses in Li, Be and Ba relative to their concentrations in the UCC. The levels of Rb and Cs are similar in both glass groups, and the concentration of Sr is less elevated in the LMLK glasses than in the HMHK glasses. Most of the elements in both groups of glasses are present in lower concentrations than they are found naturally in the earth's crust, with the noted exceptions where enrichment has occurred.

Table 5-5 Average compositions in ppb of alkali and alkaline earth elements from Pella LMLK glasses.

Pella LMLK Alkali and Alkaline Earth Elements (ppb)

		Li	Be	Rb	Sr	Cs	Ba
n=23	**Average**	45,739	351	17,623	108,686	437	94,667
	RSD	19	32	19	36	26	66
Average glass/ MUQ		1.62	0.18	0.22	0.77	0.08	0.24

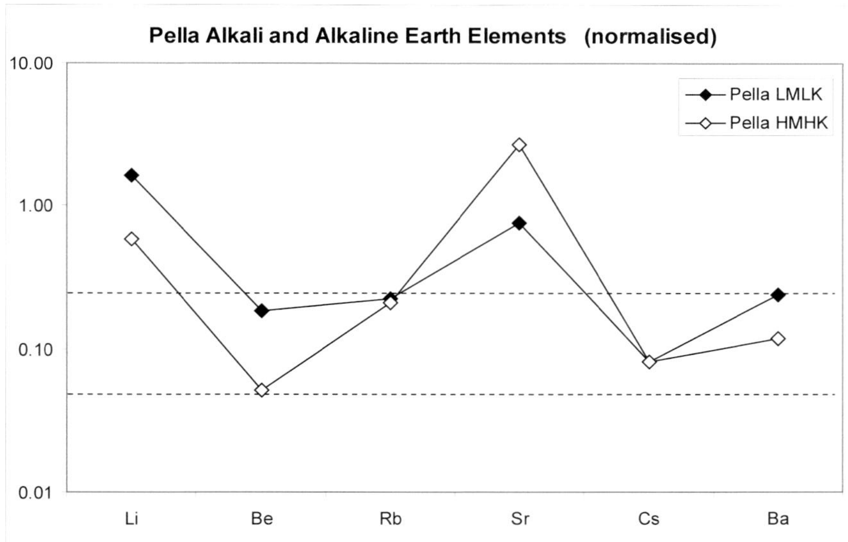

Figure 5-15 Line plot of averaged Pella alkali and alkaline earth trace elements normalised to MUQ, on a logarithmic scale. Dotted lines cross the Y-axis at 0.25 to represent the SBS of the Pella LMLK glasses, and at 0.05 for the SBS of the HMHK glasses.

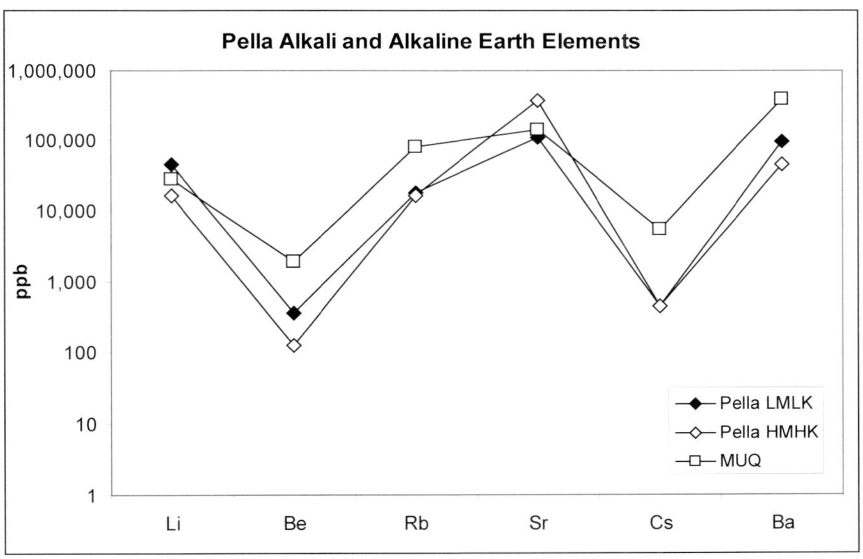

Figure 5-16 Line plot of averaged raw data for Pella LMLK alkali and alkaline earth elements compared with Pella HMHK and MUQ. A logarithmic scale is used.

In summary, the concentrations of the alkali and alkaline earth elements Be, Rb and Ba are found near SBS level, probably in association with sedimentary contamination in the LMLK glasses. Sr and Li are enriched, and are related to lime and magnesia, which if these glasses are assumed to be fluxed with mineral soda, suggests that these components could be derived from a combined silica and lime sand source.

5.4.4 Colourants and Opacifiers

Although the LMLK glasses are assumed to be coloured with iron oxide, and are not opacified or coloured blue, it is convenient to continue to treat the elements listed in Table 5-6 as a group, because it provides a means of comparison between all the groups of glass reported in the present study.

Table 5-6 Compositions in ppm of colourant and related elements from Pella LMLK glasses.*

Pella LMLK Colourant and Related Elements (ppm) *n=23*	Co	Ni	Cu	Zn	As	Cd	Sn	Sb	Pb
Average	4.79	96.0	117	66.1	110	0.11	0.94	330	28.4
RSD	35	46	80	92	105	53	52	108	256
Av.glass/MUQ	0.21	3.04	3.61	0.90			0.34		1.39

*MnO is below detection. There are no values for As, Sb or Cd in MUQ.

Table 5-6 presents the average concentrations of the elements traditionally recognised as colourants, opacifiers and the elements often associated with them, as analysed by solution ICP-MS.

By comparison with blue glasses, such as the Pella and Nuzi HMHK glasses, the Cu concentrations of the black LMLK glasses are very low (average 117 ppm), although they are elevated above the SBS at a ratio to MUQ of 3.61. Ni is found in greater concentration than in the blue glasses, and is elevated above the SBS (ratio to MUQ = 3.04), as is Zn to a lesser extent. The concentration of Co (average 4.79 ppm) is similar to that found in the blue glasses, although in the LMLK glasses it is not elevated above the SBS and appears to be an unintentional additive. The Sn content of the LMLK glasses is low (average 0.94 ppm) and is slightly elevated above the level of the SBS. In the LMLK glasses Pb and Sb concentrations are variable: 3 ppm to 356 ppm for Pb, and 5 ppm to 1170 ppm for Sb. The high RSD values for most elements are due to considerable variation in the concentrations of the elements in this group of LMLK glasses, although the glasses are recognised as a coherent group through SEM-EDS analyses of the base glass composition (section 5.3).

Figure 5-17 illustrates the correlation between the concentrations of the iron colourant and Ni in the Pella LMLK glasses (r^2 = 0.6). Iron oxide does not appear to correlate with any of the other elements.

In Figure 5-18 the averages of typical 'colourant-related' trace element concentrations for Pella LMLK and HMHK glasses are compared with the concentrations of these elements in MUQ. The LMLK glasses are coloured by high iron oxide contents and are not opacified. Their Sb

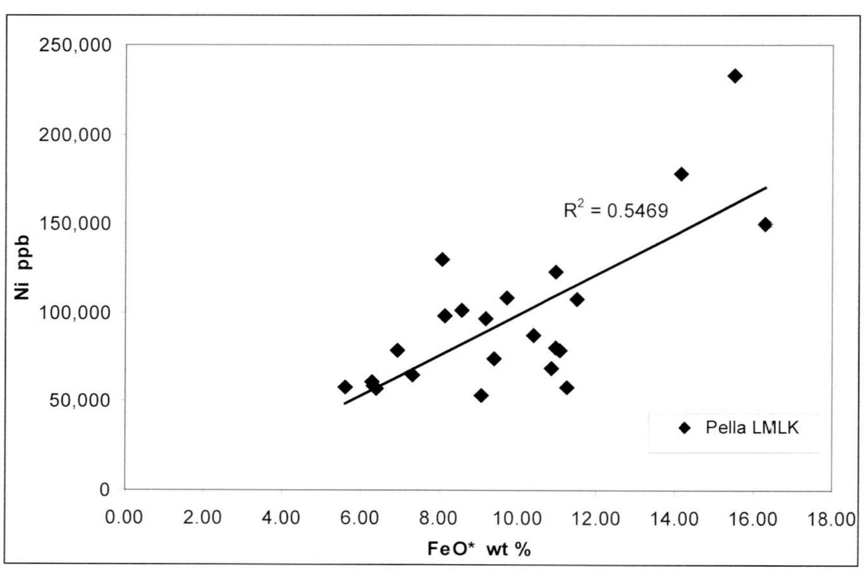

Figure 5-17 Scatter plot of iron oxide vs Ni for Pella LMLK glasses.

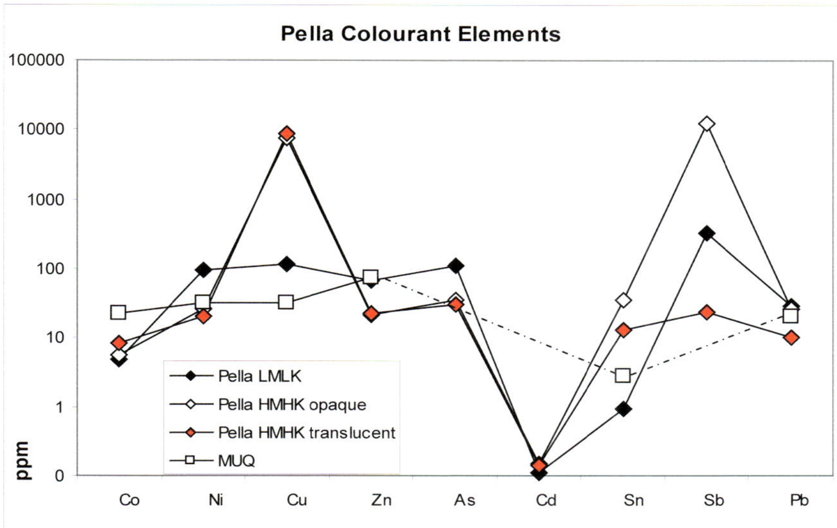

Figure 5-18 Line plot of raw averaged colourant and related elements in Pella LMLK and HMHK glasses compared with MUQ, from ICP-MS data. A logarithmic scale is used. There is no MUQ value by which to normalise As, Sb or Cd.

concentrations are higher than those of the translucent HMHK glasses, but are lower than those of the opaque HMHK glasses. The LMLK glasses contain less Sn and Cu, which are not considered to be the intended colouring agents in these black glasses. The LMLK glasses have greater concentrations of Ni, which is associated with iron oxide, Zn and As, than are present in the HMHK glasses.

5.5 General Conclusions

The IA black, high iron glasses were found to contain relatively low levels of magnesia, potash and lime which classifies them as LMLK. They represent a glass-making technology which is different from that of the HMHK Pella glasses of the LBA. The composition of the LMLK glasses could be explained if a lime-bearing sand silica source with a mineral soda flux had been used by the glass manufacturers. The higher concentrations of alumina, REE and SRE, compared with the concentrations of those elements in the HMHK glasses, supports the proposition that sand was used as the silica source. The unusual colour of these glasses appears to have been achieved by the addition of a high iron-bearing ingredient to the glass mix, perhaps in association with chromite from the sand source of silica.

Current research into different plant ash compositions allows the possibility that these glasses could have been made with a low magnesia, low potash and low lime ash with associated elevated Sr and Li as was found in HMHK ash fluxes. On balance, the evidence from the present study is insufficient to conclude that the flux used in the Pella LMLK glasses was plant ash, with the evidence more strongly supporting the proposition that the flux was derived from mineral soda, and the silica from sand. These glasses are discussed further in section 9.6.

Chapter 6

Nimrud: Iron Age Glasses Results Of Compositional Analyses Part 1

6.1 Introduction

The compositions of the 1st millennium BC LMLK glasses of IA Pella, in Jordan differ in flux and colourant technology to the typical 2nd millennium BC HMHK tradition, well represented at Nuzi and Pella in the LBA. The continuation of the LBA HMHK tradition into the 1st millennium BC at Pella is debatable, with only four possible examples surviving, which are considered the more likely to be residual LBA pieces. At 1st millennium BC Neo-Assyrian Nimrud (ancient Kalhu), in modern Iraq, there is a variety of both HMHK and LMLK glass, published in a preliminary report by Reade *et al.* (2005).

There are several key issues which need to be addressed in regard to the glasses reported in this chapter. Does the Early IA assemblage at Nimrud represent a continuity of 2nd millennium BC HMHK tradition, or do we find the beginnings of the technological innovations that have emerged so clearly by Hellenistic times, namely the use of natron as a flux? Did explcitation of 2nd millennium BC raw material sources continue, or had new sources of colourants, opacifiers and silica been adopted? Had glass-making diversified and to what extent? The glass assemblage from Nimrud, a significant proportion of which is housed at the British Museum, offered the possibility to address these questions.

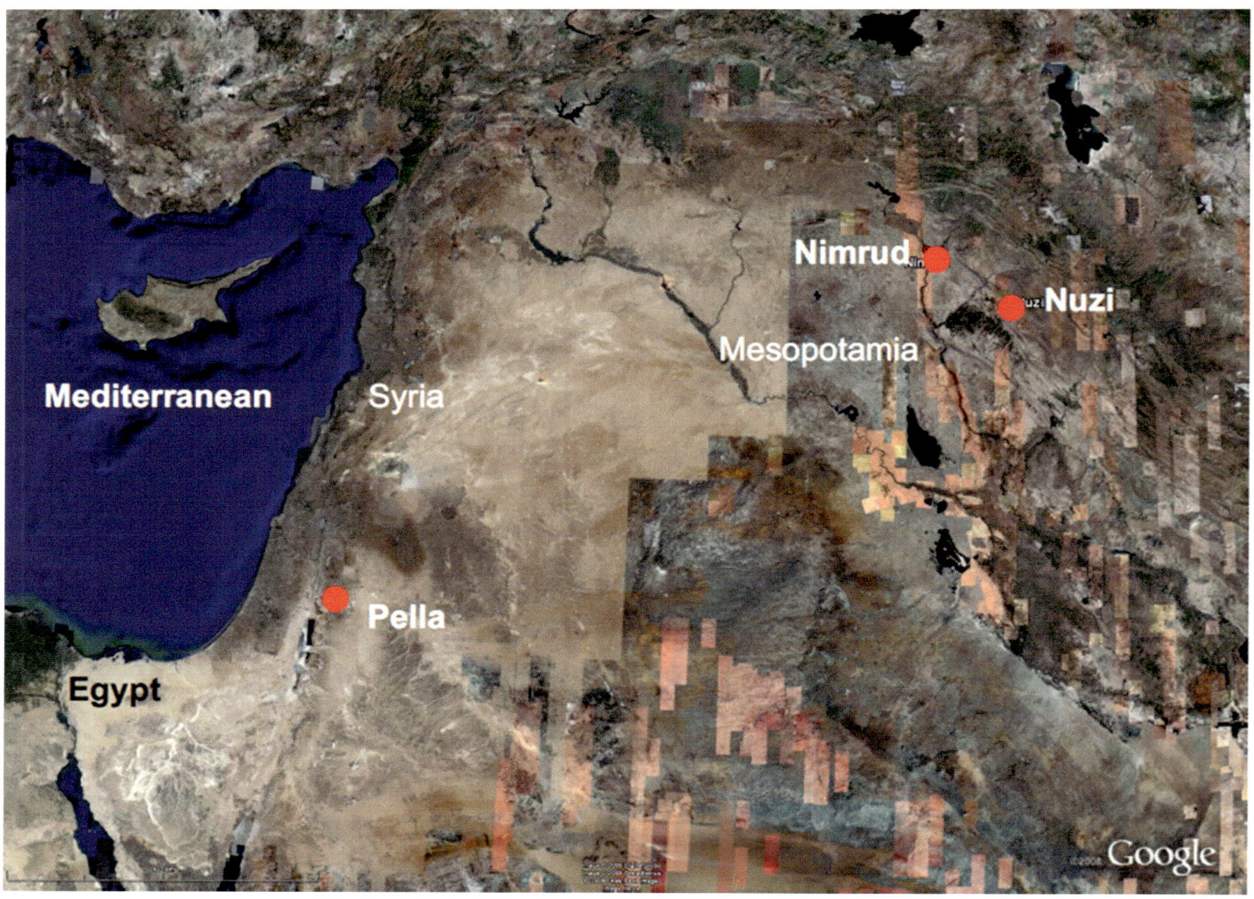

Figure 6-1 Satellite map showing location of Nimrud on the east bank of the Tigris River, in relation to Nuzi and Pella, from Google Earth.

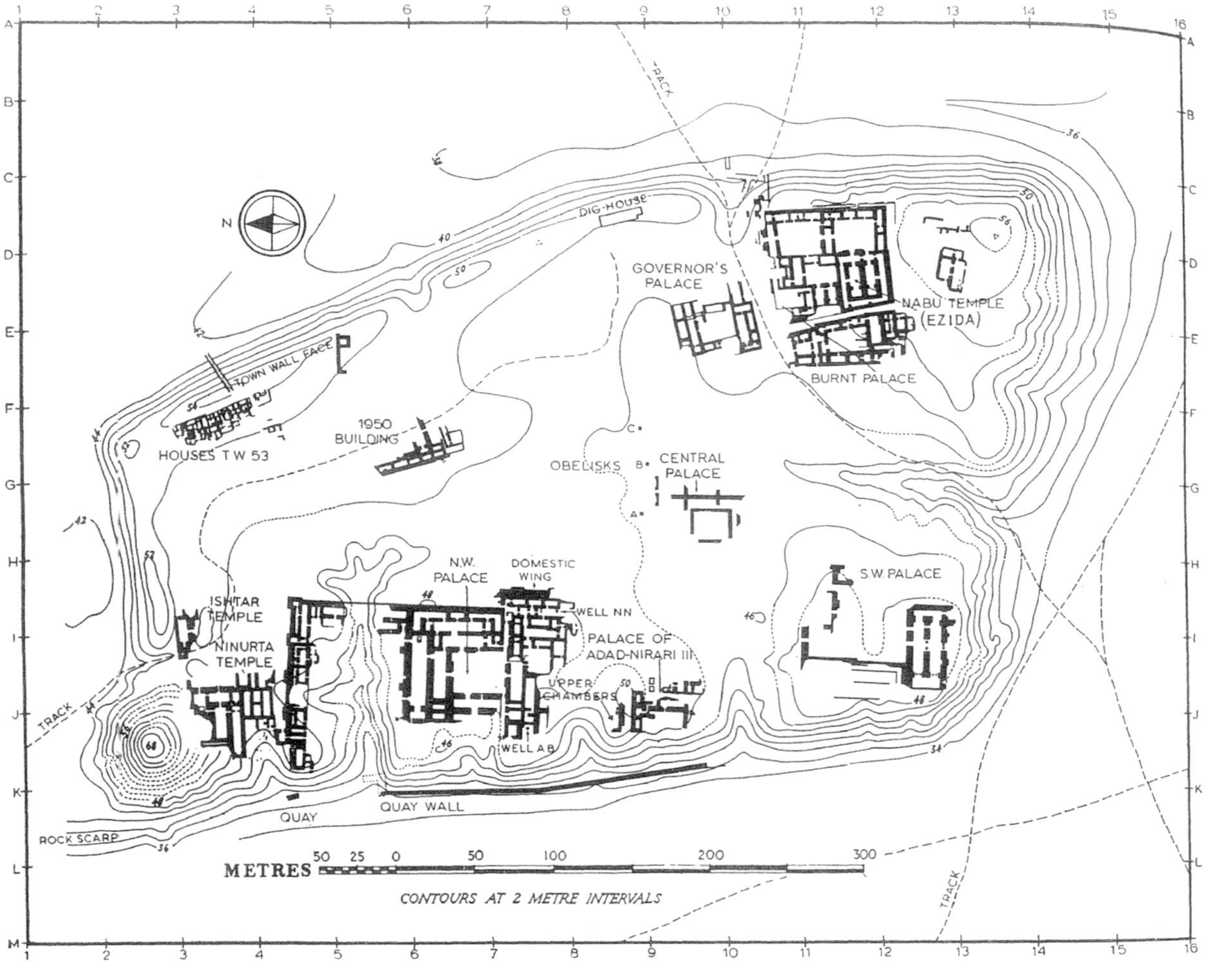

Figure 6-2 Contour map of the mound or acropolis showing the positions of excavated buildings (1957) from which much of the glass analysed in this study was unearthed. Reproduced from Mallowan (1966: 32).

6.2 The Site

Nimrud was established in 879 BC by king Ashurnasirpal II (883-859 BC) as the new capital of the Neo-Assyrian empire, situated on the fertile east bank of the Tigris river and north of the previous capital at Assur (Figure 6-1). The capital was later moved to Khorsabad and then to Nineveh, 35 km north of Nimrud, at the end of the 8th C BC, and Nimrud declined until it was finally destroyed in 612 BC by a combined Babylonian and Median army (Van De Mieroop 2004: 216, 232). As a great and relatively short-lived centre with a large corpus of glass principally dating to the 9th to 8th C BC, Nimrud provides a fascinating and important piece in the jigsaw of ancient glass research.

6.2.1 The Excavations

This large and complex site has been extensively excavated since the mid-19th century, with finds in the British Museum deriving from excavations made by Sir Austen Henry Layard at that time, and his successors, and those conducted by M.E.L. Mallowan from 1949 to 1963 on behalf of the British School of Archaeology in Iraq (Mallowan 1966; Reade 1982) (Figure 6-2). More recently, excavations were resumed by the British Museum in Fort Shalmaneser in 1989, confirming the frequency of vessels and blue glass inlays at the site (Curtis 1999; Figure 6-3). The Directorate-General of Antiquities in Iraq has carried out excavations since 1956 (Oates and Oates 2001: 10; Hussein 2002, for example). Sampling of Nimrud glasses for this study from the collection of the British Museum was generously allowed by its custodians.

The main areas of excavation at Nimrud included the North-West Palace, the Burnt Palace, the Ninurta Temple, the Ziggurat, the Ezida, or Nabu Temple, and Fort Shalmaneser.

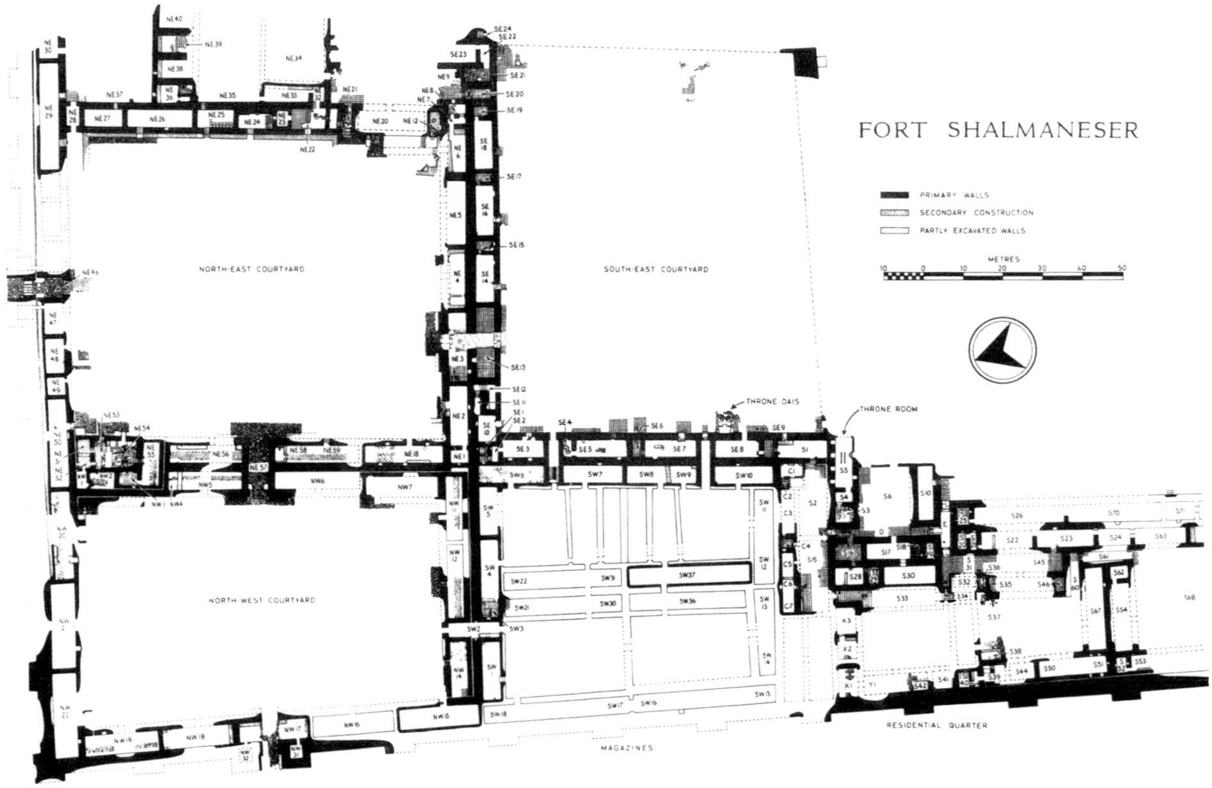

Figure 6-3 Plan of Fort Shalmaneser, located in the south-east corner of the site. Reproduced from Mallowan (1966).

6.2.2 The Glass

The glass sampled for this study consists of small Co blue inlay pieces and fragments of beards from two statuettes; translucent light blue, purple, pink and colourless hemispherical bowls and other unidentified vessels, which may well have been parts of bowls; and colourless and turquoise opaque ingots (von Saldern 1966a; Barag 1985: 51-55). Very few beads appear in the Nimrud glass assemblage, there being more made of frit, faience, shell and stone.[47] Inlays and cylinder seals were also made of lapis lazuli, and a few turquoise inlays were found, but there are many more surviving inlays of glass than of stone. Whether the relative numbers of stone and glass objects left to us are an accurate reflection of how much of each material was used in neo-Assyrian times, and which was valued more highly is difficult to determine.[48] That the glasses were in royal contexts or contexts associated with senior officials, specifically at Fort Shalmaneser, the North-West Palace and the Burnt Palace indicates that they were all luxury objects. As there are no remains of glass-making at Nimrud earlier than the Hellenistic period,[49] the mechanisms by which these items, especially the Co blue glasses, reached Nimrud would have included booty, tribute and trade, for example the tribute and treasure bearing Sargon's name found by Mallowan in the storerooms of the North-West Palace, and the inscriptions which describe its acquisition (Oates and Oates 2001: 69, 125, 240-1; see also Shortland 2001: 1; sections 1.2, 9.7.3). Remains of ingots of turquoise blue and of red glass suggest that glass-working occurred to some extent at Nimrud in Neo-Assyrian times (Moorey 1999: 203; von Saldern 1966a: 633; 1970: 223).

Previous analyses of glasses from Nimrud have been published by Turner (1955; 1956d) and Bimson and Freestone (in Barag 1985), but these studies were based on a limited number of samples. Brill (1999a: 45-47; 1999b: 47-49) has published analyses of 24 glasses, covering a range of colours including colourless, purple

[47] Stone beads were largely made from carnelian and other red stones, lapis lazuli, turquoise, agate, rock crystal (rare), and gold (collection of the British Museum, and see Hussein and Suleiman 2000).

[48] Indeed recycling, decay and removal of glass could have altered the archaeological record of any site. Some idea of the relative values of glass and stone may be gleaned from surviving texts, such as the Amarna Letters (Moran 1992) and Pliny's Natural History (Eichholz 1962).

[49] Although Moorey (1999: 202-203) preferred Mallowan's (1954: 77, 82-83) original dating of the glass-making remains to the Neo-Assyrian period.

and blue hemispherical bowls, but focusing particularly on red, yellow and blue opaque glasses, many being inlays. In this study a range of glasses is covered, mainly blue and colourless, and a further 50 analyses are added to the Nimrud corpus. In terms of both their colourant technology and base glass compositions, these glasses, and especially the opaque blues, are particularly pertinent to the question of continuity *versus* change in the LBA to Early IA transition.

6.2.3 *Summary of Samples*

Samples of the 50 glasses, referred to above, were taken from the British Museum's collection of Nimrud glasses, for chemical analysis by SEM-EDS and solution ICP-MS (sections 2.3, 2.4).

The 16 Co blue glasses analysed in the present study are extremely well preserved. Two beard fragments from statues were sampled, with the remainder having been taken from small inlay plaques, frequently square or rectangular in shape, and less commonly of other shapes. They are undecorated, or are inlaid with central white rosettes. They are generally 1mm to 2mm thick, with the decorated ones ranging in dimensions from approximately 3mm or 4mm across to approximately 8mm across. These inlays were set within glass or copper alloy frames which often were attached to ivory inlays set into furniture (von Saldern 1966: 625; Barag 1985: 71-72), while other glass inlays appear to have been hammered onto beds and/or other furniture with bronze tacks (Curtis 1999; Oates and Oates 2001: 239-240), and an 8th to 7th C BC 'Phoenician' glass bowl, also excavated at Nimrud, was decorated below the rim with plaques of this type (Barag 1985: 65-66, Fig.3, nos 40-40A; cf. also Oates 2001: 240). It has been proposed that these inlay plaques are Phoenician (Mallowan 1978). Oates and Oates (2001: 230) noted that the Phoenician ivories are recognised for their use of coloured glass inlays in a cloisonné-like technique.[50] Similar square inlays with white rosettes have also been excavated at Arslan Tash in northern Syria (Thureau-Dangin *et al.* 1931: 89-141) and Samaria in central Palestine (Crowfoot and Crowfoot 1938), and are suggested to have been the products of a south Syrian centre such as Damascus (Winter 1981: 101-130).

Furthermore, glass plaques inlaid into an alabaster cup excavated at Hasanlu in north-west Iran pose further questions over the origin and function of these inlays (de Schauensee 2001). Spaer (2005) considered these inlays, further noting the gathering opinion that they date from the 8th into the 7th C BC, although the possibility of an earlier beginning was acknowledged. Inscriptions on the palace reliefs tell of the royal hunting of elephants in Syria, but Oates and Oates (2001: 226-7) believed that a large number of the ivories found at Nimrud were acquired through plunder, tribute and royal gifts, particularly from regions west of Assyria such as Syria and Phoenicia (see also Mallowan and Hermann 1974: 38-39; section 9.7.3). It seems reasonable to assume that the glass inlays attached to these ivories would have had a similar origin.

The analyses of the Co blue glasses are compared with the analyses of other colours from Nimrud, and with those of blue glasses from Egypt and Mesopotamia of the 2nd and early 1st millennia BC in section 9.7.1.

The hemispherical bowl and unidentified vessel fragments (no complete vessels were sampled) often had thin layers of flaking opalescent weathering product on their surfaces. Decomposed glass was removed so that fresh underlying glass could be sampled.

Of the few glass beads in the collection, none was suitable for sampling. Glass 48Y, due to its fragmentary state, is of uncertain identification. Table 6-1 provides identification and descriptions of the 50 samples analysed. For convenience, they are divided into four groups, determined by SEM-EDS analysis of base glass composition. The results of these analyses are presented in section 6.3.

6.3 Base Glass Composition

SEM-EDS analysis of the glasses (section 2.3.3) defined four reasonably coherent compositional groups, with a small number of outliers or 'ungrouped' glasses. The main groups are plotted in Figure 6-9 according to their concentrations of magnesia and potash, and average reduced compositions of all groups are presented in Table 6-2. Complete raw oxide and trace element data for each sample are tabulated in Appendix V.

6.3.1 *Group 1*

Group 1 glasses form a tight compositional group with an average potash concentration of 1.5% (range of 1.3% to 1.6%) and an average magnesia concentration of 3% (range of 2.8% to 3.5%). The magnesia content more-or-less conforms to a plant ash or HMHK glass, but the potash concentration falls at the lower end of the range traditionally accepted for this type of glass (Sayre and Smith 1961: 1824-1826; section 1.5). Phosphorus levels range between 0.09% and 0.2%, averaging 0.16%. An average lime content of 8.8% (range of 8.3% to 9.2%) is consistent with a plant ash glass composition. This group has lower concentrations of magnesia and potash than typical 2nd millennium BC Near Eastern HMHK glasses, as represented at Nuzi and Pella. Their lime levels are

[50] Apart from the Phoenician style, a number of different styles of ivory carving are known, the discussion of which is beyond the scope of the present study. See Spaer (2005: 28) for further references and discussion.

Table 6-1 Description of Nimrud glass samples divided into groups determined from base glass composition by SEM-EDS analyses.*

Sample Identification	Analysis Number	Object	Colour	T or O	Comments
Group 1					
1994-11-5, 173 ND12, 541	5Y	vessel	colourless	T	large, thick walled
1994-11-5, 173 ND12, 541	7N	vessel	colourless	T	large bowl
1994-11-5, 173 ND12,541	10P	vessel	colourless	semi-T	very thin walled
1994-11-5, 173 ND 12, 541	12W	vessel- hemispherical bowl	colourless	T	rim fragment
	22L	vessel – hemispherical bowl	purple	T	
	23N	vessel – hemispherical bowl	colourless	T	large bowl, greenish tinge
WAA N821 91521 734	36M	vessel	colourless	T	
TW53 level 3	38R	vessel – hemispherical bowl	pink	T	tapered rim fragment
Group 2					
N.785 91574  Figure 6-4 Nimrud N.785 91574 dark blue opaque beard fragment	5BM	beard fragment from statue- diagonal rows of curls	dark blue	O	8th-7th C BC (Barag, 1985: 63) 11mm thick, 49 x 35mm LxW. Excavated by Sir Austen Henry Layard

Sample Identification	Analysis Number	Object	Colour	T or O	Comments
 Figure 6-5 Nimrud turquoise opaque ingot fragment	34X	ingot	turquoise	O	circular, fragmentary
TW53 level 3	40Q	vessel	turquoise	O	
ZTE or J	48Y	bead?	light blue	T	
Ungrouped					
1994-11-5,173 ND12, 541	2P	unidentified	light blue	T	
1994-11-5,173 ND12, 541	8Q	chunk unidentified	light blue	T	
ZTE or J	43K	chunk unidentified	turquoise	O	*c.* 700 BC
ZTE or J	45P	vessel	colourless	T	
Group 3					
1994-11-5, 173 ND12, 541	1M	vessel	light blue	T	
1994-11-5, 173 ND12, 541	3R	unidentified	light blue	T	
1994-11-5, 173 ND12, 541	4W	inlay	turquoise	O	rectangular shape
1994-11-5, 173 ND12, 541	6L	inlay?	colourless	T	flat fragment
1994-11-5, 173 ND12, 541	9T	vessel	light blue	T	thin walled

Sample Identification	Analysis Number	Object	Colour	T or O	Comments
1994-11-5, 234 ND12, 541  Figure 6-6 Nimrud 1994-11-5, 234 ND12, 541 light blue translucent vessel fragment	11R	vessel	light blue	T	darker shade of blue
ND5240, 5290, 5216, 5220 (as labelled on box)	19K	vessel - bowl	light green blue	T	horizontally grooved decoration
1994-11-5, 309 ND12, 541	20W	inlay	colourless	T	
	26X	vessel	light blue	T	
	27K	vessel	light blue	T	
	35K	vessel	colourless	T	
TW53 level 3	37P	vessel - bowl	light blue	T	very thin walled, *c.*630 BC
TW53 level 3	39W	lump unidentified	colourless	T	pale blue tinge, thick walled, possibly curved exterior
ZTE or J	44M	chunk unidentified	light blue	T	*c.*700 BC
ZTE or J	46R	vessel	light blue	T	
	49L	ingot	turquoise	O	circular fragment, *c.*8mm wide
Group 4  Figure 6-7 Examples of Group 4 cobalt blue inlays, various sizes					
1987-1-31, 19 140346 ND10229	3BMA	inlay	dark blue	O	white rosette decoration inlaid, square, 2.5-3mm thick, 32mm across
1987-1-31, 19 140346 ND10229	3BMB	inlay	dark blue	O	white rosette decoration inlaid, 2.5-3mm thick, 29 x 26mm LxW

Sample Identification	Analysis Number	Object	Colour	T or O	Comments
1987-1-31, 32 140359 ND8081	4BMA	inlay	dark blue	O	undecorated, narrow rectangular, 2mm thick, 32 x 4mm LxW
1987-1-31, 32 140359 ND8081	4BMB	inlay	dark blue	O	white rosette decoration inlaid, small square c.7mm L
1994-11-5, 73	13Y	inlay	smoky dark blue	O	rectangular
1994-11-5, 73 ND8157	14L	inlay	smoky mid blue	semi-T	semicircular with other edge tapered to a point, 10mm across
1994-11-5, 67 ND4138  Figure 6-8 Nimrud 1994-11-5, 67 ND4138 dark blue opaque beard fragment	15N	beard fragment from statue	dark blue	O	from burnt floor of NT3 hair implied by vertical wavy lines, 45 x 38mm LxW
	16Q	inlay	dark blue	semi-T	
2T Misc. ND3541?	17T	inlay	dark blue	O	rounded shape, 20 x 20mm LxW
1994-11-5, 293	18X	inlay	dark blue	semi-T	white rosette decoration inlaid,
ND1712	21Y	inlay	dark blue	O	white rosette decoration inlaid,
1994-11-5, 225 ND1076?	24Q	inlay	dark blue	semi-T	quarter circle shaped, 1-2mm thick, 8mm W
1994-11-5, 221 ND1029	25T	inlay	dark blue	O	rectangular, broken, 2mm thick, 8mm long
	29P	inlay	dark blue	O	
	30L	inlay	dark blue	O	
	31N	inlay	dark blue	O	
1994-11-5, 238 ZT and NWP	32Q	inlay	dark blue	O	

*Sample identification information was supplied by the British Museum. NWP = North West Palace, ZT = Ziggurat Terrace extension to NWP, NT = Nabu Temple, TW and 2T could not be found on the published plans.
Measurements are given only where they are meaningful, as all samples were taken from incomplete objects.

Table 6-2 Average reduced compositions of 50 Nimrud glasses in wt% oxide.

Samples	SiO_2*	Al_2O_3*	FeO*	MgO*	CaO*	Na_2O*	K_2O*	P_2O_5*	Total
Group 1 n=8									
Average	70.81	1.02	0.49	3.12	8.80	14.12	1.49	0.16	100.00
SD	0.50	0.11	0.07	0.19	0.36	0.48	0.09	0.04	
Group 2 n=4									
Average	66.50	1.12	0.77	5.37	6.74	16.01	3.33	0.15	100.00
SD	3.47	0.56	0.44	0.97	1.09	2.25	0.73	0.16	
Group 3 n=16									
Average	72.50	0.56	0.28	1.77	6.24	17.56	0.99	0.10	100.00
SD	1.25	0.12	0.10	0.32	1.12	1.01	0.39	0.07	
Group 4 n=17									
Average	68.06	5.79	0.84	2.90	3.14	18.81	0.42	bd	100.00
SD	3.10	1.32	0.32	0.82	1.43	1.30	0.24	0.04	
Ungrouped									
2P	67.41	0.95	0.45	3.04	8.82	17.68	1.49	0.15	100.00
45P	69.17	0.81	0.51	4.22	7.83	15.70	1.57	0.18	100.00
8Q	68.29	0.65	0.41	2.99	7.79	17.85	1.75	0.27	100.00
43K	74.61	1.92	0.74	1.70	4.39	14.16	2.21	0.27	100.00
Gp1/MUQ	1.15	0.06	0.06	1.99	8.00	17.65	0.97	0.86	
Gp2/MUQ	0.98	0.06	0.09	3.42	6.13	20.01	2.16	0.83	
Gp3/MUQ	1.07	0.03	0.03	1.13	5.67	21.95	0.64	0.55	
Gp4/MUQ	1.01	0.33	0.10	1.85	2.86	23.51	0.27	0.21	

equivalent to those of the 2nd millennium BC high-lime HMHK glasses. This group consists of colourless, weakly coloured and pink/purple hemispherical bowls.

6.3.2 Group 2 and Ungrouped

Group 2 is a small group of four standard plant ash glasses with high concentrations of potash, averaging 3.3% (range of 2.4% to 4%) and of magnesia at 5.4% (range of 4.4% to 6.7%). Phosphorus averages 0.15%. The average lime level is 6.7% (range of 5.2% to 7.7%), consistent with the concentration expected in a plant ash glass. Curiously, sample 5BM is a high Cu, high Co dark blue inlay, which it is easy to visually mistake for a Group 4 Co LMLK glass,[51] and is the only high Co HMHK glass analysed (2.4% potash and 5% magnesia, with 6.7% lime). Group 2 glasses conform to the composition of the 2nd millennium BC low-lime HMHK glasses, with regard to these elements, although ratios of base glass oxides differ to those reported for the Nuzi and Pella HMHK glasses (sections 3.2, 4.4).

There are four outliers (three turquoise opaque or translucent, and one colourless glass) that are ungrouped for the following reasons. They have potash concentrations ranging from 1.5% to 2.2%, and magnesia concentrations ranging from 1.7% to 4.2 %, although they do not conform well to HMHK Groups 1 and 2 in other respects. When all the base glass oxides are considered, glasses 2P and 8Q more nearly fit into

[51] Jackson and Nicholson (2007: 105) noted that it is impossible to differentiate Co and Cu blue glasses based on visual inspection only.

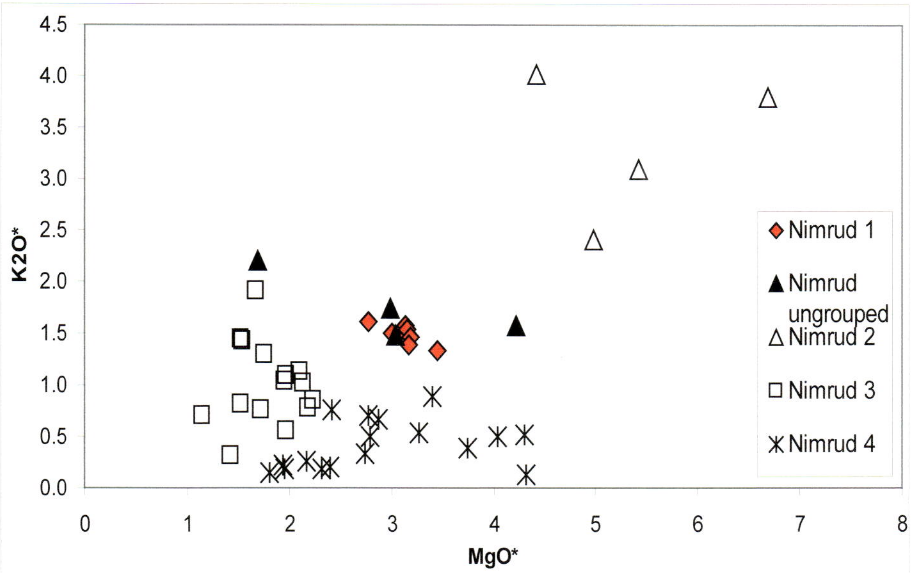

Figure 6-9 Scatter plot of magnesia *vs* potash in wt% oxide for Nimrud glasses.

Group 1, though 2P has a lower silica content and a higher soda content, and 8Q is lower in the content of most oxides except soda and potash, which are higher. Glass 45P fits somewhere between Groups 1 and 2, and closer to Group 1. Glass 43K does not fit readily into any group but is mostly an outlier to Group 3. Phosphorus levels for all ungrouped samples range from 0.15% to 0.27%, within the established plant ash range. Lime contents are between 7.8% and 8.8%, except for 43K which has 4.4% lime.

6.3.3 Group 3

Group 3 is a large group of vessels, mainly Cu blue with some being colourless. Components considered markers of plant ash are generally at relatively low levels, with a mean magnesia content of 1.8%, mean potash of 1%, and mean phosphate of 0.1%. These glasses can therefore be classified as LMLK glasses, though they have comparatively high lime contents, averaging 6.2% (range of 4.3% to 8.7%) which, if not plant ash, could reflect a limestone source, added either separately or in conjunction with a relatively pure quartz sand. This may represent an advance in the technology of glass-making from the earlier Pella LMLK glasses (Chapter 5), and LMLK glasses from the tomb of Nesikhons (Schlick-Nolte and Werthmann 2003) where lime contents alone are too low to stabilise the glass.

6.3.4 Group 4

Group 4 consists of 17 Co coloured glasses which have particularly low concentrations of potash, averaging just 0.4% (range of 0.12% to 0.9%), strongly suggesting that these are LMLK glasses. The low lime contents of these glasses (averaging 3.1% with a range of 1.1% to 5.6%) support this view, as lime is a major constituent of plant ash and the use of plant ash soda alone would typically yield a glass with 5% lime or more (section 1.7). Low phosphorus contents of less than 0.1% (i.e. below instrumental detection limits), also appear to preclude plant ash as a source of soda. These Co blue glasses are mainly inlay pieces opacified with calcium antimonate. An average magnesia level of 2.9% (range of 1.8% to 4.3%) is characteristic of an LMLK glass coloured with Co derived from alum, such as is found in the Western Desert oases of Egypt (Figure 9-13). The Egyptian alum is associated with increased magnesia levels in such glasses (section 1.10.1).

Figure 6-9 presents the groupings of the Nimrud glasses according to their magnesia and potash levels. Nimrud Group 2 plots in the region of the typical LBA or 2nd millennium BC HMHK glasses from Nuzi and Pella, as noted above, but unlike these glasses, there is no correlation between magnesia and potash in the Nimrud Group 2 glasses. Group 1 HMHK glasses are less rich in magnesia and potash than Group 2 and 2nd millennium BC HMHK glasses, but are still within the accepted plant ash range. The ungrouped glasses appear to be plant ash fluxed, and although samples 2P and 8Q group together with Group 1 glasses according to their contents of potash and magnesia, they do not always group this way when other oxide components are considered. Group 4 glasses have a wide range of magnesia concentration, dependent on the natural variation of alum composition from which they derive magnesia (section 1.10.1).

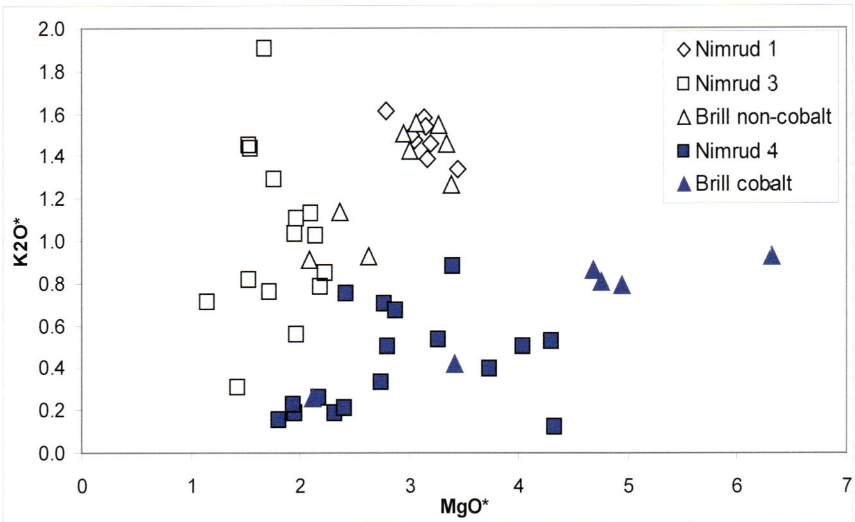

Figure 6-10 Scatter plot of magnesia vs potash in wt% oxide comparing Brill's (1999b) analyses of Nimrud glasses with those of the current study (yellow and red glasses not plotted).

The results of Brill's analyses of Nimrud glasses (excluding red and yellow glasses) (Brill 1999b: 47-49) are compared with the magnesia *versus* potash data from the current study of Nimrud glasses in Figure 6-10. Brill's samples 3235, 3236, 1712, 545, 546 and 548 fit into Group 1. Brill's samples form a similar mixture of colours, including light blue translucent inlays, a colourless plaque, and colourless and purple hemispherical bowls. None of Brill's glasses falls into Group 2. Three of his Cu blue and greenish tinged glasses (547, 3254, 1121) belong to Group 3. Co blue samples 3220, 3231, 3232, 3233, 3234 and 3229 fall into Group 4. Brill's four yellow opaque glasses (not plotted) all appear to be Group 3 LMLK. Of the five red opaque samples he analysed (not plotted), two are clearly Group 2 HMHK, one is possibly Group 1 plant ash with high magnesia and lower potash, while the remaining two fit into Group 3. This shows clearly that colour is not necessarily dependent on base glass composition.

Brill's samples do not conform as closely to these groupings when other base glass oxides are considered, highlighting the variability of composition of the Nimrud IA glasses and the difficulty of ascribing some of them to an identifiable group. It is indicative of the change and diversification in glass-making recipes and raw materials compared with the predictable, long-lived, traditional recipes of the preceding 2nd millennium BC glass-makers.

As with the other glass groups examined in the present study, magnesia, soda, lime and potash concentrations are enriched compared with the concentrations of these oxides found in the UCC, while iron oxide and alumina concentrations remain low.

The quantities of silica and soda vary relatively little between the groups of Nimrud glass as is observed in the glasses from Pella and Nuzi. Silica concentrations are similar to those found in the earth's crust. Group 1 has a well-controlled silica content ranging from 70% to 72%, average 71%; Group 2 contains less silica with an average of 67% (range of 63% to 71%); Group 3 has an average of 73% silica (range of 70% to 75%); and Group 4 has an average of 68% silica, with a range of 63% to 72%. The ungrouped samples range from 67% to 75% silica.

Soda contents range from 13.8% to 14.3% (average 14.1%) in Group 1. Group 2 ranges from 16.4% to 17.9%, averaging 16%, though this includes the Co blue plant ash glass, 5BM, which has a low soda concentration of 12.8%. Group 3 soda concentrations range from 16% to 19.5%, averaging 17.6%. Group 4 has the highest soda concentrations, ranging from 17% to 21% (average 18.8%).

Concentrations of iron oxide and alumina (Figure 6-11, 6-14) vary across the groups, with good correlations between the concentrations of these oxides in Groups 2 (r^2 = 0.67) and 3 (r^2 = 0.78). Group 1 has a weaker correlation (r^2 = 0.41) between the concentrations of these oxides. The colourless and pink to purple glasses of this group have an average alumina concentration of 1%, and an average iron oxide concentration of 0.5%, enough to impart colour to the glass (section 1.10). The glasses have, however, been decoloured (sections, 1.10.2, 6.4.4.1).

Group 2 plant ash glasses have relatively high iron oxide concentrations averaging 0.8%, with 1.1% alumina. This group contains a mixture of transparent and opaque,

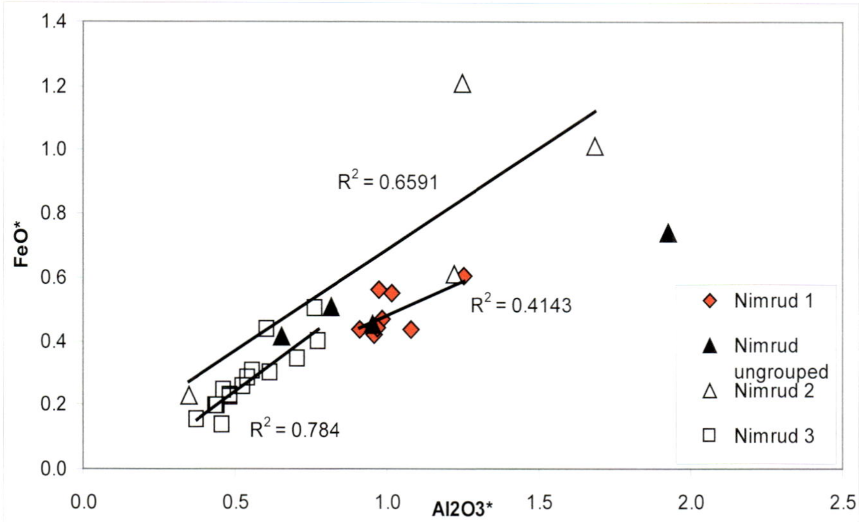

Figure 6-11 Scatter plot of alumina vs iron in wt% oxide for Nimrud Groups 1-3 and ungrouped glasses. Group 4 glasses are plotted separately as alumina is more concentrated.

light and dark blue glasses. The relatively higher oxide concentrations of iron and alumina are suggestive of a sand source of silica, rather than the purer quartz pebble source used to make 2nd millennium BC plant ash glasses.

Group 3 colourless and light blue translucent and opaque glasses have, by comparison, low contents of both oxides with 0.3% iron oxide and 0.6% alumina on average. These quantities are more typical of a purer silica source, as is found in established 2nd millennium BC HMHK recipes, but is unusual for LMLK glasses which are usually considered to have been made with mineral soda combined with sand. Could these glasses have been made instead with an LMLK high-lime plant ash in combination with a pure silica source, or mineral soda with pure silica and lime added separately? From published analyses of Near Eastern sands, the availability of a high-lime pure sand seems remote. One sand sample from Haifa was analysed by Turner (1956c: 281T) and was found to have low contents of alumina and iron oxide. The other sand samples he analysed each contained several per cent of alumina. Those analysed from near the mouth of the Belus River reported by Bezborodov (1975: 59-60) were high in lime, low in iron but with several percent of alumina. Brill (1999b: 474-477) analysed one sand (sample 686) which contained 0.35% alumina and 0.76% iron oxide, while the rest of the sands analysed contained several per cent of each of these oxides (section 1.6).

The ungrouped glasses vary in composition between 0.6% and 1.9% alumina, and 0.4% and 0.9% iron oxide. The ratio of alumina to iron oxide is 2.1:1 in both Groups 1 and 3, which approximates the ratio found naturally in the UCC, indicating that these two oxides have entered the glass as soil impurities. The ratio of alumina to iron oxide in Group 2 HMHK glasses is low at 1.6:1, due to the relatively high iron oxide content of these glasses. The Group 4 glasses have consistently high alumina contents (Figure 6-12), ranging from 4% to 7.7% (average of 5.8%). The average iron oxide content is 0.8% and there is no correlation between alumina and iron oxide, as there is in Groups 1 to 3. The high alumina concentration of the Group 4 glasses is reflected in the unusually high alumina to iron oxide ratio of 7.7:1.

The lime-alumina plot below (Figure 6-13) emphasises the distinctive character of the Co blue glasses, with high alumina and particularly low lime contents.

The high alumina contents associated with the Co blue colourants immediately bring to mind the Co blue glazes of New Kingdom Egypt, which Kaczmarczyk (1986) argued were coloured by Co-bearing alum obtained from the Kharga and Dakhla Oases of the Egyptian Western Desert, a view which has come to be widely accepted (section 1.10.1). Other characteristics of glasses coloured with this Co alum include high concentrations of magnesia, manganese oxide, iron oxide (Figure 6-12), Ni and Zn. According to Kaczmarczyk, these elements are invariably associated in New Kingdom Co blue vitreous materials, and are consequently considered to provide a reliable indicator of the alum source. As will be discussed in section 7.5 on Nimrud Group 4 colourants, the Nimrud blue glasses have elevated concentrations of the same suite of elements, and a similar Co alum source is therefore

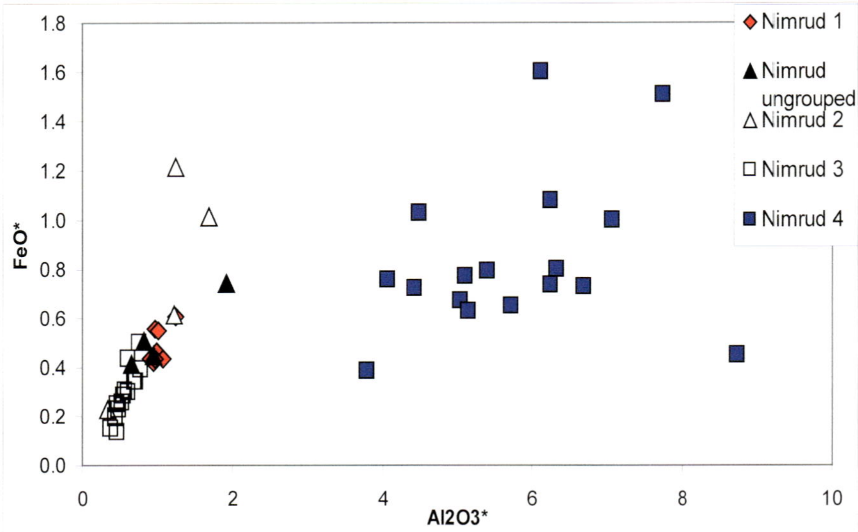

Figure 6-12 Scatter plot of alumina vs iron in wt% oxide illustrating clearly the difference between the cobalt coloured blue glasses of Group 4 and the other Nimrud glasses.

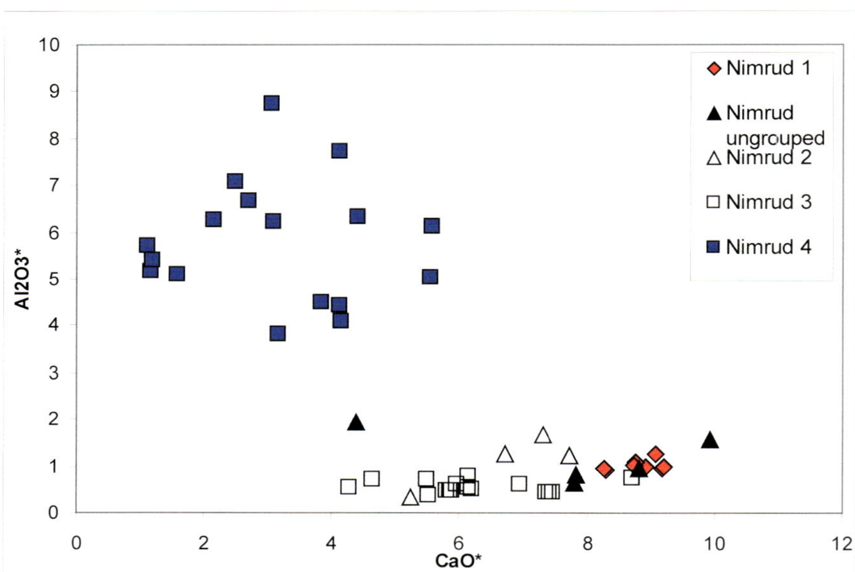

Figure 6-13 Scatter plot of lime vs alumina for all Nimrud glasses in wt% oxide.

likely. Figure 6-14 illustrates the strong correlation between alumina and magnesia in Group 4 glasses (r^2 = 0.7) that would be expected in glasses coloured with magnesia-rich Co alum from the Egyptian western oases. Group 1 glasses provide a comparison, being a tighter compositional group with low alumina content.

There is no correlation between alumina and magnesia in Groups 2 and 3 glasses. Group 1, however, has a moderately strong correlation (r^2 = 0.6) between these elements, which must be associated in one of the raw materials.

6.3.5 Summary

Groups 1 and 2 are HMHK plant ash fluxed glasses of a variety of colours with Group 2 being closest in base glass composition to 2nd millennium BC HMHK glasses. Group 3 glasses appear to be LMLK glasses with higher lime content than is seen in the earlier 1st millennium LMLK glasses from Pella and Egypt. The Group 4 glasses appear to be LMLK glasses coloured with Co alum, which introduces magnesia and alumina, and is most likely sourced from Egypt. These glasses appear to have been made in the low lime Egyptian 2nd millennium BC

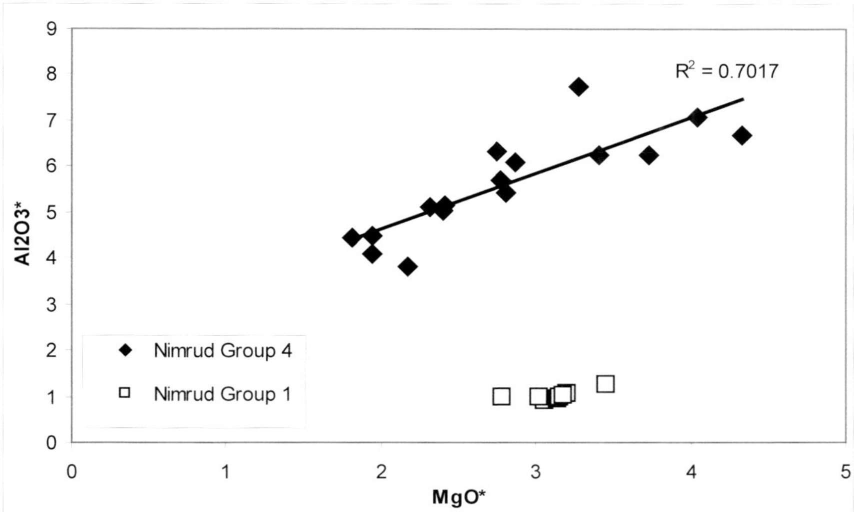

Figure 6-14 Scatter plot of magnesia vs alumina in wt% oxides for Nimrud Group 4 glasses.

tradition. The ungrouped glasses can be classified as HMHK glasses, but their compositions do not fit easily into the established groups, indicating the greater variability of 1st millennium BC glass compositions, compared with the more homogeneous compositions of the 2nd millennium BC. The earlier HMHK technology is to some extent continued into the IA, in particular Group 2, which has the closest composition to typical 2nd millennium BC Near Eastern glasses. These glasses form part of a more diversified and experimental group of glass-making technologies which might have made use of a wider range of raw materials, or possibly traditional raw materials combined in new ways. Perhaps a larger pool of craftsmen or greater number of workshops from a wider region is responsible for the diversification in glass compositions in the IA. This could have been a result of the many Assyrian conquests of foreign lands, especially to the west of Assyria, which were made throughout the period of the Neo-Assyrian empire, the time when the glass analysed in this study was made (section 9.7).

6.4 Trace Element Analyses (Groups 1 to 3)

Trace elements for glasses from Groups 1, 2 and 3 were analysed with solution ICP-MS. The results are presented below following the protocol established in Chapter 2. Group 4 trace element results will be presented separately in Chapter 7.

6.4.1 Rare Earth Elements (REE)

Averaged raw concentrations of REE from Nimrud Groups 1 to 3 and ungrouped glasses, and their ratios to MUQ, are presented in Table 6-3.

Table 6-3 Averaged raw values for Rare Earth Elements from Nimrud glasses, Groups 1 to 3 in ppb.

Nimrud Rare Earth Elements Groups 1-3 (ppb)

Samples	La	Ce	Pr	Nd	Sm	Eu	Gd	Tb	Dy	Ho	Er	Tm	Yb	Lu
Group 1 n=8														
Average	3613	6891	874	3370	674	145	651	101	634	134	391	61	397	63
RSD	13	10	13	13	13	12	13	13	14	15	16	17	16	18
Group 2 n=4														
Average	2663	5426	685	2615	548	131	542	86	511	105	295	44	267	41
RSD	32	27	29	30	30	30	28	29	31	32	33	33	36	36
Group 3 n=16														
Average	1514	2915	358	1383	290	65	271	43	250	51	144	22	136	20
RSD	19	19	19	19	18	16	16	18	17	17	18	20	20	18
Ungrouped														
8Q	2541	4188	603	2373	489	110	459	66	371	76	212	31	187	29
43K	3450	7665	885	3324	698	165	669	106	639	131	372	59	361	55
45P	2767	7749	673	2608	511	120	484	76	447	89	245	37	222	34
6BM	2863	5608	677	2581	531	113	506	81	469	99	288	45	280	44
Gp1/MUQ	0.11	0.10	0.10	0.10	0.10	0.09	0.10	0.10	0.11	0.11	0.12	0.12	0.12	0.13
Gp2/MUQ	0.08	0.08	0.08	0.08	0.08	0.08	0.09	0.09	0.09	0.09	0.09	0.09	0.08	0.08
Gp3/MUQ	0.05	0.04	0.04	0.04	0.04	0.04	0.04	0.04	0.04	0.04	0.04	0.04	0.04	0.04

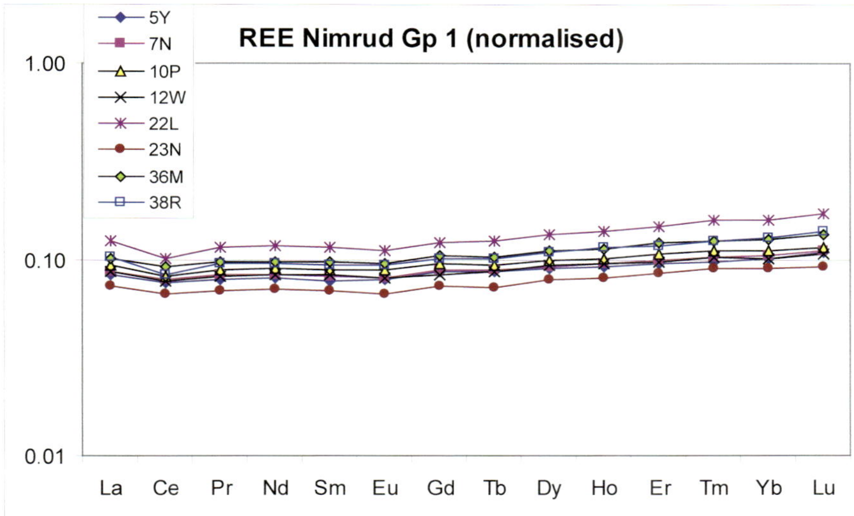

Figure 6-15 Line plot of the REE of individual Nimrud Group 1 glasses showing similarity of pattern and concentration. A logarithmic scale is used. Data are normalised to MUQ.

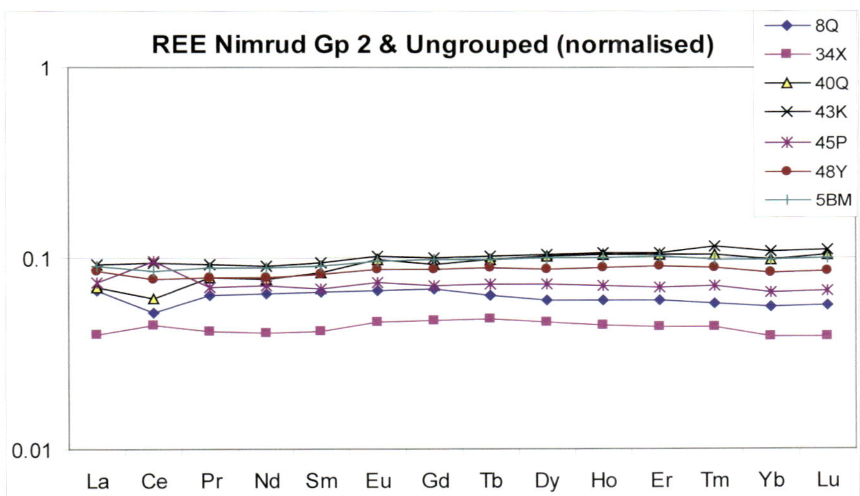

Figure 6-16 Line plot of the REE of individual Nimrud Group 2 and ungrouped glasses showing similarity of pattern and concentration. A logarithmic scale is used. Data are normalised to MUQ.

6.4.1.1 Group 1

When presented on a line plot, the REE of the Nimrud Group 1 glasses are shown to be homogeneous in pattern and concentration, as illustrated in Figure 6-15. These glasses form a tight compositional group apparently made with a consistent recipe and raw materials, as is indicated by the low RSD values for each element (Table 6-3). These samples have slight negative Ce anomalies. There is a gradual increase in the concentration of the HREE from Eu to Lu.

Group 1 is plant ash fluxed, and as a consequence it is likely, as with the 2nd millennium BC HMHK glasses, that the REE enter the glass as components of the plant ash, or in association with the ash[52] (section 1.8). It might be possible that they are derived from the silica source because these glasses have on average higher concentrations of iron oxide (0.5%) and alumina (1%) than is usual for plant ash glasses based on the 2nd millennium BC Nuzi-Pella model made with pure quartz pebbles (cf. Figure 6-21). Regardless of whether these glasses are colourless or pink/purple translucent, the REE patterns and concentrations remain the same,

[52] Although note that the analyses of Syrian plant ashes by Barkoudah and Henderson (2006: 308-310) show plant ashes to have relatively low REE content that would have added little to the glass.

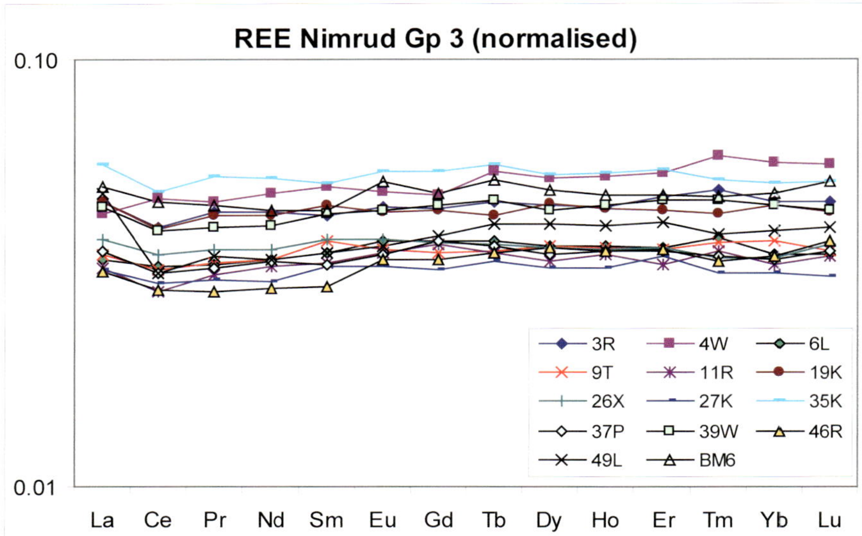

Figure 6-17 Line plot of the REE of individual Nimrud Group 3 glasses showing variability of pattern at relatively similar concentrations. A logarithmic scale is used. Data are normalised to MUQ.

indicating that the colourant or decolouring agent is not responsible for introducing the REE. The REE, as has been shown previously (section 4.5.1, 4.52, 5.4.1, 5.4.2), tend to correlate with the SRE as well as alumina and iron oxide, indicating that they are introduced with the alumina-bearing component of the glasses, either sand, mineral-bearing plant ash, ash contaminants or quartz contaminated by tools used to grind it (see Rehren and Pusch 2007: 223-225; section 1.6).

6.4.1.2 Group 2

The REE concentrations of the Group 2 and ungrouped HMHK glasses are closely similar to each other, with relatively flat patterns. Samples 40Q (Group 2) and 8Q (ungrouped) have negative Ce anomalies, and samples 43K (ungrouped) and 34X (Group 2) have positive Ce anomalies (Figure 6-16). It is possible that sample 34X is slightly different, but it fits better into this grouping than any other. La values have the largest error because of the large antimony oxide correction required, and Ce is redox sensitive, consequently the scatter on the left of the diagram is not surprising, even amongst closely similar samples.

Similar to the Group 1 glasses, the more typically 2nd millennium-type plant ash glasses of Group 2 have relatively high concentrations of iron oxide (0.8%) and alumina (1.1%). Does the concentration of alumina indicate that a sand source of silica was used to make the Group 2 glasses, or was it that a higher level of contamination from quartz or ash preparation occurred? Either possibility could account for the additional REE. The REE of Groups 1 and 2 glasses have different patterns at higher concentrations than the REE from Nuzi plant ash glasses which were most likely made with a quartz pebble silica source (Figure 6-21).

6.4.1.3 Group 3

The line plot (Figure 6-17) of individual Nimrud Group 3 glasses presents variation in the REE patterns, but at similar concentrations, which are lower than those of Group 1 and 2 glasses. The differences are not enough to justify subdivision into further groupings. These variations are not borne out in the other elemental components of these glasses, and the colour and opacity of the glasses do not influence the REE concentrations or patterns. While REE concentrations are low, they are well above instrumental detection limits (section 2.4.2) indicating that the observed variations occur naturally within this group of glasses.

6.4.1.4 Groups 1 to 3 Comparisons

Figure 6-18 clearly shows the correlations between the concentrations of the REE, represented by the HREE Yb, and alumina in Groups 1 to 3 glasses. For the Group 1 HMHK glasses $r^2 = 0.8$, for Group 2 HMHK glasses $r^2 = 0.9$ and for Group 3 LMLK glasses $r^2 = 0.4$, which indicates with some certainty that alumina and the REE were introduced from the same source for each of the Groups 1 and 2 glass mixes. The LMLK glasses exhibit a poorer correlation between these elements which are present at lower concentrations.

In Figure 6-19 the plotted Nimrud data for Groups 1, 2 and 3 are normalised to MUQ which results in relatively

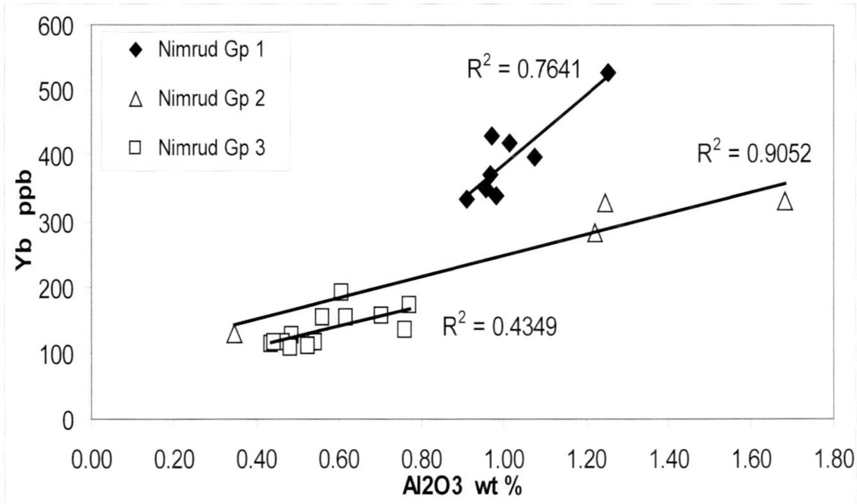

Figure 6-18 Scatter plot of alumina vs Yb for Nimrud Groups 1 to 3 glasses.

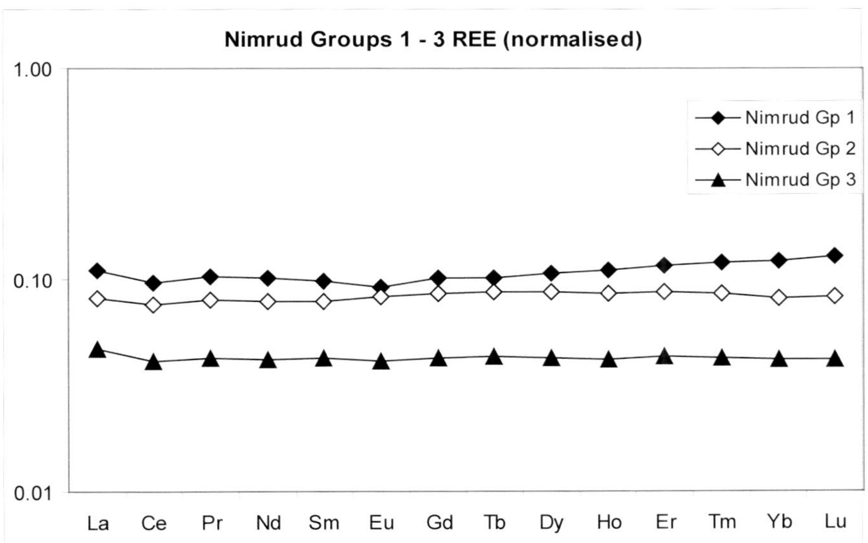

Figure 6-19 Line plot of averaged REE from Nimrud glasses. Data are normalised to MUQ. Note logarithmic scale.

flat REE patterns, resembling the pattern of sediments found in the UCC. If, as is the case in the 2nd millennium BC glasses, the ratios of the REE in the Nimrud glasses to those in the UCC (Table 6-3) are indicative of an unenriched sedimentary level or SBS, then Group 1 has a ratio of 0.09 to 0.13, being higher for the HREE; Group 2 has a ratio of 0.08 to 0.09, a flatter pattern; and Group 3 has a ratio of 0.04, which is constant across all the REE in this group. The Group 3 REE are present at a lower concentration than is found in the first two groups and the LBA HMHK glasses, indicating an even greater purity of raw materials than was used to make the LBA glasses. The Group 3 glasses are opaque and translucent and of different colours showing that the opacifiers and colourants do not introduce contaminants to these glasses.

The sedimentary pattern of the REE is highlighted in Figure 6-20 where Groups 1 to 3 are normalised to average chondrite data and plotted with MUQ. As was shown for the Pella and Nuzi data, the patterns of the REE in all the Nimrud glasses are very similar to those of the REE from the UCC, indicating their derivation from sediments, though at lower concentrations.

The REE patterns and concentrations of Nimrud Groups 1 to 3 are compared in Figure 6-21 with REE data from Nuzi glasses, representing typical 2nd millennium

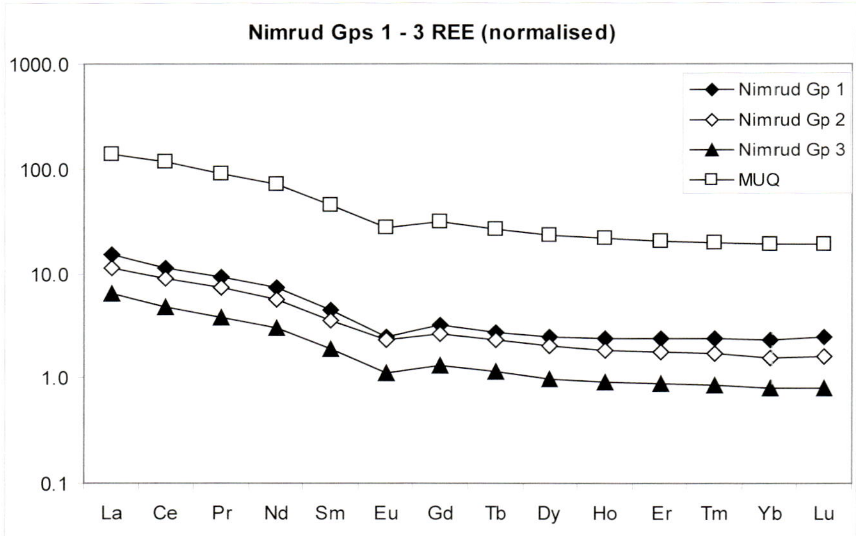

Figure 6-20 Line plot of averaged Nimrud REE data for glass Groups 1 to 3, and MUQ. Data are normalised to average chondrites for comparison (see Table 27).

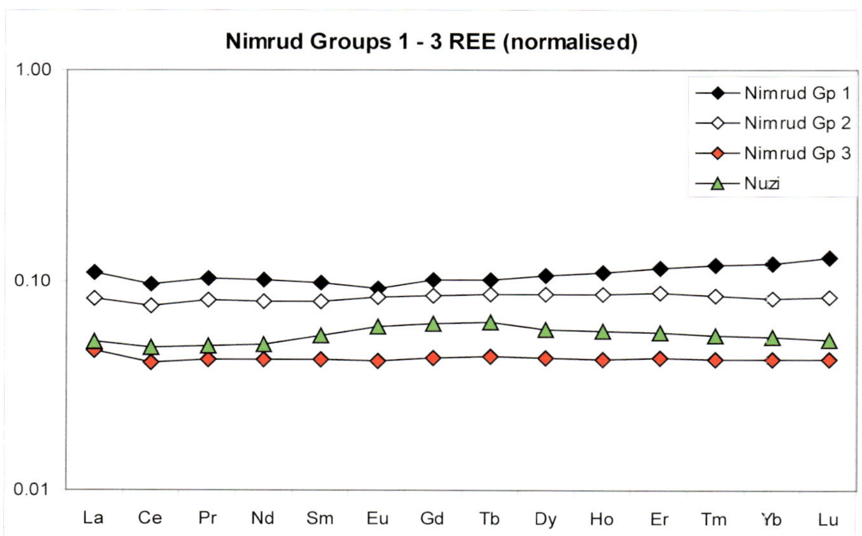

Figure 6-21 Line plot of averaged REE data from Nimrud Groups 1 to 3 glasses compared with Nuzi HMHK glass REE. Data are normalised to MUQ. Note logarithmic scale.

BC glass. None of the Nimrud groups is the same as the glasses from Nuzi, so these glasses on balance are either not likely to be 2nd millennium BC glass, or at least are not made from the same raw materials (see sections 9.9.2, 9.3). As noted previously, the recipe for Group 2 glasses in particular is comparable with that of the 2nd millennium BC glasses, but the Group 2 REE pattern, while similar, particularly to the flatter pattern of the Pella HMHK glasses, is not the same. The recipe for the Nimrud HMHK glasses, and possibly Group 3 LMLK glasses, seems to constitute a continuation of a 2nd millennium BC tradition but with raw ingredients that provide a different chemical signature. Whether this represents a change over time in the location of glass-making, or the increase in the number of glass-making sites and hence the use of a wider variety of raw materials, can only be surmised.

In summary, the REE of the Nimrud Groups 1 to 3 glasses exhibit a continental crustal pattern, though of lower concentration, due to the pure nature of the glass raw materials, especially those used to make Group 3 glasses. The ratios of the REE to the UCC are at approximately 0.1 for Group 1, 0.08 for Group 2 and 0.04

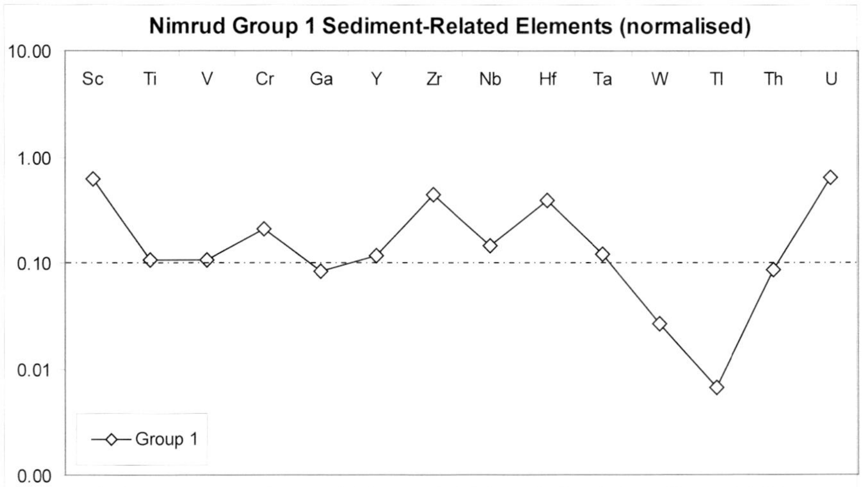

Figure 6-22 Line plot of averaged concentrations of sediment-related trace elements in Nimrud Group 1 glasses, normalised to MUQ. A logarithmic scale is used. Mo is not plotted as there is no MUQ value by which to normalise it. The dotted line crosses the Y-axis at 0.1 to represent the SBS of Group 1 glasses.

for Group 3, forming a constant base-line ratio with which to compare concentrations of other elements. The most likely source of the REE is sediment associated with plant ash for Groups 1 and 2, and possibly with a contribution from the silica source, particularly for Group 3 which has the lowest REE concentrations. The LMLK nature of Group 3 combined with its low REE levels compared with the plant ash glasses, suggest that these glasses might have been made with mineral soda. Mineral soda is known to have low contamination levels (section 1.9). The REE appear to be unrelated to colour or opacity, as was found for the glasses from Nuzi and Pella.

6.4.2 Sediment-Related Elements (SRE)

The next group of trace elements to be considered are those related to sediments, usually included as contaminants with batch ingredients. The averaged data for these elements are presented in Table 6-4. The larger RSD values indicate variability in the composition of glasses within each group, and are not indicative of unreliable measurements because the values obtained are above the instrumental detection limits (section 2.4.2).

The Group 1 colourless and purple glasses have a reasonably consistent SRE content, in keeping with their REE patterns. Sc, however, is particularly variable (range of 1197 to 36,152 ppb) and is, like Zr, Hf, U and, to a lesser extent Cr, elevated above the established SBS of approximately 0.1. W and especially Tl, are comparatively depleted (Figure 6-22). Sample 22L (a purple glass) has very low levels of Mo and W, while sample 38R (a pink glass) has low levels of Ta and W.

Group 2 consists of only four samples. Sample 34X has elemental concentrations generally inconsistent with the other three glasses in the group and has been excluded where necessary from the calculations of many of the averages and RSDs for this group. The RSD value for Ta is high as there are two samples with relatively low and two with relatively high contents (range of 14 ppb to 86 ppb). W levels vary between 33 ppb and 125 ppb and Tl between 3 ppb and 66 ppb. Sc concentrations range from 1,791 ppb to 23,679 ppb. Sc and Cr in particular, and elements from Ga to Hf, and U are enriched above the SBS of 0.08 in these glasses. Zr and Hf are not particularly elevated in Group 2 glasses compared with their concentrations in Group 1 glasses. Ta, W, Tl and Th are depleted compared with the Group 2 SBS (Figure 6-23).

In Group 3 LMLK glasses, large variations occur in the concentrations of Sc, Mo, W, Tl and U. Sample 39W has a particularly high concentration of Cr. The concentrations of Sc, V, Cr, Zr, Nb, Hf and U are elevated relative to the SBS of 0.04. The concentrations of Zr, Nb and Hf are relatively more enriched than in Group 2 glasses, but less so than in Group 1. W and Tl are depleted (Figure 6-24).

The SRE of the ungrouped glasses do not fit well into any of these groups. Samples 8Q, 43K and 45P have the greatest similarity to Group 2 glasses, but differ especially in the concentrations of the elements which are particularly variable throughout the groups. Overall, the patterns of SRE in Groups 1 to 3 are similar in their enrichment of Sc, Cr and U, and to a lesser extent of Zr and Hf (most noticeable in Group 1) compared with the SBS for these glasses. W and Tl are depleted in all

Table 6-4 Averaged concentrations in ppb of sediment-related elements from Nimrud Groups 1 to 3, and ungrouped glasses.

Nimrud Sediment-Related Elements Groups 1-3 (ppb)

Samples	**Sc**	**Ti**	**V**	**Cr**	**Ga**	**Y**	**Zr**	**Nb**	**Mo**	**Hf**	**Ta**	**W**	**Tl**	**Th**	**U**
Group 1 n=8									*n=7*		*n=7*	*n=6*			*n=7*
Average	10,175	745,696	12,867	13,623	1607	3755	86,604	2188	352	2042	133	42	3	964	1831
RSD	124	11	35	12	36	15	18	37	49	18	42	68	35	15	34
Group 2 n=4		*n=3*	*n=3*	*n=3*	*n=3*	*n=3*	*n=3*	*n=3*	*n=3*	*n=3*				*n=3*	
Average	12,537	536,999	12,352	46,593	2063	3298	23,051	1433	589	574	52	69	28	841	492
RSD	94	21	22	14	40	6	22	15	36	21	70	57	106	16	68
Group 3 n=16				*n=15*								*n=13*			
Average	7459	239,509	6387	6833	764	1459	17,260	901	479	396	56	24	4	346	1793
RSD	140	22	30	19	23	16	33	26	54	33	30	47	45	21	62
Ungrouped															
8Q	3.79	318	7.68	39.7	0.936	2.38	16.9	1.14	1.01	0.386	0.072	0.055	0.003	0.413	2.27
43K	5.02	618	12.8	46.5	2.01	3.53	31.3	1.93	0.292	0.786	0.135	0.065	0.012	1.21	0.404
45P	45.7	361	9.12	12.3	1.15	2.65	17.0	1.47	0.393	0.412	0.092	0.017	0.002	0.536	1.24
Gp1/MUQ	0.62	0.11	0.11	0.21	0.08	0.12	0.44	0.13		0.38	0.10	0.02	0.01	0.09	0.60
Gp2/MUQ	0.76	0.06	0.08	0.57	0.09	0.09	0.10	0.08		0.09	0.05	0.04	0.07	0.06	0.17
Gp3/MUQ	0.45	0.03	0.05	0.15	0.04	0.05	0.09	0.06		0.07	0.05	0.15	0.01	0.03	0.63

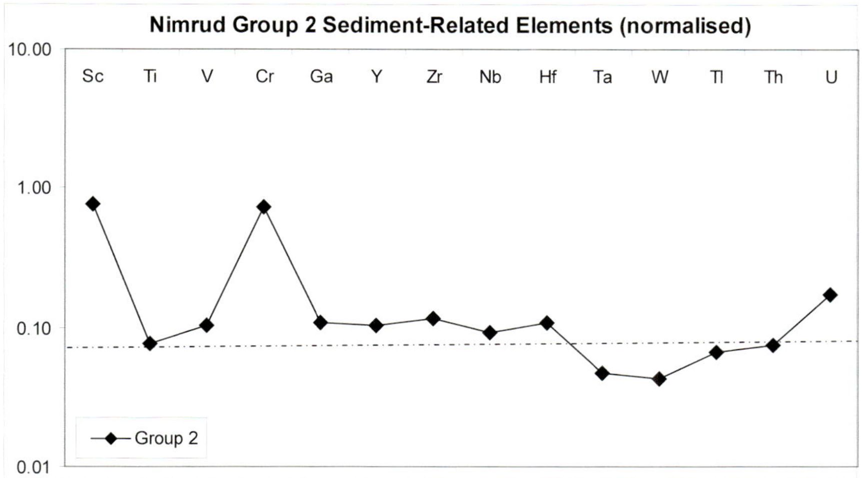

Figure 6-23 Line plot of averaged concentrations of sediment-related trace elements in Nimrud Group 2 glasses, normalised to MUQ. A logarithmic scale is used. The dotted line crosses the Y-axis at 0.08 to represent the SBS of the Group 2 glasses.

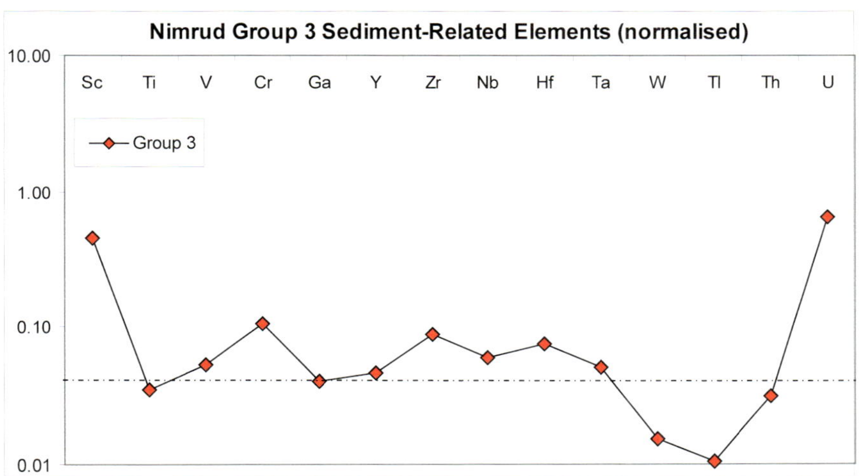

Figure 6-24 Line plot of averaged concentrations of sediment-related trace elements in Nimrud Group 3 glasses, normalised to MUQ. A logarithmic scale is used. The dotted line crosses the Y-axis at 0.04 to represent the SBS of the Group 3 glasses.

groups, but less so in Group 2. The concentrations of other elements are at approximately the SBS level of each group. Groups 1 and 3 have the greatest similarity. These relatively consistent elemental enrichments and depletions could constitute a regional if not chronological marker which is different from that established by the SRE content of glass from other sites considered in the present study (section 9.4).

Figure 6-25 is a line plot illustrating the average concentrations of SRE in Nimrud Groups 1 to 3 glasses compared with those in MUQ. Overall, the patterns are similar to each other, especially those of Groups 1 and 3, and to MUQ. Their differences occur notably in the concentrations of the elements Cr, Zr, Hf, Tl and U.

When Ti is plotted against Nb (Figure 6-26), close correlations for these elements in Groups 2 (r^2 = 0.8) and 3 (r^2 = 0.8) and a good correlation in Group 1 (r^2 = 0.5) are revealed. The different trend in Group 1 and the different REE patterns of this group suggest that it had a different REE-bearing raw material source.

Group 3 has lower contents of the SRE in general compared with Groups 1 and 2 glasses. Ti and Ta are negatively correlated in Group 1 (r^2= -0.3) and are not correlated in Group 2. In Group 3 Ti and Ta are

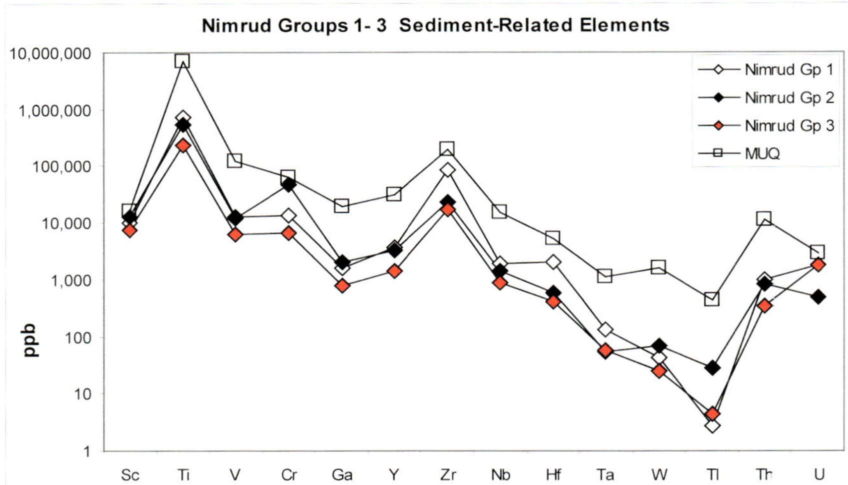

Figure 6-25 Line plot of raw Nimrud Groups 1 to 3 sediment-related element data in ppb compared with MUQ on a logarithmic scale.

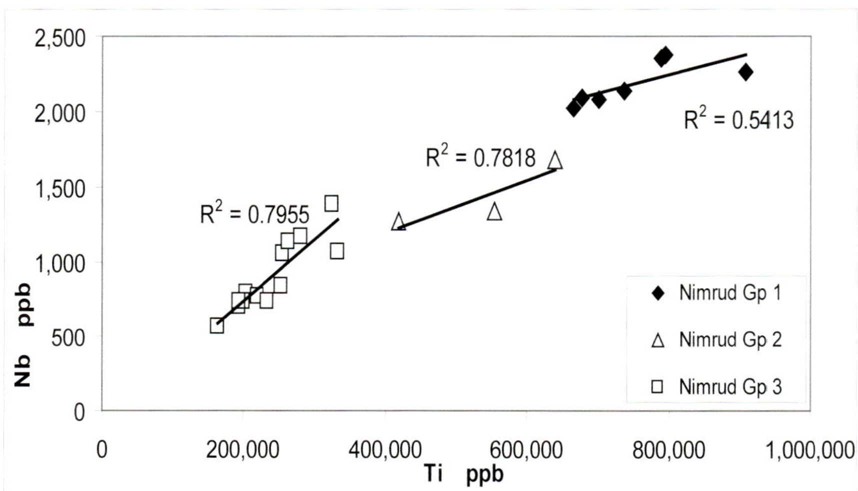

Figure 6-26 Scatter plot of Ti vs Nb in ppb for Nimrud Groups 1- 3 glasses.

positively correlated (r^2 = 0.6). The concentration of Ti is very closely correlated with those of Zr and of Hf in Group 1 (r^2 = 0.98) and they are well correlated in Group 3 (r^2 = 0.6). There is no correlation between Ti and the concentrations of these elements in Group 2. The ungrouped glasses plot with Groups 2 and 3 (Figure 6-27). Zr and Hf are very closely correlated in all glasses, with Group 1 glasses again being distinct from the others because of their higher concentrations of both of these elements (Figure 6-28).

When Groups 1 and 2 plant ash glasses are compared with the Nuzi HMHK glasses (Figure 6-29) greater similarity is found between Nuzi and Group 2 glasses, supporting previous findings that this group of Nimrud glasses, whilst not exactly the same as 2nd millennium BC HMHK glasses, is more nearly like them than are the other glasses found at Nimrud.

In summary, the SRE are associated with the REE, iron oxide and alumina in the Nimrud Groups 1 to 3 glasses. There are distinctive variations between the groups, but overall, the concentrations of Sc, Cr, Zr, Hf and U are elevated to different degrees above the SBS levels for each group. The Hf and Zr levels of Group 1 glasses are particularly elevated. These elemental anomalies may constitute a chemical signature for the glasses from this region in the 1st millennium BC. The lowest SBS is found in Group 3, which generally has the lowest levels of SRE. The plotted patterns of Groups 1 to 3

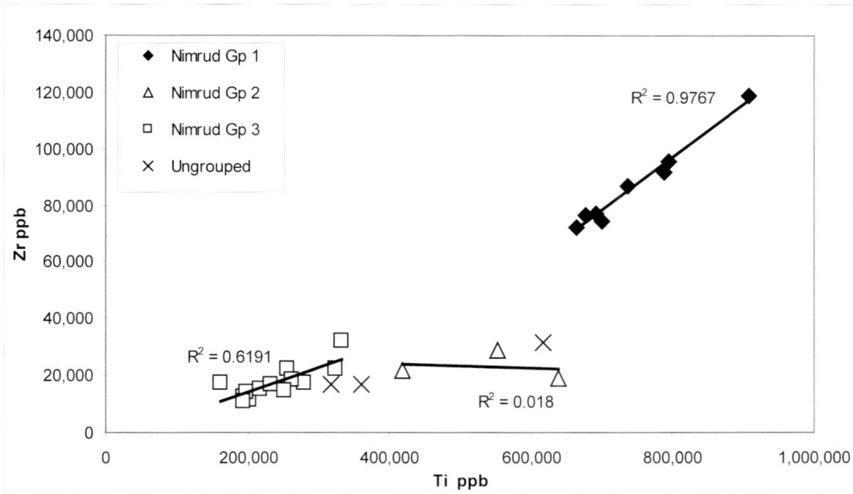

Figure 6-27 Scatter plot of Ti *vs* Zr in ppb for Nimrud Groups 1 to 3 and ungrouped glasses.

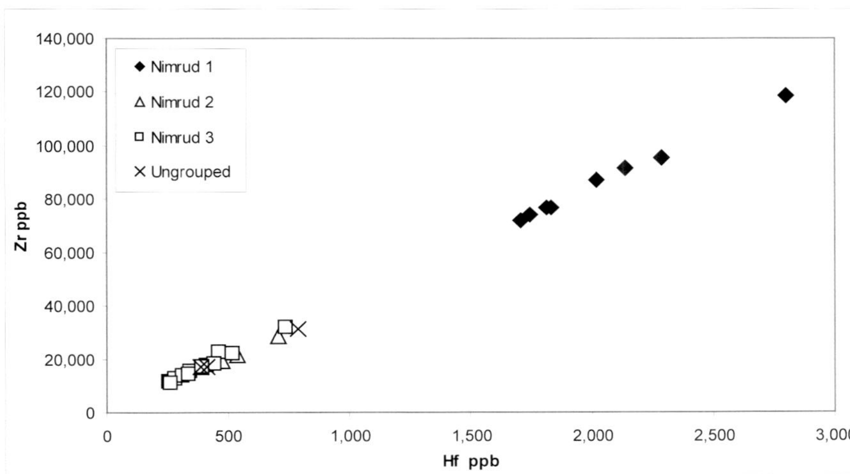

Figure 6-28 Scatter plot of Hf *vs* Zr in ppb for Nimrud Groups 1 to 3 and ungrouped glasses.

SRE are somewhat similar to that of MUQ, indicating derivation from sediments, but with variations in the concentrations of the elements Sc, Cr, W, Tl and U within groups.

6.4.3 Alkali and Alkaline Earth Elements

Nimrud Groups 1 and 2 glasses are fluxed with plant ash as established by the elevated levels of magnesia, potash, phosphorus and lime (see section 6.3). Group 3 glasses, which have low concentrations of magnesia and potash, but relatively high lime contents, appear to be mineral soda fluxed glasses. They are unusual in the apparent purity of the silica used to make them, as determined by the low REE and SRE concentrations, a feature usually associated with the quartz silica source of plant ash glasses. The averaged results of the ICP-MS analyses of alkalis and alkaline earths for Groups 1 to 3 are presented in Table 6-5, with their ratios to MUQ. Elemental concentrations in all Group 1 to 3 glasses are in general depleted relative to the content of the UCC, as observed in the REE and the SRE.

It is evident that the concentrations of alkaline and alkaline earth elements in all three Nimrud groups are almost identical (Figure 6-30), with Group 3 having slightly lower concentrations of these elements as might be expected of an LMLK glass, especially if fluxed with mineral soda. The concentrations of these elements are similar to those of the alkali and alkaline

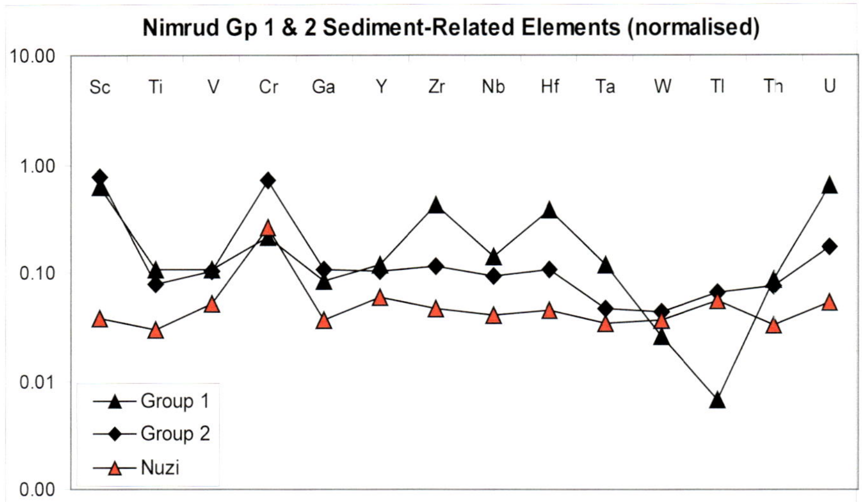

Figure 6-29 Line plot comparing Nimrud sediment-related elements in plant ash glasses (Groups 1 and 2) with typical 2nd millennium BC HMHK glasses from Nuzi. Data are normalised to MUQ. A logarithmic scale is used.

Table 6-5 Average compositions in ppb of alkali and alkaline earth elements from Nimrud glasses Groups 1 to 3, and ungrouped glasses.

Nimrud Alkali and Alkaline Earth Elements Groups 1-3 (ppb)

Samples	**Li**	**Be**	**Rb**	**Sr**	**Cs**	**Ba**
Group 1 n=8						
Average	5273	169	7434	438,390	142	112,251
RSD	29	13	4	7	12	38
Group 2 n=4	*n=3*					
Average	19,161	160	10,684	398,632	203	45,439
RSD	30	35	29	34	23	49
Group 3 n=16	*n=15*					
Average	4003	103	4982	210,354	147	49,820
RSD	25	22	27	25	42	30
Ungrouped						
8Q	5.87	0.159	7.67	342	0.145	73.8
43K	71.5	0.272	14.9	202	2.09	74.1
45P	5.89	0.171	7.06	363	0.158	54.9
Gp 1/MUQ	0.19	0.09	0.09	3.09	0.03	0.28
Gp 2/MUQ	0.54	0.08	0.13	2.81	0.04	0.11
Gp 3/MUQ	0.16	0.05	0.06	1.48	0.03	0.13

earth elements present in the earth's crust, with Sr being clearly enriched in all glass groups.

It is apparent that Group 2 glasses are consistently closer in composition to 2nd millennium BC HMHK glasses than is any other group from Nimrud. Whereas the LMLK Group 3 glasses have lower concentrations of the alkali and alkaline earth elements than the plant ash glasses, the pattern of their content as plotted in Figure 6-30 is closely similar. The patterns and concentrations of the ungrouped glasses generally conform to plant

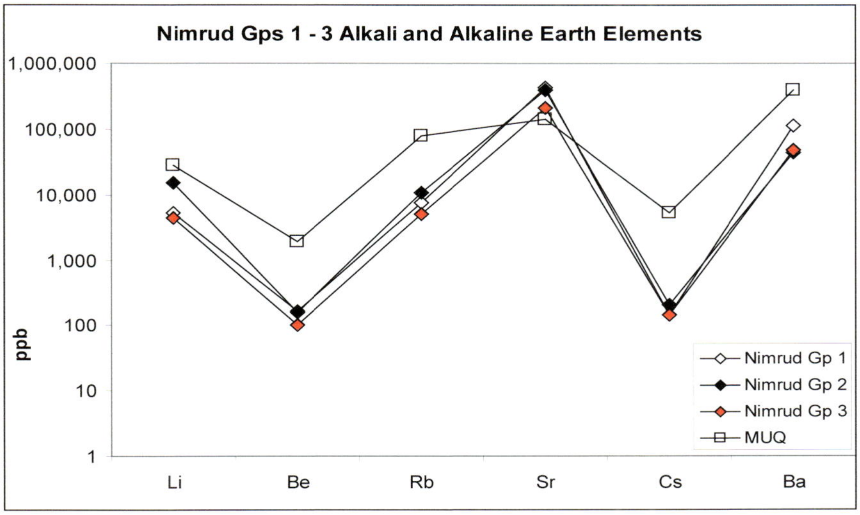

Figure 6-30 Line plot of averaged raw data for Nimrud alkali and alkaline earth elements compared with MUQ. Note the logarithmic scale.

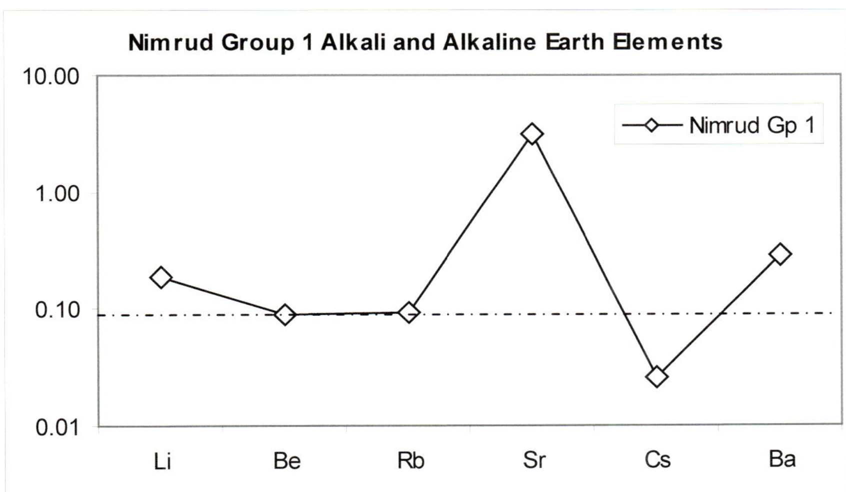

Figure 6-31 Line plot of data averaged and normalised to MUQ for Nimrud Group 1. The dotted line represents the SBS level of 0.1

ash glass compositions, although sample 43K has particularly high Li and Cs, and low Sr contents.

Figure 6-31, 6-32 and 6-33 illustrate the alkali and alkaline earth contents of each group relative to their established SBS. Be again appears not to be introduced with plant ash because it is found at SBS levels for both Groups 1 and 2, as was similarly noted for the Nuzi and Pella HMHK glasses.

The Rb concentration is slightly enriched, relative to the SBS, in Groups 2 and 3, and not at all in Group 1. Li concentrations are elevated in each group, particularly Group 2 (though sample 40Q has a comparatively low content of 3731 ppb). Ba is slightly elevated in all groups, while Cs is depleted.

It has been noted that Sr concentrations are enriched relative to that in the UCC in all of Groups 1 to 3 glasses. Sr concentrations in Groups 1 (average 438,390 ppb) and 2 (average 398,632 ppb) are consistent with levels of Sr in 2nd millennium BC plant ash glasses, whereas Group 3 has a lower average Sr level (210,354 ppb) more typical of mineral soda fluxed glasses but higher than that of Pella LMLK glasses (average 108,686 ppb) and Nimrud Group 4 glasses (average 105,777 ppb).

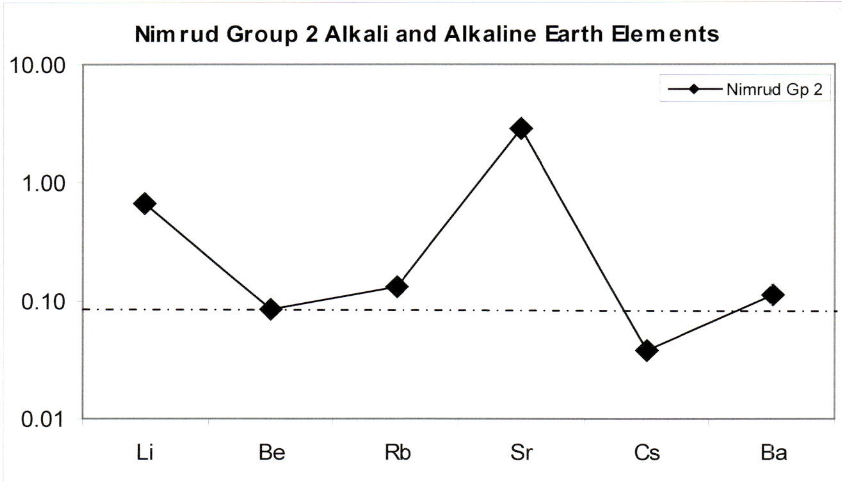

Figure 6-32 Line plot of data averaged and normalised to MUQ for Nimrud Group 2. The dotted line represents the SBS level of 0.08.

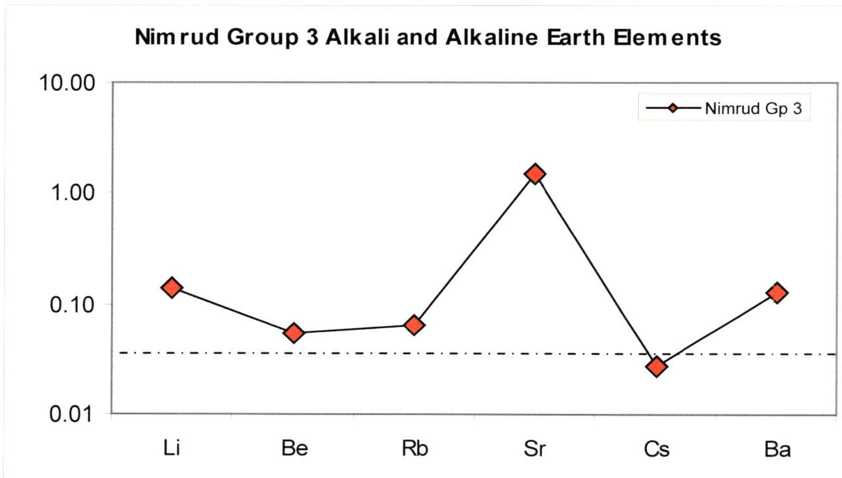

Figure 6-33 Line plot of data averaged and normalised to MUQ for Nimrud Group 3. The dotted line represents the SBS level of 0.04.

Figure 6-34 and 6-35 provide a comparison for the Nimrud glasses with the HMHK glasses from 2nd millennium BC Nuzi and Pella. Pella and Nuzi glasses are closely similar to each other, but differ from the Nimrud glasses in their greater enrichment in Li, Rb, Sr and Cs, and their lower SBS level: 0.05 for 2nd millennium BC glasses compared with 0.1 and 0.08 for Groups 1 and 2 1st millennium BC Nimrud glasses. The purity of the 2nd millennium BC raw materials approaches that of the Group 3 LMLK glasses (SBS of 0.04).

The alkali and alkaline earth patterns of the Pella HMHK and LMLK glasses are also broadly similar to each other and to MUQ. Of the three Nimrud glass groups under consideration here, the pattern of the Group 3 LMLK glasses bears greatest resemblance to that of the Pella LMLK glasses, although with lower elemental concentrations. Sr is the exception, being twice as concentrated in the Nimrud Group 3 glasses. The explanation for Sr enrichment, even in the Group 3 LMLK glasses, is not clear. One explanation might be that it is associated with a sand source of silica, which is low in REE and SRE contaminants. Alternatively, it could be the result of a Sr-bearing calcium ingredient being added to the glass batch to stabilise the glass mix. This might represent an improved understanding by the glass-makers of the role of lime as a glass stabiliser, leading to an advance in mineral soda-fluxed glass-making technology since the manufacture of the unstable, low lime LMLK glasses of Nesikhons,

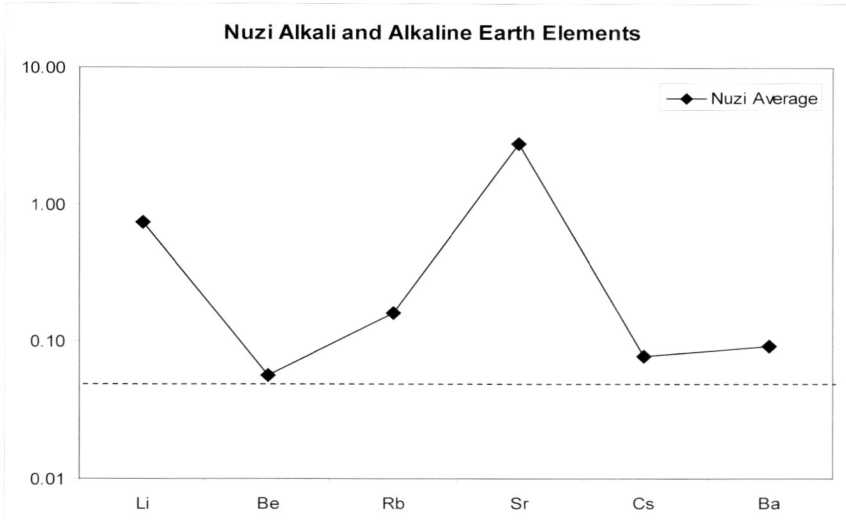

Figure 6-34 Line plot of data averaged and normalised to MUQ for Nuzi glasses. The dotted line represents the SBS level of 0.05.

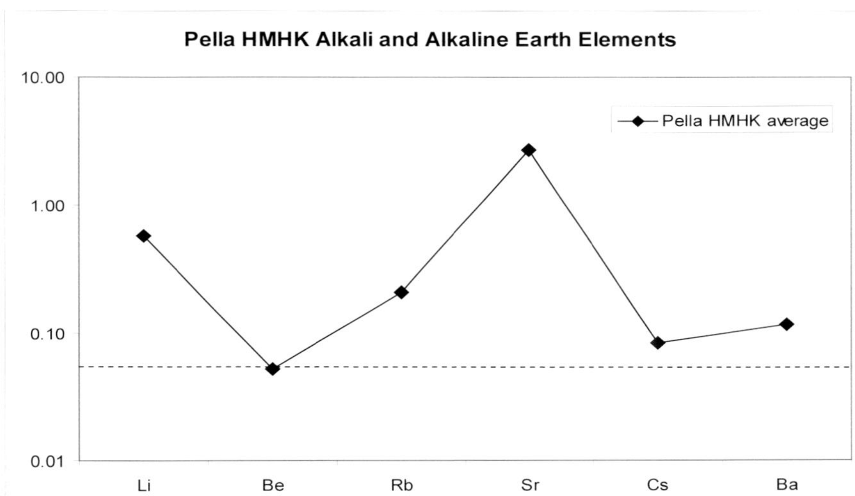

Figure 6-35 Line plot of data averaged and normalised to MUQ for Pella HMHK glasses. The dotted line represents the SBS level of 0.05.

approximately two centuries earlier (section 1.7). Finally, there remains the as yet unconvincing proposition that the Sr is introduced as part of an LMLK plant ash, which would have contained unusually low levels of REE and SRE contamination unlike the ashes used to make the recognised HMHK plant ash glasses. There is no correlation between lime and Sr in Group 3 glasses, while there is a weak correlation of $r^2 \approx 0.3$ to 0.4 in Groups 1 and 2 glasses by comparison.

In summary, Nimrud glass Groups 1 to 3 are enriched in their concentrations of Li, and Sr, and to a lesser extent, Ba. Be is not enriched and does not seem to be associated with plant ashes, as is observed in glasses from Nuzi and Pella. Cs is depleted in each group. Groups 1 to 3 patterns are broadly similar to each other and to that of the UCC. With respect to the concentrations of alkaline and alkaline earth elements, Nimrud Group 2 glasses again bear the closest similarity to 2nd millennium BC HMHK glasses, while Nimrud Group 3 is not dissimilar to the LMLK Pella glasses, but at generally lower concentrations, with higher Sr.

6.4.4 Colourants, Decolourants and Opacifiers

A range of colours is found in Nimrud Groups 1 to 3 glasses. Group 1 contains purple, pink and colourless glasses; Group 2 consists of light blue translucent, turquoise opaque, dark blue and colourless glasses; the ungrouped glasses are a mixture of colours too; and

Table 6-6 Compositions in ppm of colourant, opacifier and related elements from Nimrud Groups 1 to 3 glasses.*

	Nimrud Colourants, Opacifiers and Related Elements							**(ppm)**		
Group	**Colour**	**MnO (wt%)**	**Co**	**Ni**	**Cu**	**Zn**	**Cd**	**Sn**	**Sb**	**Pb**
	Purple T									
1	22L	1.53	4.67	18.0	14.3	24.7	0.100	0.429	105	8.42
1	38R	1.54	6.36	14.1	30.4	28.8	0.062	1.55	1,731	13.6
	Average	**1.54**	**5.52**	**16.0**	**22.4**	**26.7**	**0.081**	**0.988**	**918**	**11.0**
	RSD	**0.33**	**22**	**17**	**51**	**11**	**33**	**80**	**125**	**33**
	Colourless									
1	5Y	bd	1.95	6.04	12.0	22.0	0.053	0.234	1961	2.88
3	6L	bd	0.45	3.22	10.4	14.2	0.018	0.104	5129	11.2
1	7N	bd	1.95	7.78	12.0	22.5	0.058	0.278	2706	11.7
1	10P	bd	1.89	6.08	15.3	20.8	0.065	0.237	5212	13.7
1	12W	bd	1.80	6.25	10.9	21.6	0.059	0.282	2446	10.8
1	23N	bd	1.97	6.50	13.2	26.5	0.062	0.588	1582	111
3	35K	0.11	1.72	6.36	25.2	18.7	0.024	0.192	2701	14.7
1	36M	0.19	2.58	7.25	14.5	22.9	0.070	0.489	2322	202
3	39W	bd	1.35	9.02	19.4	14.1	0.024	0.486	6648	3.21
-	45P	bd	1.83	11.0	24.0	32.8	0.037	0.236	4142	8.85
	Average	**0.15**	**1.90**	**7.36**	**15.7**	**20.6**	**0.045**	**0.312**	**3485**	**9.63**
	RSD	**37**	**16**	**32**	**34**	**30**	**43**	**47**	**49**	**46**
	Blue T									
3	3R	bd	2.79	6.55	12,710	21.2	0.021	111	207	421
-	8Q	bd	2.03	26.2	13,624	52.9	0.023	6.28	83.6	89.8
3	9T	bd	1.54	9.93	9492	21.4	0.024	63.7	9.66	3.09
3	11R	bd	3.57	9.95	12,360	14.7	0.016	174	65.5	15.6
3	19K	bd	4.54	9.22	12,274	22.4	0.019	511	61.5	60.9
3	26X	bd	2.17	7.85	11,216	21.1	0.021	52.1	12.4	2.63
3	27K	bd	2.90	7.34	11,669	17.6	0.018	87.4	31.2	3.52
3	37P	bd	8.71	12.2	14,195	23.8	0.017	319	2150	9.24
3	46R	0.42	2.87	11.0	10,627	25.0	0.017	59.7	351	39.1
2	48Y	bd	8.77	27.2	13,695	559	0.047	62.7	32.0	369
	Average		**2.80**	**9.25**	**12,186**	**24.5**	**0.020**	**87.2**	**94.8**	**101**
	RSD		**34**	**21**	**12**	**45**	**15**	**50**	**119**	**156**
	Turquoise O									
3	4W	bd	3.94	9.30	15,615	80.7	0.043	141	8206	222
2	34X	bd	1.21	8.72	20,530	180	0.176	20.8	10,618	862
2	40Q	bd	8.65	39.8	37,438	60.4	0.037	2455	11,511	492
-	43K	0.16	3.93	30.9	14,703	96.5	0.070	5.2	9735	290
3	49L	bd	9.58	23.3	36,277	20.2	0.042	559	21,806	208
	Average		**3.02**	**22.4**	**16,949**	**64.5**	**0.048**	**636**	**12,375**	**303**
	RSD		**52**	**61**	**18**	**51**	**32**	**164**	**44**	**43**

Nimrud Colourants, Opacifiers and Related Elements								(ppm)		
Group	Colour	MnO (wt%)	Co	Ni	Cu	Zn	Cd	Sn	Sb	Pb
Purple T/MUQ		11	0.25	0.51	0.69	0.36		0.36		0.54
Colourless/MUQ		1.07	0.08	0.23	0.49	0.28		0.11		0.47
Blue T/MUQ			0.13	0.29	377	0.33		31		5.0
TurquoiseO/MUQ			0.14	0.71	524	0.88		230		15

*MnO is determined by SEM-EDS and measured in wt%. There are no values for Sb or Cd in MUQ. Data have been grouped according to colour and translucency. Inconsistently large or small values omitted from average and RSD calculations are underlined.

Group 3 glasses are colourless, light blue translucent and turquoise opaque. The dark blue glass (5BM) is included in Chapter 7 with the Nimrud Group 4 Co glass results.

Results of ICP-MS analyses of colourant and opacifying elements, and those elements associated with them, are presented in Table 6-6 in ppm. The samples are grouped according to colour, and data for individual samples are presented because there is often considerable variation in the concentrations within groups, reflected by large RSD values. It is immediately obvious that these glasses are much less consistent in colourant content than their predecessors from 2nd millennium BC Nuzi. Greater variation in element concentrations occurs in the Pella Cu blue glasses by comparison with those glasses from Nuzi.

The removal of inconsistently large or small values from the calculation of averages and RSD values is possible for some elements, but for other elements there is so much variation in the data that it is not possible to remove extreme values to calculate averages and RSDs, e.g. Sn and Sb in most groups, Pb in light blue translucent glasses, and Co, Ni and Zn in the turquoise opaque glasses. In these cases, the RSD values are large.

It is clear from these results that the initial grouping of glasses by base glass composition bears no relation to colour, except perhaps for Group 1. HMHK Group 2 and LMLK Group 3 glasses consist of the same variety of colours, which suggests that there might have been more than one workshop which used different fluxes to produce the same range of colours.

6.4.4.1 Pink/Purple and Colourless Glasses

The one pink and one purple glass (samples 38R and 22L of Group 1) are coloured with manganese which is elevated to 1.5% in both glasses. The difference in depth of colour is most likely to be related to the kiln conditions. The technology of colouring with manganese is known from 18th Dynasty Egypt and becomes much more common in the Classical and Hellenistic periods of the later part of the 1st millennium BC. The pink/purple glasses have higher concentrations of all other colourant and related elements, except Sb, by comparison with the concentrations found in the colourless glasses.

The ten colourless glasses (from Groups 1, 3 and ungrouped) are generally low in manganese, with samples 35K and 36M having the highest concentrations at 0.1% and 0.2% respectively. Purple glass 22L is lower in Sb (105 ppm) than pink glass 38R which has a higher concentration of this element (1731 ppm), falling at the lower end of the range of concentrations found in the colourless glasses. The colourless Nimrud glasses have Sb levels of 1600 ppm up to 6600 ppm. Sb does not appear to correlate with other elements in any of the Nimrud glasses suggesting that it was added as a pure raw material, a practice inferred from the analyses of opaque LBA glasses from Pella and Nuzi (sections 3.3.4, 4.5.4).

Sb is used both to opacify, in combination with calcium or lead, and to decolour glass, which is not naturally colourless unless the raw ingredients are pure. The use of Sb as an intentionally added decolourant has not been recorded before the mid-1st millennium BC, but the Nimrud colourless glasses represent the earliest instances of this glass-making practice so far recorded in Near Eastern glasses, nearly half a millennium earlier than previously recognised (section 1.10.2).

Colourless glasses 23N and 36M have higher than average Pb content, but it is still low at 100 ppm to 200 ppm. All glasses have otherwise relatively low levels of colourant and related trace elements, not significantly higher than the levels of their SBS, and likely to reflect their introduction to the glass mix as contaminants. The low concentrations of these elements are a good indication that these glasses are not made from recycled glass, and are deliberately decoloured with Sb. The raw data for the individual glasses are displayed in a line

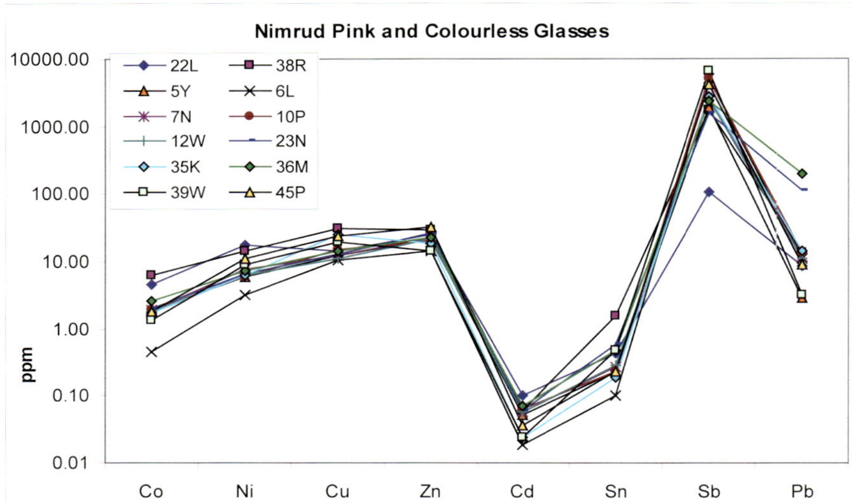

Figure 6-36 Line plot of raw data for colourant and related elements for individual Nimrud pink/purple and colourless glasses. A logarithmic scale is used.

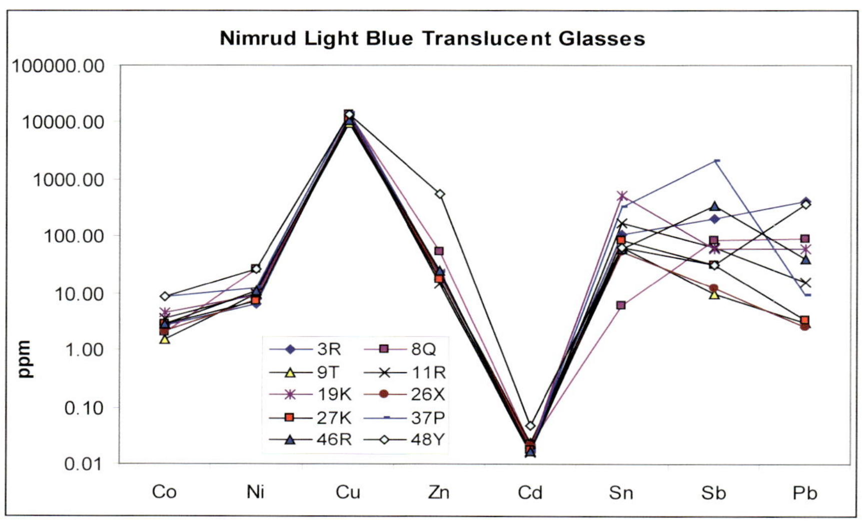

Figure 6-37 Line plot of raw data for colourant and related elements for individual Nimrud light blue translucent glasses. A logarithmic scale is used

plot (Figure 6-36), suggesting their overall similarity, with the noted exceptions.

6.4.4.2 Light Blue Translucent Glasses

Of the ten light blue translucent glasses analysed, eight belong to Group 3, one to Group 2, and one is ungrouped. All have relatively low concentrations of Co (average 3 ppm), Ni (average 9 ppm) and Zn (average 24 ppm).

Cu levels for all glasses in this group are in the range of 9500 ppm to 14,000 ppm, averaging 12,000 ppm. This is higher than the average Cu content for 2nd millennium BC translucent Cu coloured glasses from Nuzi (approximately 8000 ppm), and from Pella (approximately 9500 ppm). The individual glasses are compared in Figure 6-37, showing some variation in the levels of Co, Ni and Zn. These particular elements remain at low levels by comparison with the levels in the Co coloured blue glasses. Cu contents are reasonably consistent (RSD of 12). The concentrations of Sn, Sb and Pb are particularly inconsistent.

In all Cu blue glasses, opaque or translucent, the proportion of Sn to Cu is generally less than 1:100. Samples 19K, 37P and 40Q have higher ratios than this, but are still less than 7:100. As similar ratios are to be found in the colourless glasses, this does not suggest Cu sourced from bronze scrap, but rather it reflects naturally occurring proportions in the batch ingredients. The 2nd millennium BC glasses similarly do not make use of Cu from scrap. This practice is thought to have occurred in Egypt, but not in Mesopotamia in the 2nd millennium BC (Shortland 2005:1; section 1.10). The data presented in the current study for 2nd and 1st millennia BC Near Eastern glasses support this view.

The range in concentrations of Pb (3 ppm to 369 ppm) and Sb (10 ppm to 2150 ppm) are not easy to explain. They are not correlated and do not appear to be related. Sample 37P has a high level of Sb (2150 ppm), as though it were decoloured before it was coloured blue. The observations of the use of Sb as a decolourant in the glasses analysed in the present study will be discussed in section 9.8.

6.4.4.3 Turquoise Opaque Glasses

Of the five turquoise opaque glasses analysed, samples 4W and 49L are from Group 3; samples 34X and 40Q are from Group 2; and 43K is ungrouped. They are coloured with Cu and opacified with calcium antimonate. Cu levels are relatively consistent (RSD of 18), averaging 17,000 ppm (range of 15,000 ppm to 37,000 ppm), higher than in the translucent blue glasses. Sb concentrations have an approximate average of 12,300 ppm (range of 8000 ppm to 22,000 ppm), and are at a much more consistent level than in the translucent blue glasses. By comparison, blue opacified 2nd millennium BC glasses from both Pella and Nuzi have similar Sb levels of approximately 11,000 ppm.

The concentration of Pb in the Nimrud Groups 1 to 3 glasses is much less variable (average 303 ppm) in turquoise opaque than in blue translucent glasses, and is correlated with the concentration of Zn in this group only (r^2 = 0.7). Sn levels are again found to be inconsistent. Co concentrations are low at an average of 3 ppm, similar to the concentrations observed in Nuzi Cu blue glasses. Co levels in Pella blue glasses are marginally higher on average: 5 ppm in opaque and 9 ppm in translucent glasses. The individual Nimrud turquoise opaque samples are compared in Figure 6-38 below.

In Figure 6-39 the averaged colourant-related trace element concentrations of opaque and translucent Nimrud glasses of all colours from Groups 1 to 3 and ungrouped are compared. It is evident that the Nimrud glasses are enriched in these elements when compared with the average concentrations found in the UCC. Cu and Sn are particularly enriched. There are no MUQ values for Sb and Cd with which to make a comparison.

Both opaque and translucent Cu blue glasses have similar colourant composition, apart from higher contents of Sn, Sb and Pb in the opaque glasses. The pink/purple and colourless glasses have similar patterns to each other. The colourless glasses have higher concentrations of Sb, and the pink/purple glasses have greater manganese content, while sample 38R, a purple coloured glass, has a relatively high Sb content also.

Figure 6-40 is a line plot of the averaged raw data for Nimrud and Pella opaque and translucent Cu blue glasses. It is suggested that while the 1st millennium BC glass raw materials might be sourced from different places and combined following different recipes to those used in the 2nd millennium BC, as is evidenced

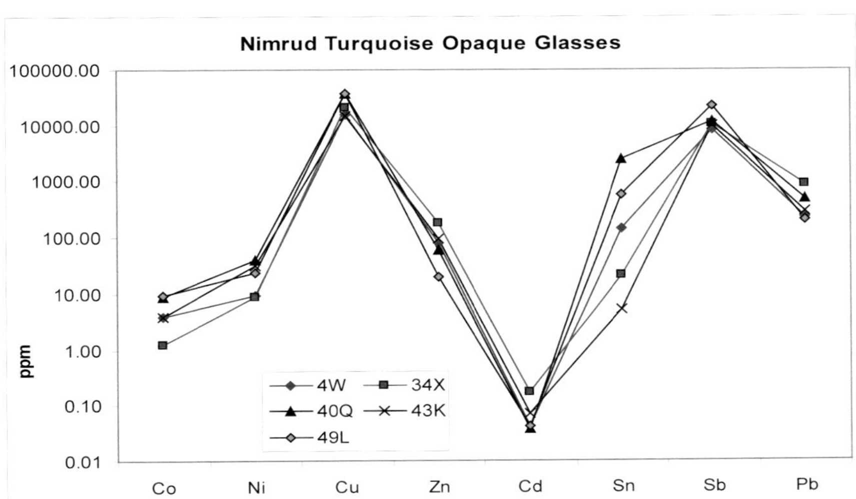

Figure 6-38 Line plot of raw data for colourant and related elements for individual Nimrud turquoise opaque glasses. A logarithmic scale is used.

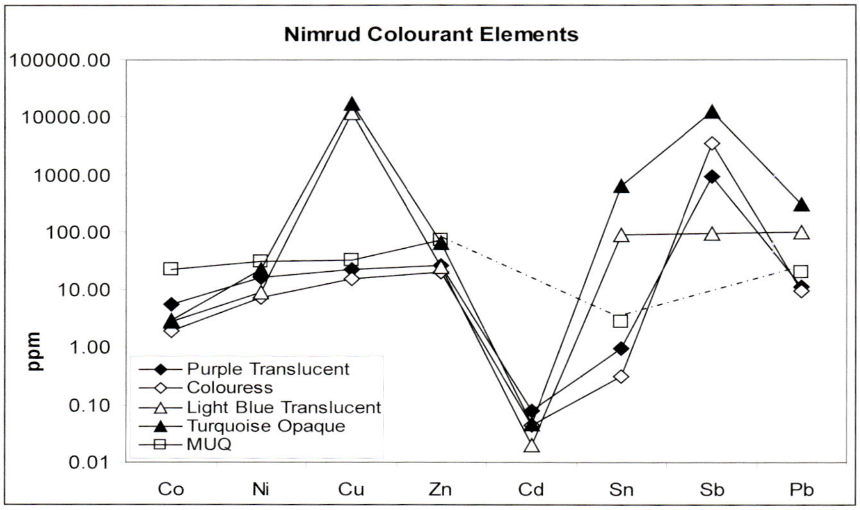

Figure 6-39 Line plot of averaged raw data for Nimrud colourant and related elements and MUQ, from ICP-MS data. A logarithmic scale is used. There is no MUQ value for Sb and Cd.

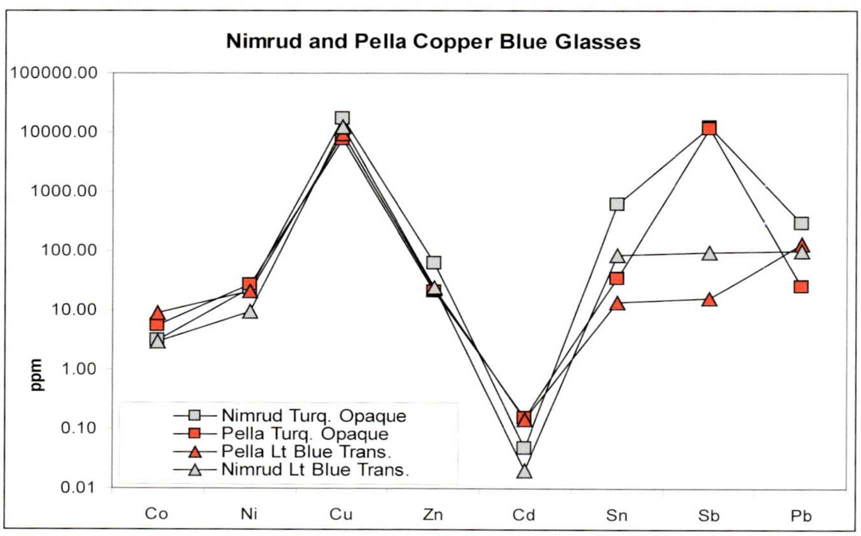

Figure 6-40 Line plot of averaged raw Nimrud Cu blue glass colourant and opacifier related trace element data compared with similarly coloured Pella glasses. A logarithmic scale is used.

by their distinctive base glass compositions, and the variations in their REE patterns, the method of colouring the blue glasses is in essence little changed over time and location. The comparison of LBA Pella opaque and translucent Cu blue glasses with those from IA Nimrud shows surprising similarities of composition and elemental concentration. The higher Sn, Pb and to a lesser extent Zn contents in the Nimrud opaque glasses are clearly shown. The Nimrud translucent blue glasses also have higher Sn and Sb concentrations than their LBA counterparts.

It is more difficult to compare colourless glass production of the LBA and IA when there is only one earlier sample from LBA/IA Pella, but colourless examples of LBA glasses are recorded from other sites in the Near East and Egypt and none is decoloured with Sb (section 1.10.2). Intentional decolouring of glass using Sb was clearly practised in the early 1st millennium BC (see section 9.8).

In summary, Group 1 glasses from Nimrud are colourless and pink/purple. The colourless glasses are decoloured with Sb, and the two purple glasses are intentionally coloured with manganese. Their trace element signatures are otherwise very similar, as are their base glass compositions, suggesting that the manganese does not add other elements to the trace element signature of these glasses. Groups 2 and 3 glasses consist of colourless glasses, and both opaque and translucent blue glasses. These blue glasses are coloured with Cu, and when opacified it is with calcium antimonate.

It is interesting to note that Group 2 glasses are made with a plant ash flux while Group 3 glasses are LMLK, as a result of the use of a mineral soda flux or less likely an LMLK plant ash. Both groups include the same range of colours, which suggests the existence of more than one source of glass where different fluxes were employed to produce glasses of similar appearance.

6.5 General Conclusions

Base glass analysis of the Nimrud glasses of the early 1st millennium BC IA has identified four distinct compositional groups and several ungrouped outliers of the typical soda-lime-silica type. Group 1 colourless and purple glasses are of the HMHK type. The colourless glasses are the earliest examples of decolouring with Sb. The pink/purple glasses are coloured with manganese. Group 2 glasses are HMHK glasses similar in composition to the HMHK glasses from 2nd millennium LBA Pella and Nuzi. They form a small group of translucent and opaque Cu blue glasses opacified with calcium antimonate. The ungrouped glasses are of the HMHK type but do not fit well into any group. Group 3 glasses are LMLK with low trace element concentrations. This group contains glasses of the same colours (except purple) represented in Groups 1 and 2 HMHK glasses, suggesting the existence of at least two workshops which produced the same range of colours but used different fluxes. Group 4 glasses form a distinct group of high magnesia, low potash Co blue glasses with high concentrations of alumina and low lime contents. They appear to be LMLK glasses with additional magnesia introduced with the Co alum colourant. This group likely represents the products of another distinct workshop which made this glass in the 2nd millennium BC Egyptian Co blue glass tradition. The trace element analyses for these glasses are presented in Chapter 7.

The patterns and concentrations of the REE for Groups 1 to 3 glasses differ and confirm the groupings defined by the base glass compositions. The two HMHK groups have higher concentrations of the REE than Group 3 which suggests that the raw materials used to make Group 3 glasses were relatively uncontaminated. All glasses have sedimentary crustal REE patterns generally similar to those observed in the LBA Pella and Nuzi glasses. The variations between the REE patterns and concentrations of each group, however, show that the REE-bearing raw materials have distinct differences which can be ascribe to raw material source, or type or method of preparation. For example, ashes from the same plants grown in different soils, or ashes of different plants or of mixtures of plants, or of ashes or silica prepared with tools that introduce contaminants or contact with other REE-bearing materials like soil, which introduce contaminants during the glass-making process. Despite all of these possible variations in the sources and amounts of REE which could be incorporated in the glasses, the compositional distinctions which define each group of glass are clear. The relative uniformity of each glass group at trace element level suggests that each group was made in a particular way in a particular place, giving weight to the proposition that several glass-making workshops were operating during the early 1st millennium BC in the Near East (section 9. 5).

The SRE appear to be associated with the REE, iron oxide and alumina. While there are variations between the groups in the concentrations of these elements, with Group 3 again having the lowest concentrations, there are distinctive features which apply to all groups: the elevated concentrations relative to the content of the earth's crust of the elements Sc, Cr, Zr, Hf and U. Comparison with glass compositions from other regions will help to determine whether these elemental enrichments form a chemical signature for glasses from this region in the early 1st millennium BC.

The concentrations of alkali and alkaline earth elements in each of the three groups follow a similar pattern with small variations between the groups. Sr and Li are enriched in all groups above their defined SBS levels regardless of whether they are classed as having HMHK or LMLK compositions. The concentrations of these elements in the Group 3 glasses are again comparatively low. The Pella LMLK low lime IA glasses have somewhat elevated levels of Sr, but not to the extent that they are in the Nimrud Group 3 glasses which have higher lime contents as well. It is likely that the Sr and lime components of the Group 3 glasses are associated, and while the source of the alkalis and alkaline earth elements including lime is less likely to be an LMLK plant ash with high lime, it remains an unproven possibility. Current knowledge assumes that these glasses were made with a mineral soda flux and a clean silica source, perhaps with added lime (section 1.9).

The Nimrud IA Groups 1 to 3 and ungrouped glasses reveal an increased diversity in 1st millennium BC glass manufacturing by comparison with the 2nd millennium BC tradition. Glasses other than those coloured with Co have LMLK base compositions, and the same colours of glasses are being produced using HMHK and LMLK fluxes. Whether this reflects the use of different plant ashes, or combinations of ashes, or the concurrent production of plant ash and mineral soda fluxed glasses with high lime for non-Co colours, is difficult to determine in the light of recent research on plant ash compositions. The colours and physical forms of the IA glasses are unrelated, which has been observed previously in the LBA glasses (sections 3.1, 4.2).

Chapter 7

Nimrud:
Iron Age Glasses
Results Of Compositional Analyses
Part 2

7.1 Introduction: Cobalt Blue Glasses

SEM-EDS analyses reported in Chapter 6 identified the Co blue glasses (Group 4) from Nimrud as a distinct and coherent group of low lime, low potash glasses coloured with a high alumina, high magnesia Co alum, reminiscent of the Egyptian 2nd millennium BC Co blue glasses. This chapter presents the results of trace element analyses of the Nimrud Group 4 glasses by solution ICP-MS (section 2.4.2).

7.2 Rare Earth Elements (REE)

Averaged raw concentrations of REE from Nimrud Group 4 Co blue glasses are given in Table 7-1. When presented on a line plot, the REE concentrations of these glasses form two sub-patterns, labelled 4A and 4B (Figures 7-1, 7-2), which appear to be geologically closely related and form part of a continuum of concentrations. All other data for these glasses suggest that they belong to a single group which is how they will be treated in the following sections. Sample 18X belongs to REE Group 4B, but has unusually high HREE content and is therefore reported separately in Table 7-1.

The REE contents of the Group 4 glasses are within the same range of concentrations in both REE sub-Group 4A and 4B. The Group 4A pattern consists of an increase in LREE concentration from La to Eu-Gd and then a gradual decrease in the HREE from Gd to Yb which forms a rise in the graphed lines. In Group 4B there is a gradual increase in concentration of the LREE from La to Gd and then a flattening out of the pattern at higher concentrations for the HREE.

Table 7-1 Averaged raw values for Rare Earth Elements from Nimrud Group 4 glasses in ppb.

Nimrud Group 4 Cobalt LMLK Rare Earth Elements (ppb)

n=15

	Group A n=9		*Group B n=5*		*All*		Av. All/	
	Averages	RSD	Averages	RSD	Averages	18X	MUQ	18X/MUQ
La	2736	38	3342	20	2953	6,939	32	0.21
Ce	7934	47	7865	17	7909	17,829	38	0.25
Pr	1332	50	1317	20	1327	2677	41	0.32
Nd	6404	49	6213	22	6336	12,946	40	0.39
Sm	1687	46	1618	23	1663	4049	38	0.59
Eu	438	42	442	24	440	1423	36	0.91
Gd	1727	40	1928	23	1798	8634	34	1.36
Tb	233	38	292	26	254	1530	34	1.55
Dy	1240	38	1722	27	1412	9183	36	1.56
Ho	231	37	356	27	276	1833	38	1.50
Er	581	36	960	26	716	4738	40	1.41
Tm	82	36	144	27	104	661	42	1.30
Yb	475	36	843	26	606	3784	43	1.16
Lu	70	36	128	26	91	576	44	1.18

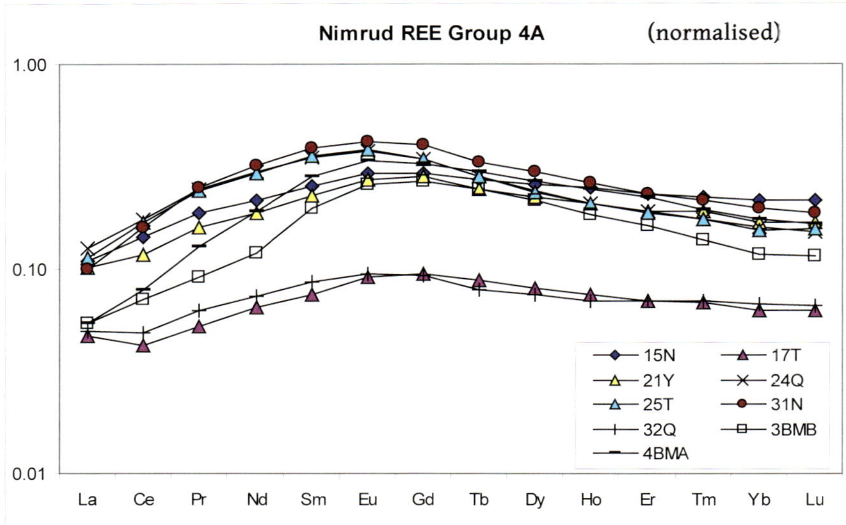

Figure 7-1 Line plot of the REE of individual Nimrud Group 4A glasses. A logarithmic scale is used. Data are normalised to MUQ.

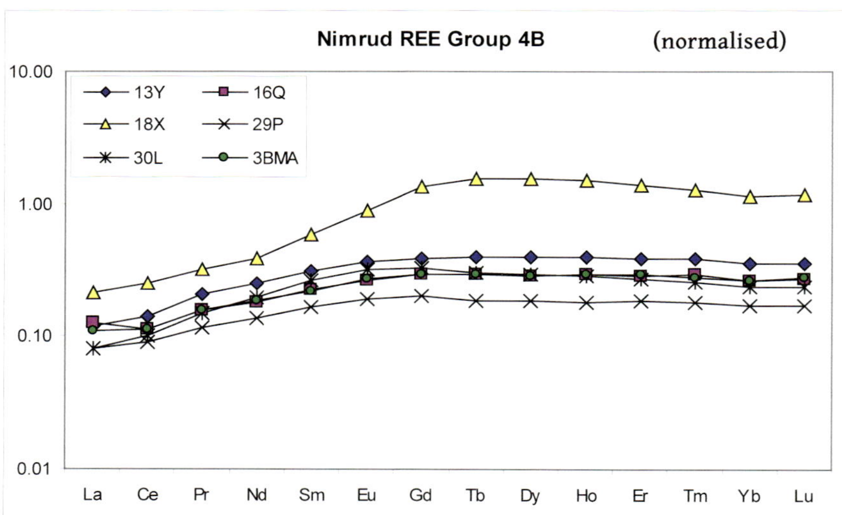

Figure 7-2 Line plot of the REE of individual Nimrud Group 4B glasses. A logarithmic scale is used. Data are normalised to MUQ.

In Group 4A, samples 24Q and 25T have identical REE patterns which indicates that they are from the same batch of glass, but are taken from two separate pieces of inlay. Samples 17T and 32Q are closely similar to each other in REE pattern and concentration, and although they have lower concentrations of the REE than the other Group 4A glasses, this is not geologically significant, and they belong to the same geochemical group (Kamber, personal communication). Their inclusion in the calculation of averages for Group 4A has resulted in higher RSD values than for Group 4B.

The REE patterns of the Group 4 glasses are different from the sedimentary patterns of the other glasses examined in the present study, and it is proposed that some of the REE are entering the glass together with the Co alum colourant.

The correlation ($r^2 = 0.5$) of HREE represented by Eu/Yb *vs* Co (Figure 7-3) shows that the alum is responsible for adding REE to the glasses. The pattern of the REE from the alum overlies the REE pattern already present from the base glass ingredients, before the addition of the Co

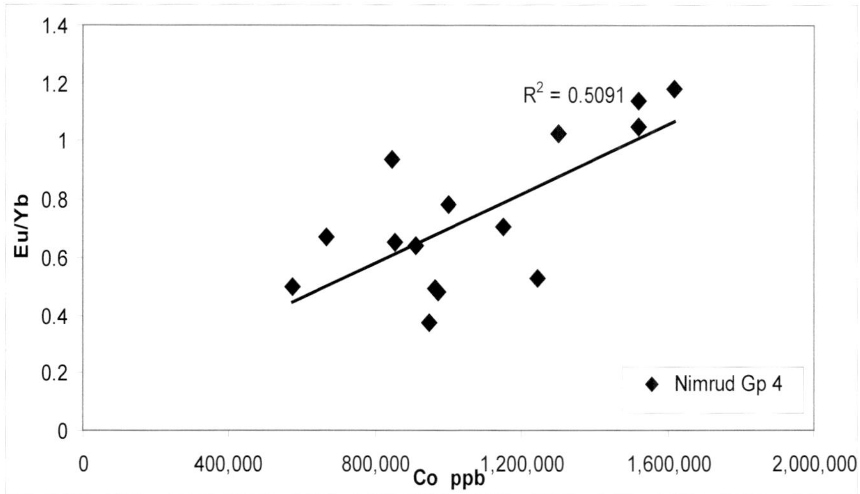

Figure 7-3 Scatter plot of Eu/Yb vs Co in ppb for Nimrud Group 4 glasses.

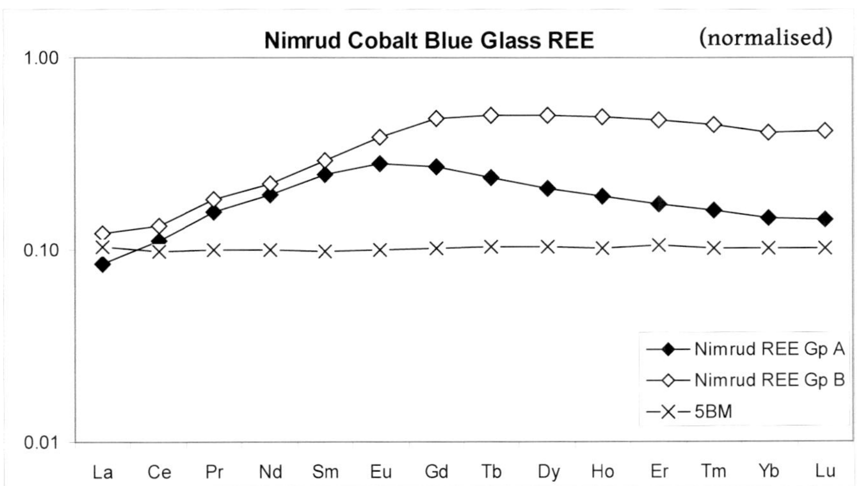

Figure 7-4 Line plot of averaged REE from Nimrud Groups 4A and 4B glasses compared with Group 2 Co blue glass 5BM. Data are normalised to MUQ. A logarithmic scale is used.

alum colourant, thereby modifying and complicating the REE patterns for this group. The REE from the silica source would have a flat sedimentary pattern, which is found in all other glasses from IA Nimrud analysed in this study. The REE from the alum give the graph the greater rise at Eu and increased HREE concentrations.

The existence of two REE patterns from different parts of the Group 4 continuum indicates that two sources of alum were utilised in the manufacture of these glasses. It is known that Co alum from the Western Oases of Egypt, the only source known in the Middle Eastern region, has a naturally varying composition (section 1.10.1), which means that it is likely that the alum colourant was sourced from different deposits in the same geographic region. This is discussed further in section 7.5.

Figure 7-4 summarises the differences between the REE patterns of the Nimrud Group 4A and 4B glasses compared with the REE pattern of sample 5BM, which is the only Co coloured HMHK glass analysed in this study. This glass is a typical 2nd millennium BC plant ash glass with a flat crustal REE pattern closely similar to those of the Pella HMHK glasses. It appears to have been made with a non-alum, non-REE bearing source of Co, which explains the dissimilarity of the REE patterns between glass 5BM and the Co alum coloured Nimrud glasses, and supports the proposition that the Co alum introduces REE to the glasses. In addition, the REE are

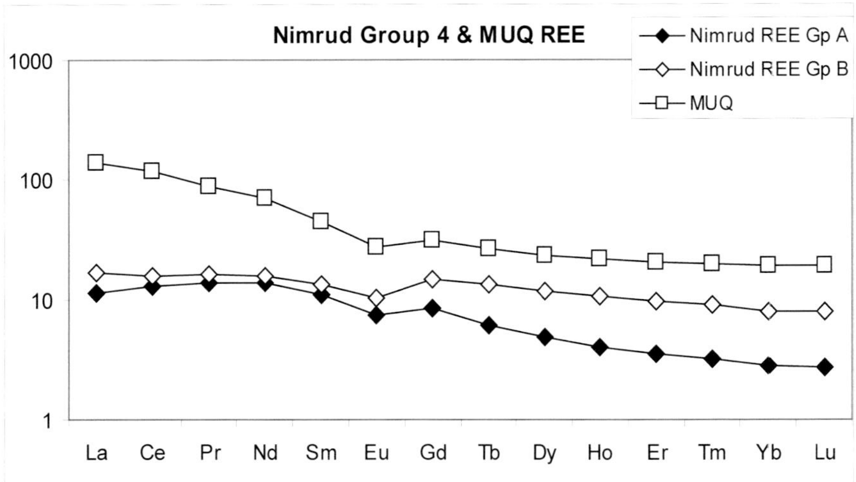

Figure 7-5 Line plot of averaged Nimrud REE data for Groups 4A and 4B glasses. Data are normalised to average chondrites for comparison (Table 27).

correlated with the elements Be ($r^2 \approx 0.9$), Sr ($r^2 \approx 0.6$), Y ($r^2 \approx 0.9$) and Cr ($r^2 \approx 0.7$), which indicates an association with the alum.

When compared with the REE content of the earth's crust (Figure 7-5), the Nimrud REE Groups 4A and 4B have LREE with a flatter pattern, and at lower concentrations. Group 4A has a lesser REE concentration on average for the HREE than Group 4B, though their patterns are similar. Like the crustal pattern, both Nimrud groups have a small negative Eu anomaly.

The HREE patterns of the glasses diverge: Group 4B has a gently downward sloping pattern, while the slope of the HREE pattern for the Group 4A glasses is steeper and the difference in the concentrations of the two groups is greater here. The slope of the Group 4B HREE pattern more closely parallels that of the UCC.

Nimrud Group 4 glasses may be divided into two geochemical sub-groups based upon their REE patterns. The patterns when normalised to average chondrites and compared with the pattern of the REE from the earth's crust, are similar to it but less so than the other glasses analysed in the present study. The Group 4 REE patterns diverge in the region of the HREE. It is believed that the crustal REE patterns, observed in the 2nd and other 1st millennia BC glasses, are modified by REE introduced with the Co alum colourant. Co coloured sample 5BM exhibits a 2nd millennium BC plant ash glass REE pattern and appears to be coloured with a different, non-alum source of Co which, like other colourants investigated in the present study, does not introduce REE.

7.3 Sediment-Related Elements (SRE)

The next group of trace elements to be considered are those related to sediments. The averaged data for these elements are presented in Table 7-2. The larger RSD values indicate variability in the composition of the glasses within Group 4, and are not a result of data collected being at or below instrumental detection limits (section 2.4.2). In order to average data and calculate RSD values, inconsistently high or low values have been removed from the data-set. For Sc, W, Tl and U, this was not possible because there is variation across the data for all samples, resulting in high RSD values for these elements.

The SRE have probably been introduced to the glasses by ingredients such as silica, and not the Co alum, because many of the SRE, including Sc, Cr, Ti, V, Nb, Zr, Hf and Th, are insoluble and hence could not have formed part of the alum, which is a precipitated salt. There is no correlation between the concentrations of the SRE and Co. There are, however, close correlations between the concentrations of Ti and each of Nb, Zr and Hf ($r^2 = 0.88$ to 0.99), with weaker correlations occurring between the concentrations of Ti and V ($r^2 = 0.6$), and Ti and Th ($r^2 = 0.6$). As was found in the other Nimrud glass groups, the SRE are introduced to Group 4 glasses as sedimentary contaminants. The particular enrichment in the concentration of Y parallels that of the REE because this element behaves in a similar manner to a REE. The concentration of U correlates with that of Co ($r^2 = 0.5$) indicating that it is one of the elements carried by the alum. U concentrations are particularly elevated in Group 4 glasses (Figure 7-7).

The Group 4 averaged SRE data are plotted in Figure 7-6 together with the SRE data for MUQ. The patterns are generally similar, with observable enrichment in Sc, Cr, Y, Zr, Hf and U. This enrichment is reflected in the ratios of the SRE data for Group 4 glasses to the MUQ SRE data (Table 7-2).

Table 7-2 Averaged concentration in ppb of sediment-related elements from Nimrud Group 4 glasses.

Nimrud Group 4 Sediment-Related Elements (ppb)				
		Average	RSD	Gp 4/MUQ
Sc	n=14	4380	93	0.27
Ti	n=15	353,633	41	0.05
V	n=15	7166	24	0.06
Cr	n=15	13,141	42	0.20
Ga	n=15	1321	23	0.07
Y	n=14	7664	41	0.24
Zr	n=13	29,440	49	0.15
Nb	n=13	1207	37	0.08
Hf	n=13	647	49	0.12
Ta	n=13	70	39	0.06
W	n=15	75	122	0.05
Tl	n=15	10	179	0.02
Th	n=15	1027	32	0.09
U	n=15	4475	72	1.58
Mo	n=15	179	58	

The SBS level (see Table 7-2) is less easy to determine in Group 4 glasses because the addition of REE from the alum has resulted in enrichment of these elements. Seven values for SRE ratios to MUQ fall between 0.05 and 0.09 (average 0.07 for Ti, V, Ga, Nb, Ta, W and Th) probably representing a reasonable SBS range, based on SBS levels of the other Nimrud glass groups. Tl has a ratio to MUQ of 0.02 and appears to be depleted relative to the content of Tl in the UCC as it is in Nimrud Groups 1 and 3 glasses. The average normalised concentrations of the SRE from Group 4 glasses are plotted in Figure 7-7 using an SBS of 0.07. The resulting picture of enrichment and depletion of these elements is not dissimilar to that observed in the other Nimrud glass groups, especially Group 3. This suggests a common geological origin for the Nimrud glasses.

Figure 7-8 provides a comparison of the SRE patterns or chemical signatures for all four Nimrud glass groups. While their patterns are broadly similar, their differences enable them to be distinguished clearly from the others. While Groups 1 and 2 are both plant ash glasses, their SRE patterns vary noticeably at Cr, Zr, Hf, W, Tl and U. Group 3 has, on the whole, the lowest concentrations of these elements and of REE and base glass oxide concentrations (other than silica and soda), which reflects the use of purer raw materials in the manufacture of Group 3 glasses. Group 4 glasses have greater concentrations of Y, as this element is closely related to the REE, and has somewhat elevated concentrations of Zr, Hf and U by comparison with the other groups.

Figure 7-9 provides a comparison of Nimrud Group 4 glasses with the other LMLK glasses examined, i.e. Pella black glasses and Nimrud Group 3 glasses. The Pella

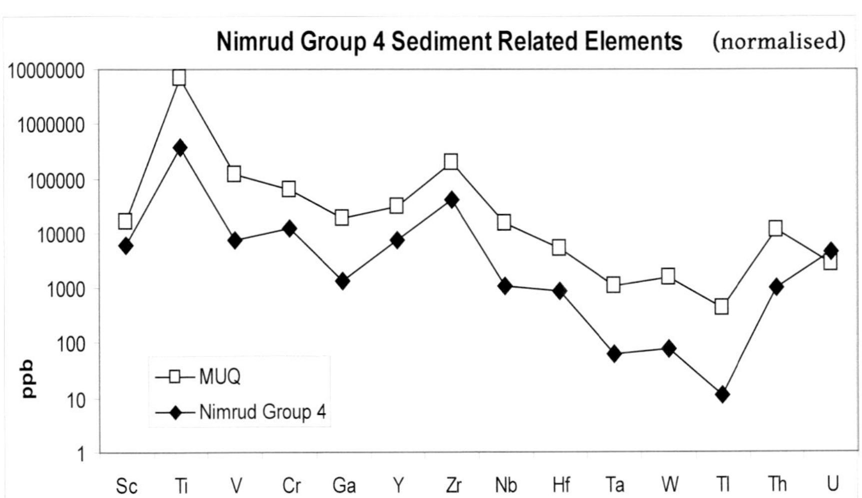

Figure 7-6 Line plot of raw Nimrud Group 4 sediment-related element data in ppb compared with MUQ on a logarithmic scale.

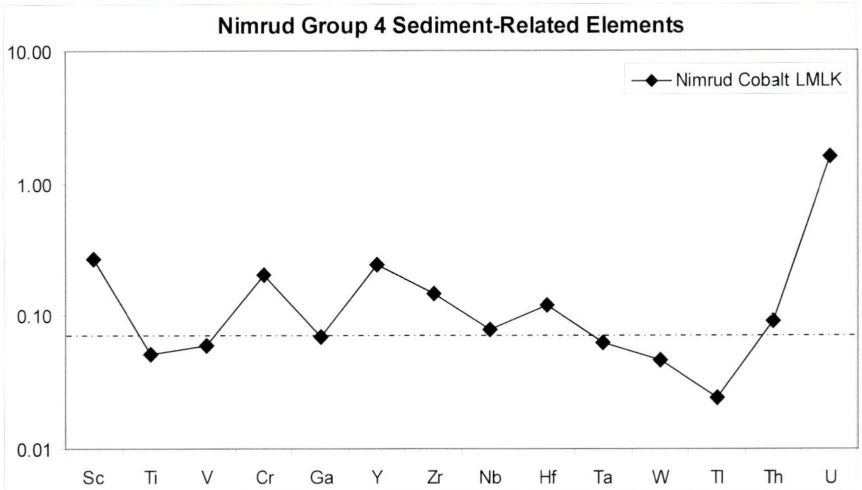

Figure 7-7 Line plot of averaged concentrations of sediment-related trace elements in Nimrud Group 4 glasses, normalised to MUQ. A logarithmic scale is used. The dotted line crosses the Y-axis at 0.07 to represent the approximate SBS of the Group 4 glasses.

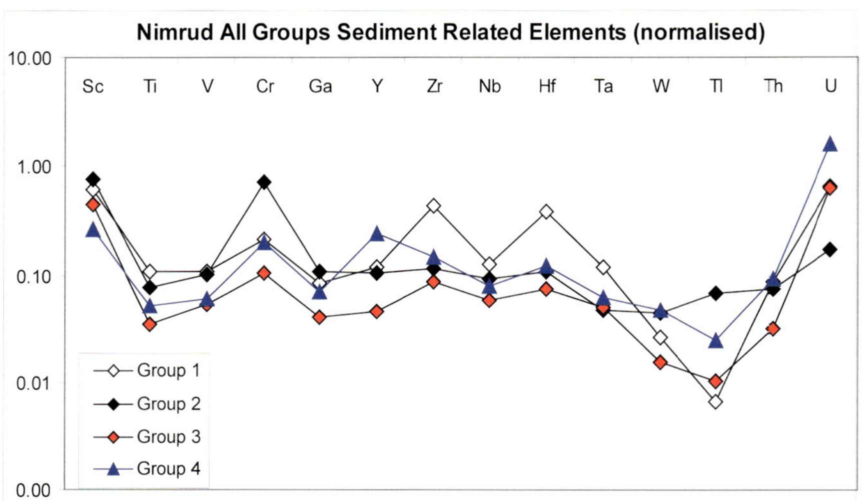

Figure 7-8 Line plot of averaged sediment-related element data for Nimrud Groups 1 – 4 glasses. Data are normalised to MUQ. A logarithmic scale is used.

glasses are chemically distinct from the Nimrud glasses, suggesting that different SRE contaminant-bearing raw materials were used to make them. The compositional differences are also reflected in their REE patterns, strongly suggesting that the Pella and Nimrud LMLK glass groups are made by different workshops in different locations, or at least in the same workshop using differently sourced raw materials, not surprising given the chronological differences between these glass groups.

In summary, there is a general similarity of pattern between the Pella and Nimrud LMLK glasses with notable differences in elemental concentrations from Sc to V, and for W. The Pella glasses overall have higher concentrations of the SRE than the Nimrud glasses, with the exceptions of Sc and U. Nimrud Groups 3 and 4 glasses have a similar SRE pattern except for the enriched Y content of the Group 4 glasses which is influenced by the addition of REE and Y-bearing alum. The Group 4 glasses have higher SRE concentrations than Group 3 glasses, except for Sc. On the basis of the trace element data presented so far, it is proposed that the Nimrud Groups 3 and 4 glasses could share a similar origin, while a different origin is proposed for the Pella LMLK glasses.

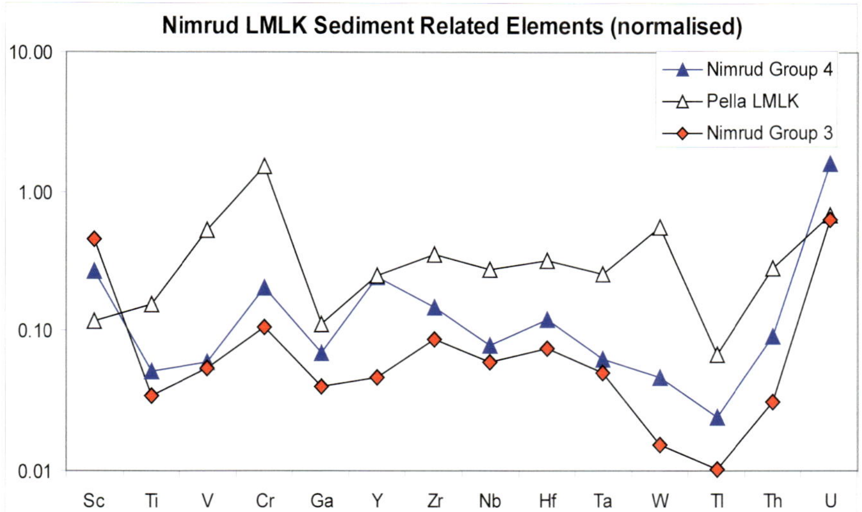

Figure 7-9 Line plot of averaged sediment-related trace element data for Nimrud LMLK glasses compared with Pella LMLK glasses. Data are normalised to MUQ. A logarithmic scale is used.

The SBS level of the glasses is approximately 0.07 as described by the averaged ratio of Group 4 SRE to MUQ. As with all glasses examined here, the pattern of elemental concentrations is similar to that found in the earth's crust, but at lower concentrations.

7.4 Alkali and Alkaline Earth Elements

Nimrud Group 4 glasses appear to be fluxed with mineral soda as established by the low levels of lime, potash and phosphorus detected by SEM-EDS analysis (section 6.3.4). The division of the Group 4 glasses into two groups based on their REE patterns is not evident when the alkali and alkaline earth elements are examined. The data for Group 4 glasses as a single group are averaged and presented in Table 7-3. Inconsistent values have been omitted in the calculation of averages and RSD values for Li, Be and Rb.

The concentration of Li is relatively high in the Group 4 glasses, ranging between 5000 ppb and 12,600 ppb, with three samples (24Q, 25T and 15N) having levels between 21,000 ppb and 26,000 ppb. The levels of Li in the plant ash glasses from Groups 1 and 2 vary from an average of 5000 ppb for Group 1 to 15,000 ppb for Group 2. The nature of the flux can not be determined from the Li content, unless all glasses contain plant ash, or there is another as yet unidentified raw material which contains Li if the LMLK glasses are considered to be fluxed with mineral soda.

The concentration of Be is particularly enriched in the Group 4 Co glasses (between 1000 ppb and 3400 ppb), more so in the REE Group 4B than in Group 4A, compared with other glass groups analysed. There is no discernible correlation, however, between the concentrations of Co and Be. Sample 30L has 5000 ppb of Be, and sample 18X has an unusually high Be content of 10,000 ppb, which might be related to this sample's exceptionally high level of HREE, despite the lack of correlation. Where the REE concentrations of other glass groups define the SBS levels of those glasses, Be is generally at the SBS level. Where the HREE are enriched in Group 4 glasses, there is a consequent rise in the content of Be. Be is probably derived from ground-water, which has evaporated leaving the Be to crystallise as a salt, and become an unintentional contaminant in the Co alum salt (Kamber, personal communication).

The level of Rb (average 4500 ppb) is similar to that observed in Group 3 glasses, both being lower than the levels found in the plant ash fluxed Groups 1 and 2. The ratio of Rb to MUQ (0.06) in Group 4 indicates that it is at approximately SBS level and is strongly correlated with Li ($r^2 = 0.8$) (Figure 7-10). There is no correlation between these elements in the Groups 1 to 3 Nimrud glasses. There might be a high and a low Rb group, but the significance of this is difficult to determine.

The concentration of Sr (average 106,000 ppb) is lower in Group 4 than in other glasses, including Group 3 LMLK glasses, and correlates moderately well with lime ($r^2 = 0.5$), which is present also at low levels in the Group 4 glasses (section 6.3.4) (Figure 7-11). The use of a calcareous sand in their manufacture may be responsible for the relationship of lime and Sr. These elements do not correlate in the other Nimrud glass groups. Lime concentrations correlate moderately well with those

Table 7-3 Average compositions in ppm of alkali and alkaline earth elements from Nimrud Group 4 glasses.

Nimrud Group 4 Alkali and Alkaline Earth Elements (ppb)

	Li	Be	Rb	Sr	Cs	Ba
	n=12	*n=11*	*n=14*	*n=15*	*n=15*	*n=15*
Average	8076	1881	4501	105,777	126	35,602
RSD	42	41	46	33	56	38
Gp 4/ MUQ	0.29	0.98	0.06	0.74	0.02	0.09

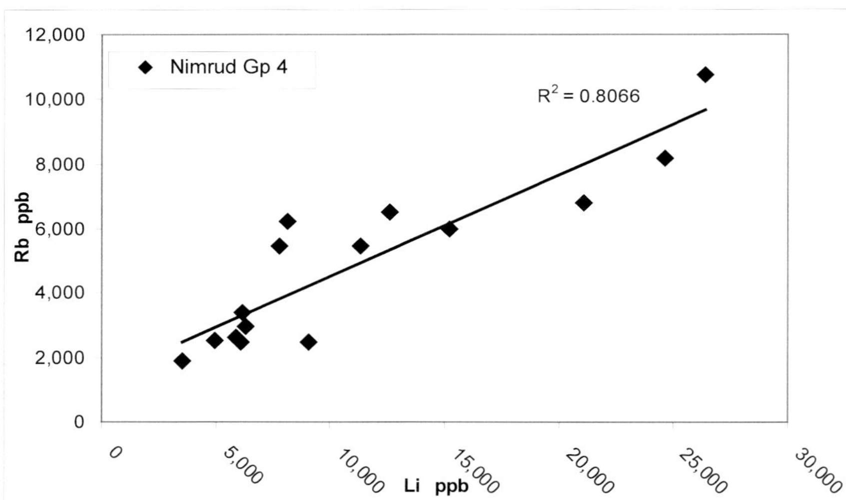

Figure 7-10 Scatter plot of Li *vs* Rb for Nimrud Group 4 glasses in ppb.

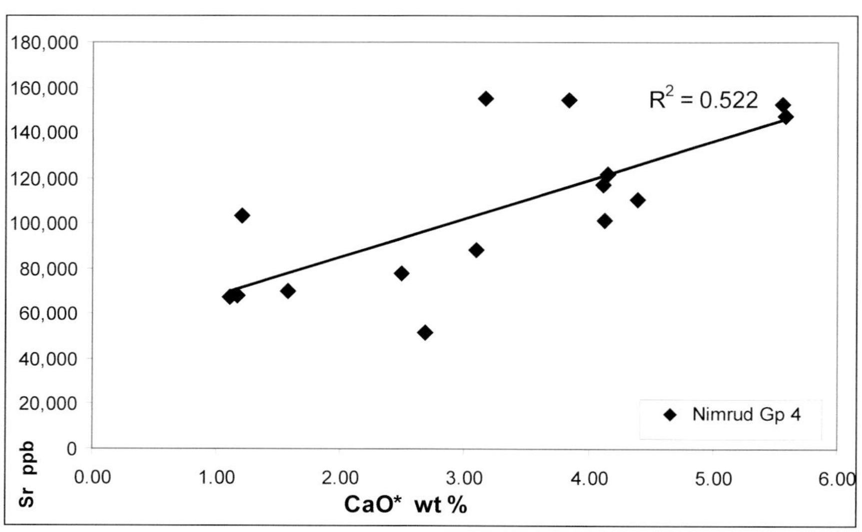

Figure 7-11 Scatter plot of lime *vs* Sr for Nimrud Group 4 glasses.

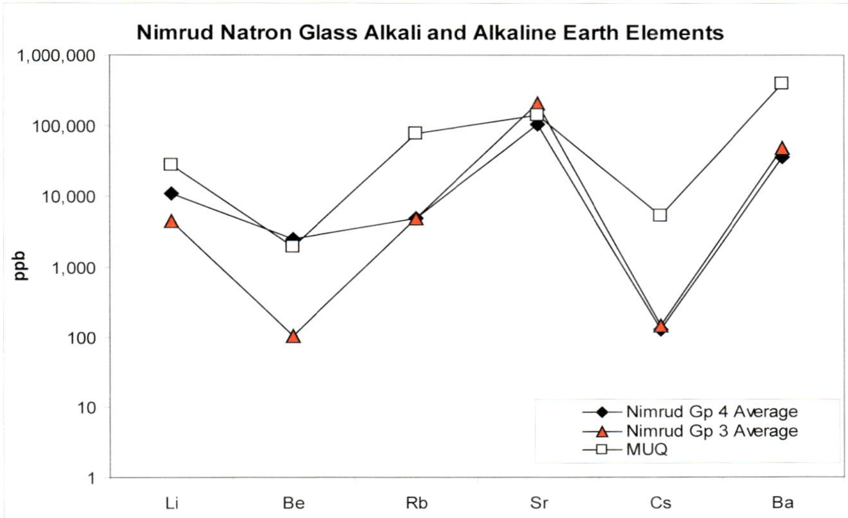

Figure 7-12 Line plot of raw alkali and alkaline earth element data for Nimrud Groups 3 and 4 mineral soda glasses and MUQ. A logarithmic scale is used.

of Ba (r^2 = 0.4) and this relationship, considered with the lime-Sr correlation, could indicate the use of a silica source containing lime from limestone.

In Figure 7-12, the raw data patterns for alkali and alkaline earth elements are plotted for Groups 3 and 4 glasses and MUQ. As observed previously (section 6.4.3), the patterns for these elements in the Group 3 glasses generally follow the pattern of the elemental concentrations found in the earth's crust, but at lower concentrations, except for the particular enrichment of Sr. The Group 4 glasses have similar patterns and concentrations to those exhibited in Group 3 glasses, with the exception of the greater enrichment of the concentrations of Be and Li which are probably associated with Co in Group 4 glasses. The Nimrud Groups 1 and 2 plant ash glasses follow the same alkali and alkaline earth pattern as Group 3 glasses, but the Groups 1 and 2 glasses have higher concentrations of most of these elements, presumably because they are introduced with the plant ash component of the glasses. Cs and Ba concentrations are both low, and are similar to the levels found in Groups 2 and 3 glasses, and Rb and Sr concentrations are little different between the HMHK and LMLK groups, which indicates that the addition of the Co colourant to Group 4 glasses has not influenced the concentrations of any of these elements.

When the alkali and alkaline earth data for Groups 3 and 4 and Pella LMLK glasses are normalised to MUQ and plotted together (Figure 7-13), the differences between these groups are clear. The alkali and alkaline earth elements are more concentrated in the Pella LMLK glasses than in the Nimrud glasses, except for Sr, which is elevated in all three groups, and Be which is more concentrated in Nimrud Group 4 glasses, due to their alum component. The Pella LMLK glasses also have higher concentrations of REE, SRE and some of the alkali and alkaline earth elements than are found in the HMHK and Nimrud LMLK glasses.

Figure 7-14 illustrates the enrichment of Li, Be and Sr for Group 4 glasses when compared with their approximate SBS of 0.07. Cs is depleted, while Rb and Ba remain near SBS level, meaning that they approximate the natural level of contamination in the glasses relative to the level of these elements found in the earth's crust.

In summary, Nimrud Group 4 glasses appear to be fluxed with mineral soda, and are depleted in the alkali and alkaline earth elements relative to the content of the earth's crust, except for Be and Sr which are enriched to almost crustal levels. Li is less enriched relative to the SBS, while Cs is depleted. The pattern of the line plot reveals the modifications made by the glass-makers, either intentionally or inadvertently, to the glass composition relative to the composition of the UCC. In Group 4 glasses, these modifications are unrelated to the use of plant ash as a flux. Plant ash glasses are more concentrated in these elements, but with lower levels of Be. The increased Be content, peculiar to Group 4 glasses, appears to be related to the addition of the Co alum colourant.

7.5 Colourant and Related Elements

The Nimrud Group 4 glasses are coloured dark blue mainly with Co from alum which imparts high levels of magnesia and alumina to these glasses (section 6.3.4). One Group 2 HMHK glass (sample 5BM) is similarly

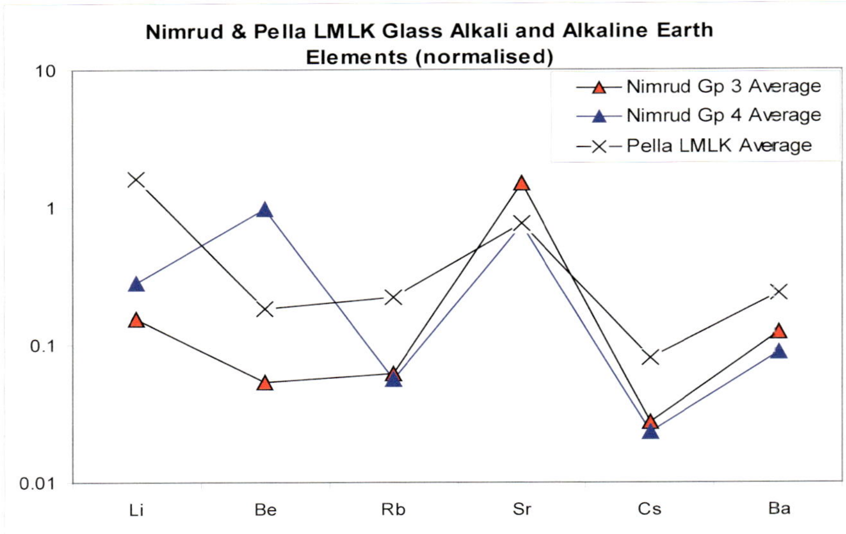

Figure 7-13 Line plot of averaged alkali and alkaline earth element data for Nimrud Groups 3 and 4, and Pella LMLK glasses. Data are normalised to MUQ. A logarithmic scale is used.

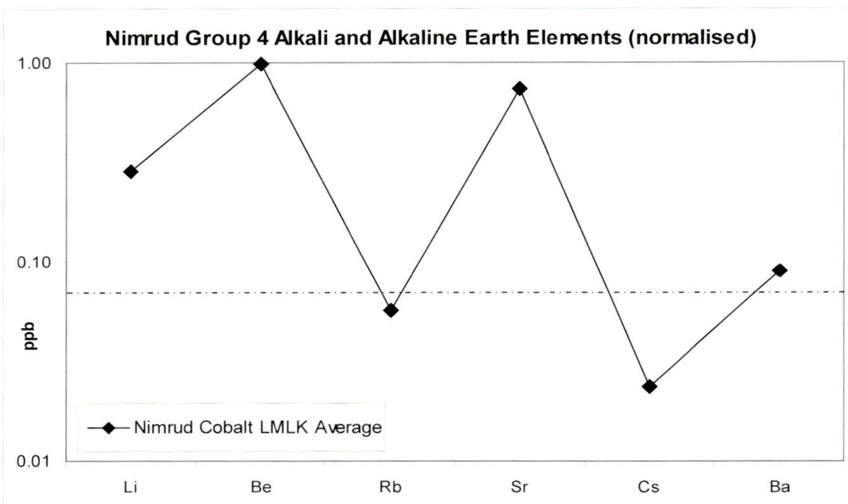

Figure 7-14 Line plot of data averaged and normalised to MUQ for Nimrud Group 4. The dotted line represents the SBS of approximately 0.07.

coloured and will be examined in this section with the other Co glasses. Averaged results of ICP-MS analyses of colourant and related trace elements for these glasses are presented in ppm, in two groups: low and high Cu glasses (Table 7-4). Sample 5BM is presented separately. Extreme values are excluded from the calculation of averages and RSD values.

All Group 4 glasses are coloured dark blue with Co. There is a sub-group of 11 glasses with lower Cu content, averaging 97 ppm, and a second sub-group of four glasses with higher Cu content, averaging 989 ppm (Table 7-5, Figure 7-15). The average Co content for the two groups varies similarly. The low Cu group has a lesser Co content with an average of 981 ppm, and the high Cu glasses have on average a greater Co content averaging 1324 ppm. Plant ash glass 5BM has 808 ppm of Co, a little lower than the average of the low Cu - low Co (LCLC) glasses, but has a comparatively high Cu content of 13,860 ppm (and hence is not plotted). This is considerably higher than the average Cu content of the 2nd millennium BC Cu blue glasses (approximately 8000 ppm), and it has a greater concentration of Sn (28 ppm) and a high concentration of Sb (7800 ppm).

Table 7-4 Average compositions in ppm of colourants and related trace elements from Nimrud cobalt coloured glasses*.

Nimrud Group 4 Colourant and Related Elements						(ppm)			
n=11	MnO wt%	Co	Ni	Cu	Zn	Cd	Sn	Sb	Pb
Average Low Cu	0.40	981	743	97.1	994	0.054	0.77	15574	45.4
RSD	42	27	46	52	56	37	30	30	64
n=4									
Average high Cu	0.34	1324	732	989	1074	0.047	4.23	144	223
RSD	7	26	13	14	26	29	39	20	50
n=1 5BM	0.76	808	76.6	13,860	102	0.234	27.9	7772	91.0
Av. low Cu/MUQ	2.9	44	24	3.0	14		0.28		2.2
Av. high Cu/MUQ	2.4	59	23	31	15		1.5		11
5BM/MUQ	5.5	36	2.4	428	1.4		10		4.5

*MnO is determined by SEM-EDS and measured in wt%. There are no values for Sb or Cd in MUQ. For low Cu glass averages and RSD values ***n***=10 for Cu, Zn, Cd and Sn, and ***n***=9 for Pb. For high Cu glass averages ***n***=3 for Sn, Sb and Pb.

Table 7-5 Cobalt blue Group 4 glasses classified as LCLC or HCHC.

Low Cu Low Co (LCLC)	High Cu High Co (HCHC)
13Y, 16Q, 17T, 18X, 21Y, 29P, 30L, 32Q, 3BMA, 3BMB, 4BMA	15N, 24Q, 25T, 31N

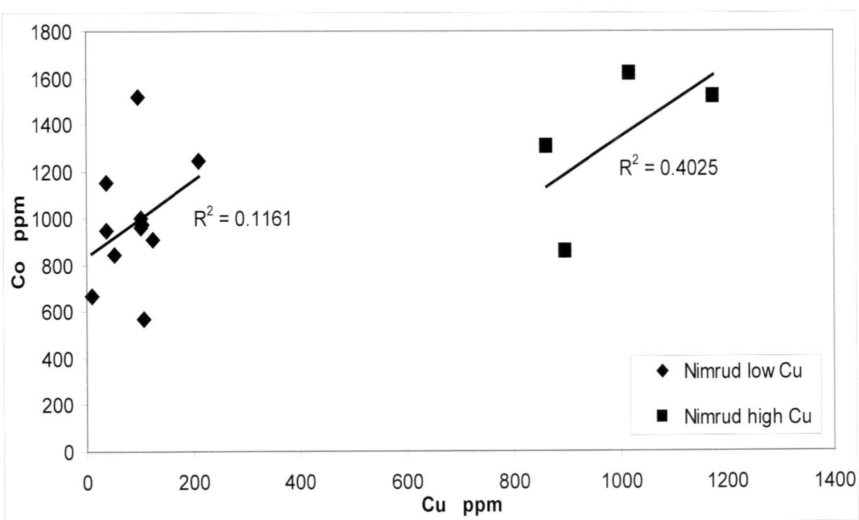

Figure 7-15 Scatter plot of Cu vs Co for Group 4 blue glasses.

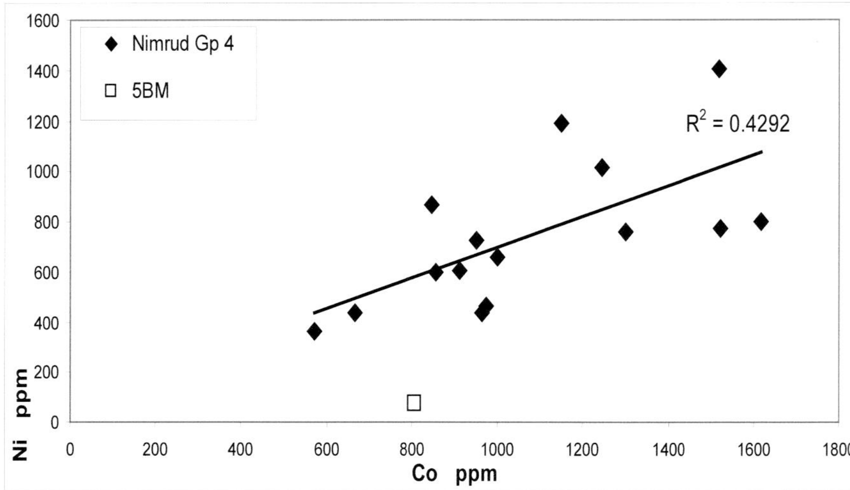

Figure 7-16 Scatter plot of Co *vs* Ni in ppm for Group 4 blue glasses.

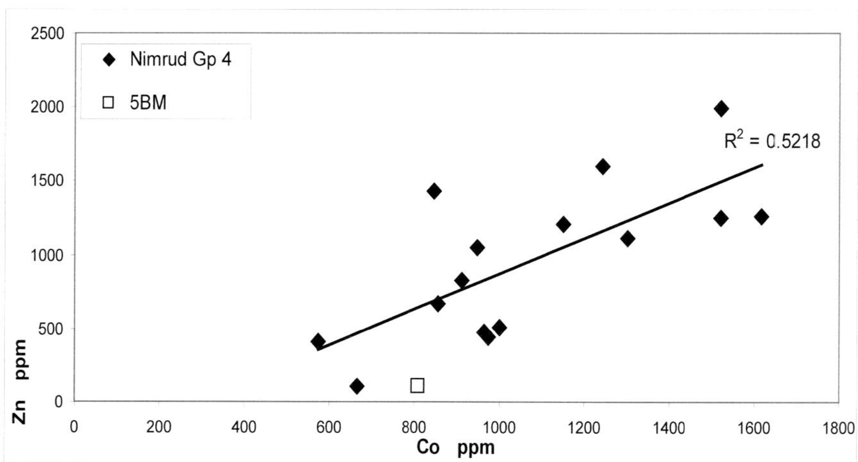

Figure 7-17 Scatter plot of Co *vs* Zn in ppm for Group 4 blue glasses.

Mn, Ni and Zn levels are comparable between the high Cu-high Co (HCHC) and LCLC groups, and all are highly elevated relative to non-Co coloured glasses. The average content of manganese is 0.4%, average Ni content is 740 ppm and the concentration of Zn averages approximately 1000 ppm. The elevated levels of these elements are consistent with the derivation of the Co colourant from cobaltiferous alum from the oases of the Western Desert of Egypt, the only source of this material now known in the Middle East. The implications of this information and comparisons with Egyptian Co alum coloured glasses are addressed in section 9.7. In the Group 4 glasses, the Co concentration is moderately well correlated with that of Ni (r^2 = 0.4) (Figure 7-16) and of Zn (r^2 = 0.5) (Figure 7-17), as would be expected. Surprisingly, the positive correlation between the concentrations of manganese and Co observed in 2nd millennium BC Egyptian Co coloured glasses by Kaczmarczyk and Hedges (1983; section 1.10.1), does not appear to occur in the 1st millennium BC Nimrud Co glasses. Nor is there a correlation between Co and alumina, or Co and magnesia (Figures 7-18, 7-19, 7-20).

Sample 5BM, on the other hand, has low concentrations of Ni, Zn and alumina. Ni and Zn are strongly correlated (r^2 = 0.8) in Group 4 glasses. It is likely that the Co in sample 5BM was derived from a non-alum source, such as Iran (section 1.10.1). In the HCHC group, the ratios of the concentrations of Co to each of Ni and Zn are 1.8:1 and 1.3:1, and in the LCLC group the ratios of these elements are 1.2:1 and 1.1:1 respectively. In glass 5BM, the ratios are much higher at 10.5:1 for Co to Ni and 7.9:1 for Co to Zn due to the lower levels of both Zn and Ni in the glass by comparison with the high level of Co.

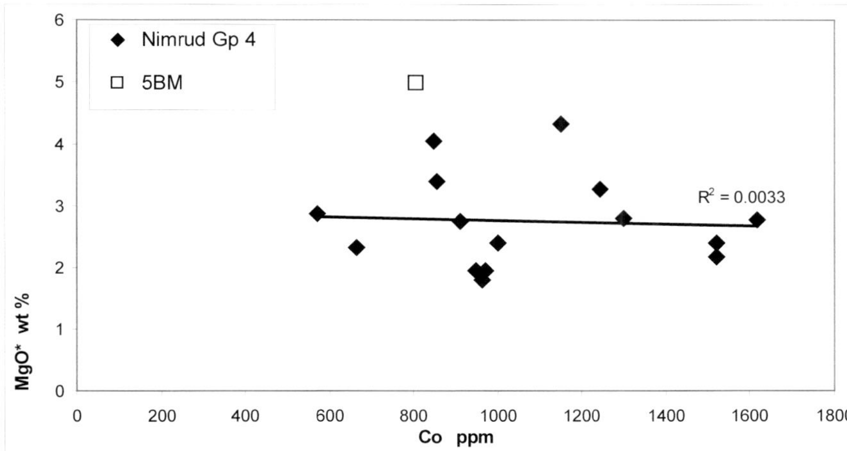

Figure 7-18 Scatter plot of Co *vs* magnesia for Group 4 blue glasses.

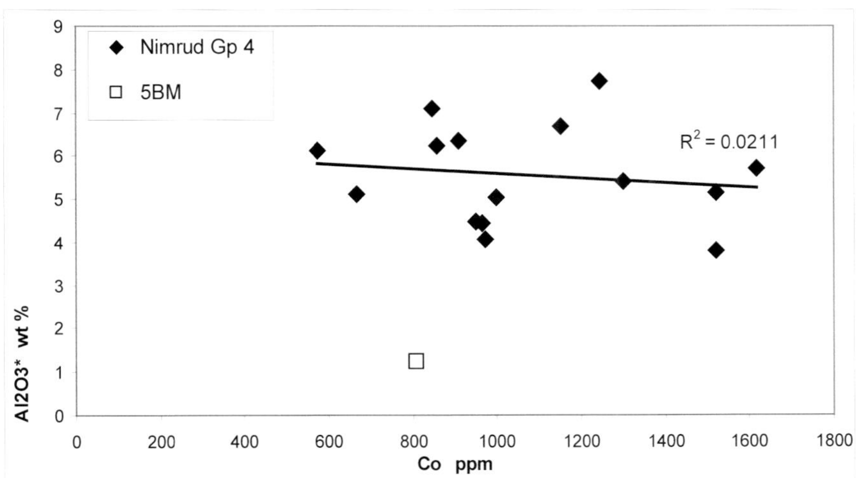

Figure 7-19 Scatter plot of Co *vs* alumina for Group 4 blue glasses.

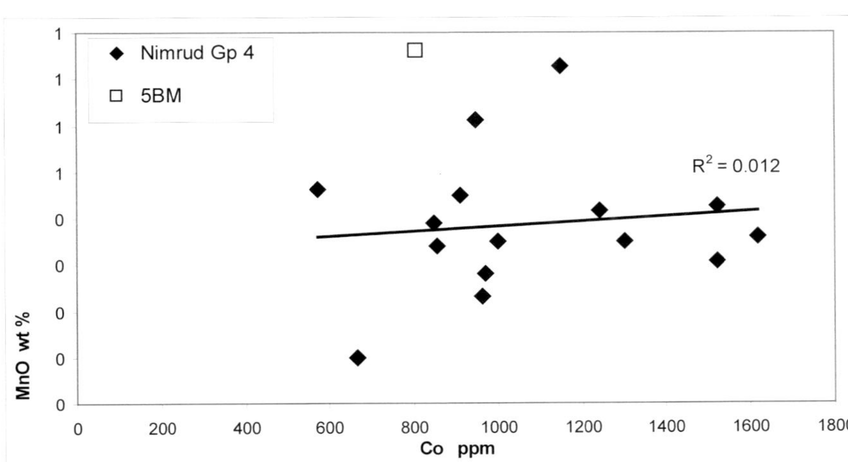

Figure 7-20 Scatter plot of Co *vs* manganese for Group 4 blue glasses.

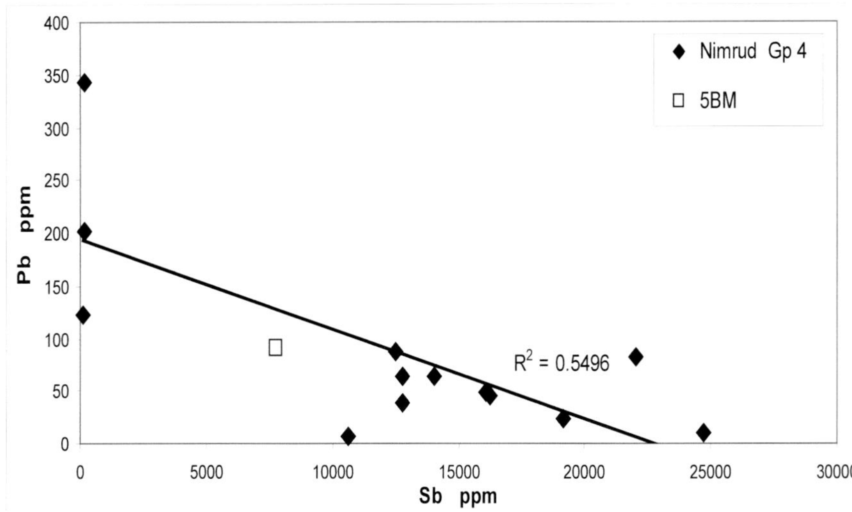

Figure 7-21 Scatter plot of Sb *vs* Pb in ppm for Group 4 blue glasses. Samples 29P and 4BMA are omitted as they have extremely high levels of Pb.

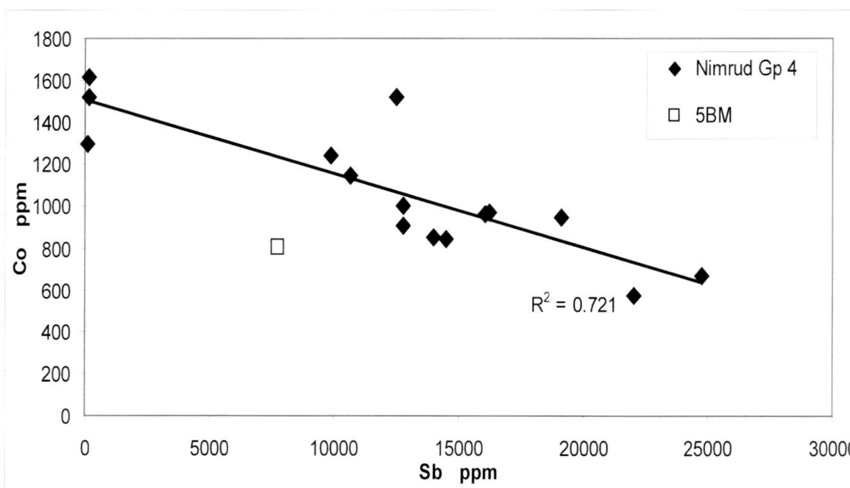

Figure 7-22 Scatter plot of Sb *vs* Co in ppm for Group 4 blue glasses.

Glass 5BM has almost twice the level of MnO (0.76%) than that found in the Group 4 glasses.

The Sb content of the Co glasses varies markedly between the groups. The HCHC group has low Sb concentrations averaging 144 ppm and a Pb content of 223 ppm on average[53]. Conversely, the LCLC group has much higher levels of Sb averaging 15,600 ppm with 45 ppm of Pb, and a higher Sn content (average 4 ppm). As a single group Sb and Pb concentrations of the Group 4 glasses have a negative correlation of r^2 = -0.6 (Figure 7-21).

[53] One glass from this Group (15N) has higher Sb (14,000 ppm) and low Pb (64 ppm), approximating more closely the levels found in the LCLC Group.

There are several individual glasses which have unexpectedly high or low contents of some elements. Sample 32Q has comparatively low values for Cu (9 ppm), Zn (108 ppm) and MnO (0.1%) with Co (666 ppm) at the lower end of the range for these glasses. Sample 29P has unusually high Sn (8 ppm) and extremely high Pb (23,334 ppm) for this group. Samples 24Q, 25T and 31N have high Co and Cu with low Sb contents. There are other glasses that are high in one or more of these elements, but not all three. The explanation for these variations is uncertain at present.

Figure 7-22 reveals a strong anti-correlation (r^2 = -0.7) between Sb and Co for Group 4 glasses. A similar anti-correlation exists between Sb and Cu.

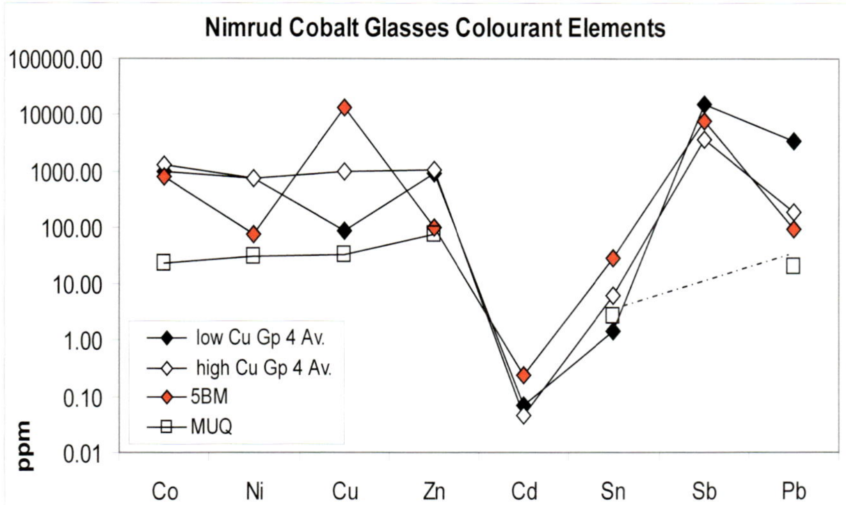

Figure 7-23 Line plot of raw colourant-related trace element data for Nimrud Group 4 blue glasses compared with MUQ. A logarithmic scale is used. There are no MUQ values for Cd and Sb.

These relationships show that when the levels of Pb, Cu and Co are lower in the Group 4 glasses, the content of Sb increases and *vice versa.* It is unclear why this has occurred, unless Sb in some way helps to achieve the desired colour when the Co and Cu colourants were added at low concentrations.

Figure 7-23 is a line plot of the raw colourant-related trace element concentrations for Nimrud Co blue glasses compared with levels of these elements found in the earth's crust. In these glasses the elemental concentrations are elevated above those of the UCC indicating deliberate addition to the glass to achieve colour. The high Cu group has higher Sn, but lower Sb and Pb. The low levels of Ni and Zn and the unusually high Cu concentration in plant ash glass 5BM mark this as having been made with differently sourced Co to produce the dark blue colour. This glass has more enriched Cd and Sn contents but has a lower level of Pb.

In summary, the Nimrud Group 4 glasses are coloured with Co alum with associated Ni, Zn, alumina and magnesia levels similar to those found in Egyptian Co blue glasses. The Group 4 glasses may be divided into two groups based on Cu and Co content: HCHC and LCLC. HCHC glasses also have higher Sn concentrations. LCLC glasses are lower in Sn, but higher in their Sb and Pb content. Plant ash glass 5BM employs a different non-alum Co colourant source, with low Ni and Zn contents, and has an extremely high Cu content with relatively higher Sn. Low ratios of Sn to Cu (0.07:1 to 0.002:1) for all Co coloured glasses indicate that Cu was not derived from bronze scrap, in keeping with observations made in this study for 2nd millennium BC glasses (sections 1.10, 3.3.4, 4.5.4), and noted by Shortland in his analyses of Near Eastern glasses (2005: 4).

7.6 General Conclusions

The trace element analyses of the Co blue Group 4 glasses from Nimrud reveal that the Co alum colourant is associated with REE, which give the glasses distinctive REE patterns at higher concentrations than observed in the non-Co coloured glasses analysed. It appears that two related sources of alum were used to make the Nimrud Group 4 glasses, probably from different deposits in the same geological region. The SRE content, except for Sc, of the Group 4 glasses is higher than in the LMLK Group 3 glasses. Group 4 glasses appear to have been fluxed with mineral soda. Levels of Li, Be and Sr are elevated relative to the SBS of the glasses, but this enrichment appears to be unrelated to plant ash in these glasses. Be is associated with the Co alum colourant, which is associated with elevated levels of magnesia, alumina, Ni, Zn and manganese, similar to the chemical characteristics of Co alum sourced from the Western Desert oases of Egypt. There are two groups of glass based on relatively high Cu and Co with lower Sb, or low Cu and Co with higher Sb contents. Plant ash glass 5BM is coloured by a non-alum sourced Co colourant.

Chapter 8

Late Bronze Age Glasses - Comparisons and Interpretations

8.1 Introduction

In this chapter, the nature of the LBA glasses from Pella and Nuzi is discussed and compared with compositional data from the published analyses of other LBA glasses from the Near Eastern region, principally from Tell Brak, and from Egypt (pre-Malkata, Malkata, Amarna, Lisht and Timna). The compositional characteristics identified in this study are explored further, and the results are placed in the broader context of glass compositions from the whole region, as far as they are currently known. Differences and similarities of composition between glasses from the Near East and Egypt are identified. The synthesis of information from comparable studies is intended to provide a more comprehensive understanding of glass production from these regions in the LBA.

8.2 LBA Glasses

8.2.1 Composition

The LBA glasses from Nuzi and Pella have been identified by base glass analyses as soda-lime-silica glasses of the typical 2nd millennium BC HMHK or plant-ash type, which exhibit closely similar base glass compositions with elevated levels of magnesia, potash, lime and phosphorus. The LBA HMHK glasses form a distinct and generally homogeneous compositional group except, surprisingly, with respect to their lime content. It is interesting and unexpected to find that LBA HMHK glasses from the Near East can be divided into groups of high and low lime concentration (sections 3.2, 4.4, 8.4).

The trace element data reveal that these LBA glasses have nearly identical sedimentary REE patterns, similar to those of the REE from the UCC, but at lower concentrations (0.04 to 0.1) (sections 3.3.1, 4.5.1). The SRE are also present at lower concentrations than are found in the UCC (sections 3.3.2, 4.5.2), which is an indication of the use of relatively pure sediment-bearing raw materials, such as silica from quartz, with plant ash and associated soil contaminants probably adding a small amount of REE and SRE.

There are small differences between the concentrations of the REE (section 4.5.1) and of the SRE (section 4.5.2) in high- and low-lime glasses from Pella, where the lime concentrations correspond closely to the opacity (high-lime) or translucency (low-lime) of the glasses. These differences are not apparent between the high- and low-lime groups from Nuzi where only two opaque samples were analysed, and translucent glasses were of both high- and low-lime content (sections 3.3.1, 3.3.2).

When Nuzi glasses were divided into groups based on opacity (not lime content), their REE concentrations corresponded with those of the opaque and translucent glasses from Pella (Figure 4-27). It appears from these observations that the differences in REE and SRE concentrations occur between opaque and translucent glasses, not between high- and low- lime glasses, although the scatter in the data is too great to be certain.

The REE and SRE trace element concentrations of the opacified glasses are lower than in the translucent glasses, showing that the opacifier is not adding REE or SRE to the glasses, and that these differences in trace element composition are not related to the lime levels, or to the source of lime. The observed differences in REE and SRE concentrations might reflect the use of different proportions of a raw material, unrelated to lime, which was used to make the opaque and translucent glasses, but differences between opaque and translucent glasses are not apparent in the base glass compositions from either the Near East or Egypt. There is not sufficient published trace element data for opaque Egyptian glasses to allow comparison with their REE or SRE levels, as has been done for Near Eastern glasses in the present study. The reason for the small differences in SRE and REE concentrations between opaque and translucent glasses remains uncertain and their significance is tentative. The alkali and alkaline earth contents show almost no variation between glasses from Nuzi and Pella, or between high- and low-lime groups from either site (sections 3.3.3, 4.5.3).

Colourant concentrations in LBA glasses from Nuzi and Pella indicated that Cu was the blue colourant and Sb, as calcium antimonate, was responsible for opacification (sections 3.3.4, 4.5.4). In opacified blue glasses from Nuzi, elevated Sb contents were associated with elevated levels of Sn and Pb. This was not so for blue opacified glasses from Pella, suggesting that the raw material sources were possibly different. The concentrations of Sn, Sb and Pb were higher in the low-lime translucent than the high-lime translucent glasses from Nuzi, the reason for which is unclear, but might be an indication of the use of different raw materials, and perhaps of manufacture at more than one workshop. Overall, the HMHK glasses from LBA Nuzi and Pella are of closely similar composition, with minor subordinate variations.

8.2.2 The LBA Context: Chronological Considerations

The glasses from Pella Tomb 62 are the earliest analysed in the present study and are some of the earliest glasses known to have been made in quantity as the products of deliberate manufacture. They are dated to *c.*1550-1525 BC, whereas the glasses from the Pella temple and Area IVE are dated between *c.*1300-1200 BC (section 4.1). The glasses from Nuzi have a *terminus ante quem* of *c.*1350-1330 BC, the probable date of the destruction of Nuzi, though their date of manufacture is uncertain and could well be earlier (section 3.1.2). The dating of the contexts in which the glasses were found at Pella confirms that this type of glass was produced as early as the 16th C BC, i.e. 200 years earlier than the date generally given for the glass from Nuzi. The proposition of an earlier date of manufacture for the glasses from Nuzi is supported by the similarities of form and chemical composition of these glasses to the earlier glasses from Pella, and would fit with the period of the decoration and architecture of the Nuzi temple, in which some of the glasses from Nuzi were incorporated.

The glasses from the temple and Area IVE at Pella confirm that this type of glass continued to be made in the 13th -12th C BC, after the destruction of Nuzi, and might have continued into the IA, dating down to the 9th C BC, though in much reduced numbers. The appearance of a small number of LBA-type glasses in IA contexts at Pella could be explained by a number of possibilities: that this type of glass was going out of production by the early IA, or that the pieces in question were heirlooms handed down from the LBA, or that these are LBA objects in disturbed archaeological contexts (section 4.2, Table 4.2). On balance, the chronological and contextual evidence presented in section 4.2.1, and the typically LBA HMHK compositions of the glasses, it is probable that the four HMHK glasses found in IA contexts are remnants from the LBA and do not represent continuity of this glass-type into the IA at Pella. Barag (1985: 51) commented that some glass objects were kept as 'precious possessions and heirlooms' during the centuries between the end of the 2nd millennium BC glass industry and the appearance of Neo-Assyrian glass. The 13th C BC mosaic-glass found in the ruins of the citadel of Hasanlu, which was destroyed in the late 9th C BC, and the fragments of early glass vessels found in a ritual deposit in the Ashuritu temple at Assur, dating from the 11th to the early 9th C BC, were cited by Barag as examples of this practice.

The HMHK glass found at Pella and Nuzi appears to have been derived from a long and steady tradition of glass-making in which consistent behaviour of the artisans witnessed the use of particular recipes for each of opaque and translucent blue glasses, using established sources of raw materials.[54] It is possible that the earliest experimental glasses have not survived in the archaeological record, as discussed in section 1.7, leaving only the perfected products of an established and well-regulated industry.

The number of workshops involved in glass production is difficult to determine (section 1.3). The consistency of glass composition from a broad geographic region from the Jordan valley to northern Mesopotamia, over a span of at least 300 years, together with the archaeological evidence for state and/or temple control, suggest that this was an organised, far-reaching industry that made use of the established trade routes of the time. If the decreased incidence of the HMHK glass in the IA is attributable to the waning of this industry at the end of the LBA, it would not be surprising considering the political upheavals and widespread destruction of that time (sections 1.2, 3.1, 4.1). The stable political framework that enabled glass-making to continue so steadily in the Near East from the 16th to 13th C BC consisted of an integrated inter-regional system which extended from western Iran to the Aegean Sea, and from Anatolia to Nubia (section 1.2). The balance of political power in the entire region at this time provided an environment in which the production and the distribution of goods flourished before the changes that ended the LBA and disrupted the manufacture of glass.

8.3 Inter-Site Comparisons

8.3.1 Near Eastern Glasses

A total of 35 glasses from the Mitannian site of Tell Brak in northern Syria, which dates from the 15th C down to 1300 BC, a slightly later end date than the accepted date for Nuzi glass of *c.*1350-1330 BC, have been analysed for composition by Brill and Shirahata (*n*=6) (1997: 89-91; 1999b: 39), Henderson and Velde (*n*=18) (in Henderson 1997: 94-100), Shortland and Eremin (2006: 595) and Shortland *et al.* (*n*=11) (2007: 784).[55] Many glasses were unearthed from the Mitanni Palace at the top of the tell, and all samples are dated to the late 14th C down to 1300 BC, except for sample Br9 from the street between the Mitanni Palace and the temple, which probably dates to the 15th C BC (Henderson 1997: 95). Glass types

[54] Rehren (2000b: 14) believed that glass-making centres in LBA Egypt strictly followed standard recipes.

[55] The studies of Shortland and Eremin (2006) and Shortland *et al.* (2007) form two parts of a single study, with the first publication concentrating on EPMA analyses of glass oxides, and the second on trace element analyses of a selection of the same glasses. The total number of glasses analysed from Tell Brak is 11, and the total number of glasses analysed from Nuzi by EPMA is 17, with 14 of these having been analysed for trace elements.

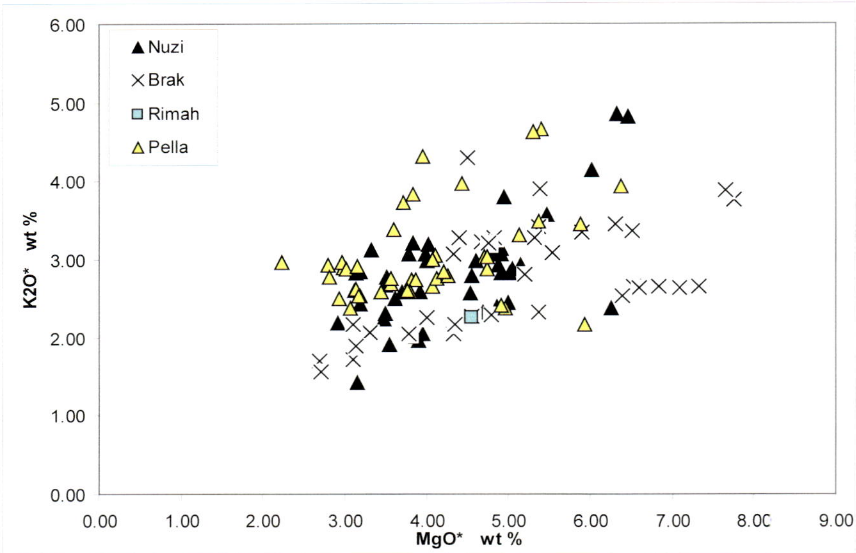

Figure 8-1 Scatter plot of magnesia *vs* potash for LBA HMHK glasses from Nuzi (*n*=46), Pella (*n*=43), Tell al-Rimah (*n*=1), and Tell Brak (*n*=35). Sources cited in text.

analysed are similar to those from Nuzi and Pella, and include a large number of ingots, vessels, beads and pendants, but in a wider range of colours, including translucent blue, amber and purple, and opaque blue, white and black. This suggests that Tell Brak stands apart from Nuzi and Pella, perhaps being a centre for glass-working, if not production, for which there is circumstantial, but no direct evidence (Oates *et al.* 1997: 86).

Compositions of the glasses from Tell Brak are compared with those of LBA glasses analysed from Nuzi in the present study, by Vandiver (*n*=2) (1983: 243), Brill (*n*=11) (1999b: 40-41), Shortland and Eremin (*n*=17) (2006: 595) and Shortland *et al.* (2007: 784); and the 43 glasses from Pella included in the present study. One opaque blue glass fragment of a tile dated to *c.*1400-1200 BC[56] (sample 721), analysed by Brill (1999b: 42) from the Mesopotamian site of Tell al-Rimah in Iraq, is included.

The base glass composition of all glasses was found to be of a closely similar soda-lime-silica HMHK type, with magnesia levels between 2.2% and 7.8% and potash levels between 1.4% and 4.9% (Figure 8-1).

When the data for magnesia and lime are plotted for the LBA Near Eastern glasses (Figure 8-2), the division based on lime concentration, observed in the glasses analysed from Nuzi and Pella in the present study (sections 3.2, 4.2), is largely supported. One opaque blue sample (30-2-7-1) from Nuzi, analysed by Vandiver (1983: 243), has a low lime content of approximately 5.3%. The specific issue of the lime content of the LBA glasses is addressed in section 8.4.

Comparisons of available data for 2nd millennium BC Near Eastern HMHK glasses confirm that they are of a relatively consistent composition over a geographic range from the Levant to northern Syria and northern Iraq, which concurs with observations by previous researchers who have noted in individual studies that, between 1500-1200 BC in Mesopotamia, the same basic raw materials were used in the manufacture of glass by an industry that was sophisticated and apparently well-regulated (McGovern *et al.* 1991: 401; Henderson 1997: 95, for example).

8.3.2 Egyptian New Kingdom Glasses

The origin of Egyptian and Near Eastern glasses has been much discussed (section 1.3), but a detailed investigation of the similarity or dissimilarity of their composition has not been made until now (sections 1.12, 1.13). In this section, the consistent base glass compositions of the Near Eastern glasses are compared with compositions of 265 HMHK (non-Co[57]) glasses from pre-Malkata, Malkata, Amarna, Lisht, Wadi Qirud and Timna in Egypt (Lilyquist *et al.* 1993: 36-39; Brill 1999b:

[56] This was the only sample from Tell al-Rimah analysed by Brill that was suitable for comparison. The rest were weathered. For typological studies of the Tell el-Rimah glass, which are beyond the scope of this dissertation, see von Saldern (1966b: 1970). An older chemical study of Tell al-Rimah glasses was made by Pollard and Moorey (1982), however these analyses have compositional features which do not seem to be consistent with other compositional studies. LBA glasses from Tchoga Zanbil in Iran analysed by Brill (1999a: 45) were weathered and not included as comparative material.

[57] Co coloured glasses have an unusual base glass composition which has low potash and high magnesia and alumina contents and hence is not of the typical HMHK type. The source of the soda is controversial (section 1.10.1).

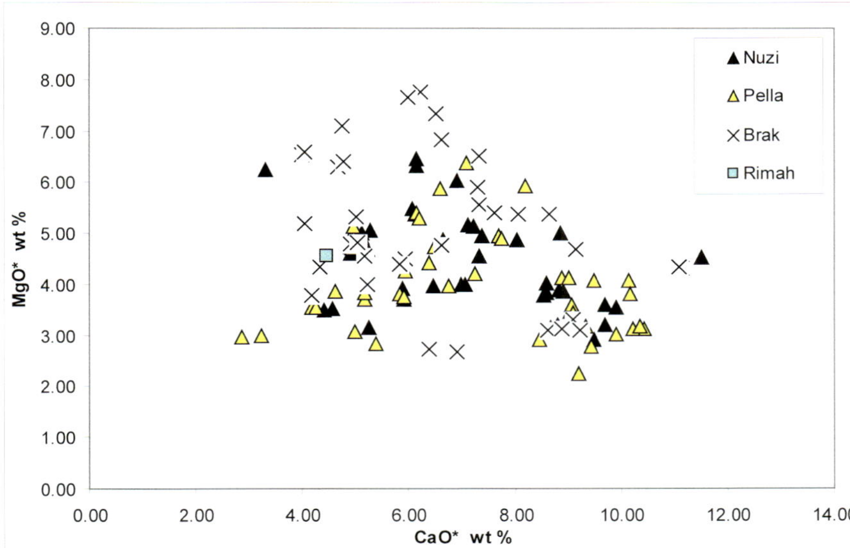

Figure 8-2 Scatter plot of lime *vs* magnesia for LBA HMHK glasses from Nuzi (*n*=46), Pella (*n*=43), Tell al-Rimah (*n*=1), and Tell Brak (*n*=35). Sources cited in the text.

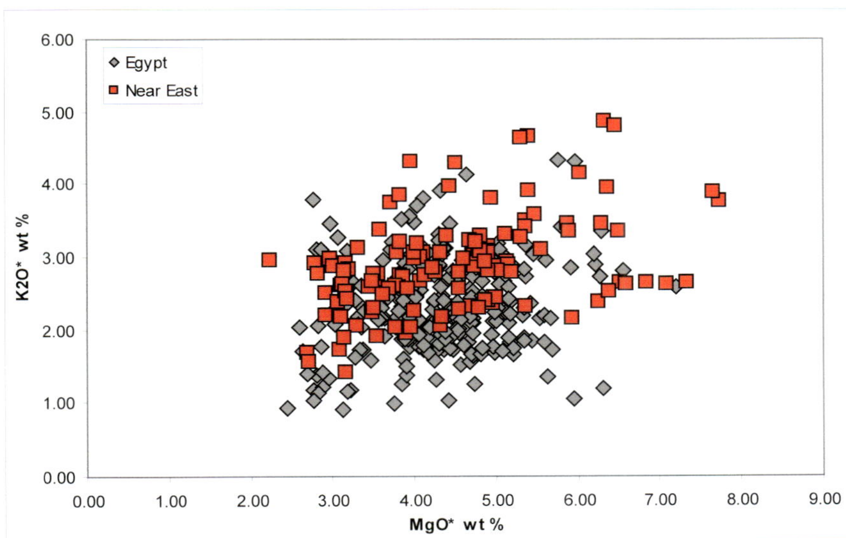

Figure 8-3 Scatter plot of magnesia vs potash for LBA HMHK Egyptian and Near Eastern glasses. Sources cited in text. Co blue glasses are excluded.

27-35; Nicholson and Henderson 2000: 215; Shortland and Eremin 2006: 584-588; Jackson and Nicholson 2007: 181-182; Freestone unpublished).

Shortland and Eremin (2006: 583) found that 14th C BC Egyptian glasses from Amarna and Malkata had been made with closely similar, if not identical raw materials, and the similarity of the composition of pre-Malkata glasses with those from Nuzi, analysed by Brill, was noted by Lilyquist *et al.* (1993: 43). The relative homogeneity of LBA HMHK glass compositions of Near Eastern and Egyptian glasses, illustrated in Figure 8-3 and the following plots, suggest the use of similar, but not identical, proportions of raw materials in both regions. The average potash content of the Near Eastern glasses is higher than that of their Egyptian counterparts, which have a wider range of potash concentrations, but the range of magnesia is similar in both.

It is evident from Figure 8-4 that the distribution of lime in LBA Egyptian glasses is different from that in the LBA Near Eastern glasses, where the two lime-based groups are clearly visible, and the high-lime group has a more restricted magnesia range at lower average concentration. The division into two lime-based groups

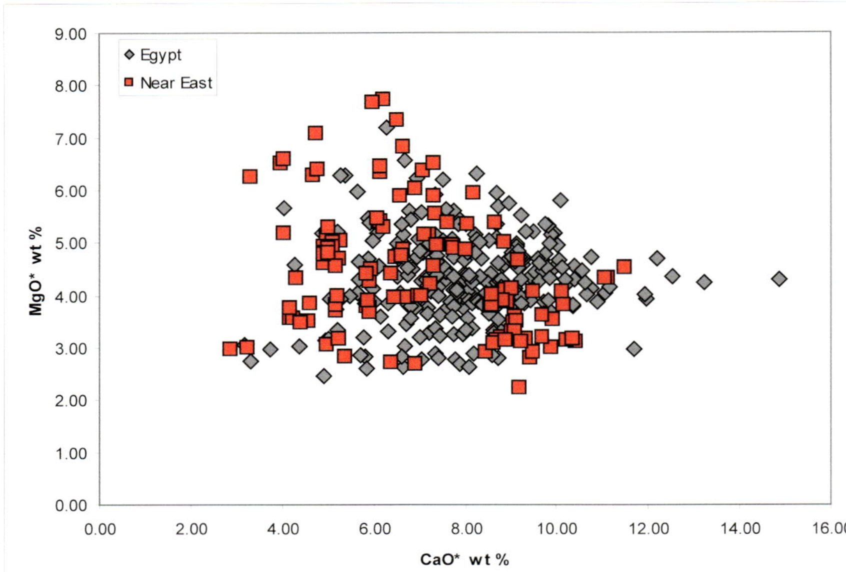

Figure 8-4 Scatter plot of lime *vs* magnesia for LBA HMHK Egyptian and Near Eastern glasses. Sources cited in text. Co blue glasses are excluded.

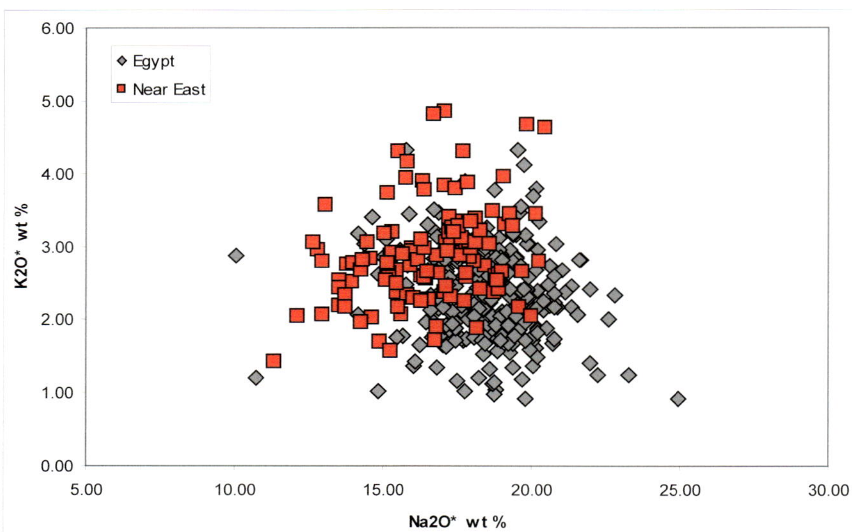

Figure 8-5 Scatter plot of soda *vs* potash for LBA HMHK Egyptian and Near Eastern glasses. Sources cited in text. Co glasses are excluded.

is not a feature of Egyptian glass compositions, although the range of total lime concentrations is similar between the regions, with a few Egyptian samples containing higher lime contents of up to almost 15%.

Figures 8-5 to 8-7 illustrate further subtle differences between the base compositions of Egyptian and Near Eastern HMHK glasses. Egyptian glasses have, on average, a relatively higher soda content and lower silica content, which might be due to the addition of silica and plant ashes in different proportions to those of Near Eastern glasses. There are more Egyptian glasses with alumina contents over 1% and iron oxide contents over 0.6%, which is indicative of the use of raw materials of lesser purity than was used in Near Eastern glasses.

Overall, the base glass compositions of Near Eastern and Egyptian HMHK glasses are similar, but with clear indications of the regional differences in the nature and proportions of raw materials. It is impossible to say, using the base glass data, whether individual glasses in the areas of overlap on the graphs were made in Egypt or the Near Eastern region, but it is acknowledged that

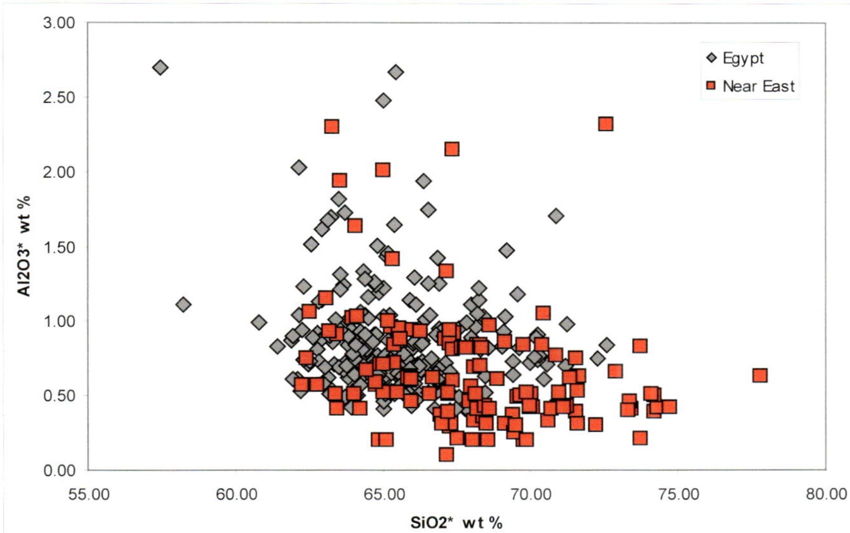

Figure 8-6 Scatter plot of silica *vs* alumina for LBA HMHK Egyptian and Near Eastern glasses. Sources cited in text. Co glasses are excluded.

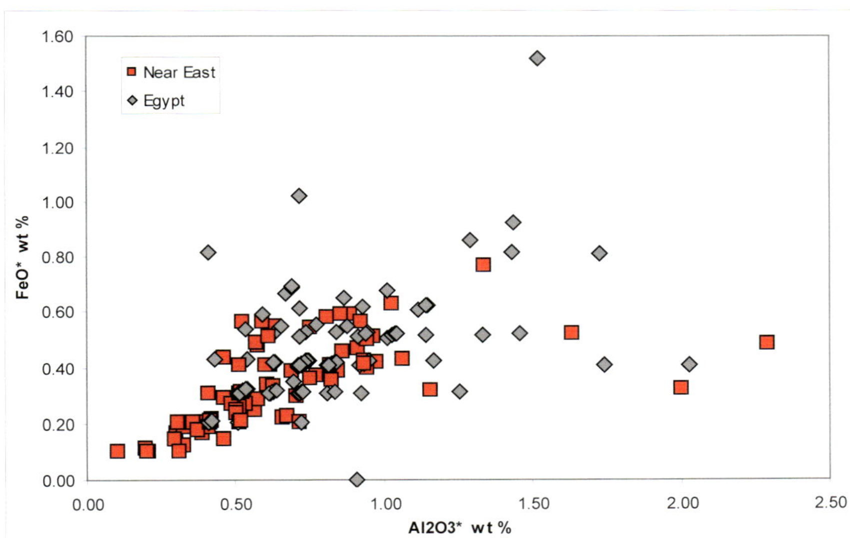

Figure 8-7 Scatter plot of alumina *vs* iron oxide for LBA HMHK Egyptian and Near Eastern glasses. Egyptian glasses are from Amarna and Malkata analysed by Shortland and Eremin (2006: 584-588). Nuzi and Tell Brak glasses analysed by Shortland and Eremin (2006: 594-595) are combined with analyses of Pella and Nuzi glasses from the present study. Only analyses which give data for FeO are included. Co blue glasses are excluded.

HMHK glasses must have been made in both places (section 1.3), presumably with locally available silica and soda sources. By comparison with Near Eastern glasses, the lower potash and higher soda contents of Egyptian glasses is probably related to the use of a different plant ash, even though both groups contain similar magnesia and lime contents, on average. Alternatively, ashes could have been used from locally grown plants of the same kind that were used in the Near East, but which exhibit compositional variations due to having been grown in different soils (section 1.8).

8.3.3 Comparisons of Trace Element Compositions

It has become possible with the availability of high-quality trace element data, such as those provided by Shortland *et al.* (2007: 784-785), to search more closely for indicators of raw material origin which might be

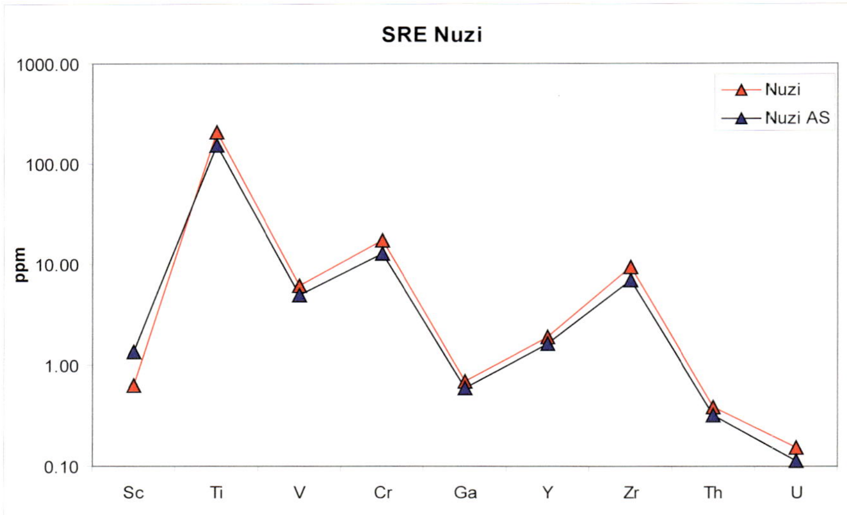

Figure 8-8 Line and dot plot of SRE data for glasses from Nuzi analysed by Shortland *et al.* (2007: 784) (Nuzi AS) and analysed in the present study (Nuzi).

imprinted in the concentrations of elements from ancient glasses.

The following comparisons of LBA Near Eastern and Egyptian glass trace element data aim to identify further distinctions and similarities between glasses from the two regions. The full suite of REE has not been analysed for Egyptian glasses, or for Near Eastern glasses other than in the present study. The Egyptian glass data used for comparisons in this section are from the analyses of 24 colourless and Cu blue HMHK glasses from Amarna and Malkata. The Near Eastern comparative data are from the analyses of 14 HMHK glasses from Nuzi and 11 HMHK glasses from Tell Brak, all Cu blue, by Shortland *et al.* (2007: 784-785). The trace elements are grouped as SRE and alkalis and alkaline earths. Shortland and co-workers have not analysed the SRE Nb, Hf, Ta, W or Tl, which were analysed in the present study.

The colourant data were not graphed in a line and dot plot because the Egyptian data are varied such that little was gained from averaging them for this kind of comparison. It is of interest that a much wider range of glass colours was produced, or at least worked, in Egypt than in the Near East. This expanded colour range included colourless, translucent brown and purple, opaque red, translucent and opaque white, yellow, green, Co blue, Cu blue, and a category described by Shortland and Eremin (2006: 586, 591) as Co blue with added Cu, possibly to deepen the colour when relatively low levels of Co were used[58].

Initially, the Nuzi glass SRE data from the ICP-MS trace element analyses by Shortland *et al.* (2007: 784) were plotted with the Nuzi glass SRE data from the present study (Figure 8-8). The result shows that these data accord well, but with some difference observed for Sc.

When the SRE data for glasses from Nuzi, Tell Brak, Pella and Egypt are plotted together (Figure 8-9), the concentrations of these elements are closely similar for glasses from Nuzi and Pella. Concentrations of the SRE in the glasses from Tell Brak are close, but less similar to those from Pella and Nuzi glasses. There is a divergence in the concentrations of U from each of the three Near Eastern sites. The Egyptian glasses have higher concentrations of the SRE except for Cr, which is higher in the Near Eastern glasses, the elevated levels of which have been noted in sections 3.3.2 and 4.3.3.

The findings of the present study for Near Eastern glasses concur with those of Shortland *et al.* (2007: 787) who noted that the lower average concentration in Mesopotamian glasses of these contaminant elements (except for Cr) implied that the Mesopotamian glass raw materials were purer than those used to make Egyptian glasses. The latter glasses were compositionally more varied, supporting the findings of the base glass analyses.

The alkali and alkaline earth data for Nuzi glasses from the ICP-MS trace element analyses by Shortland *et al.* (2007: 784) have been plotted with the alkali and alkaline earth data for Nuzi glasses from the present study (Figure 8-10) to show that these data accord well.

When the alkali and alkaline earth data for glasses from Nuzi, Tell Brak and Pella are plotted together

[58] Tite and Shortland (2003) define a Cu blue glass as one which has less than 0.05% CoO and Co glass as one which has more than 0.05% CoO.

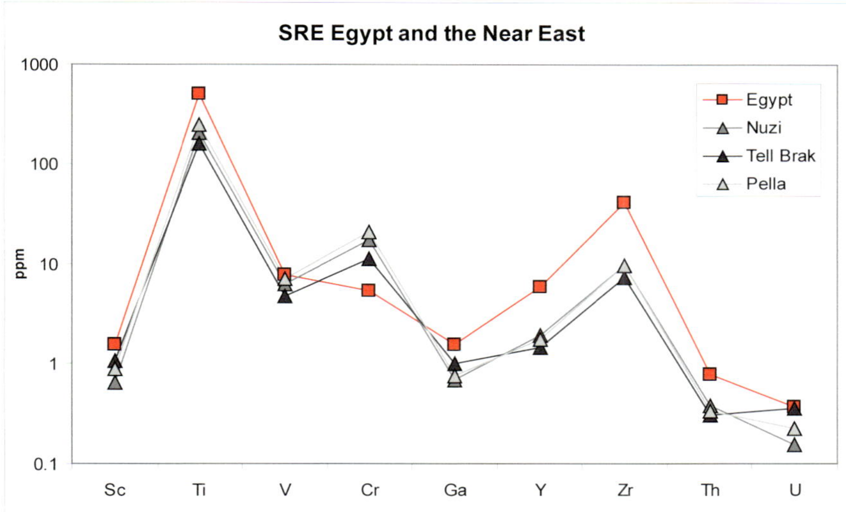

Figure 8-9 Line and dot plot of SRE data for glasses from Egypt and the Near East. Data for Egypt and Tell Brak from Shortland *et al.* (2007: 784-785) and for Nuzi and Pella from the present study.

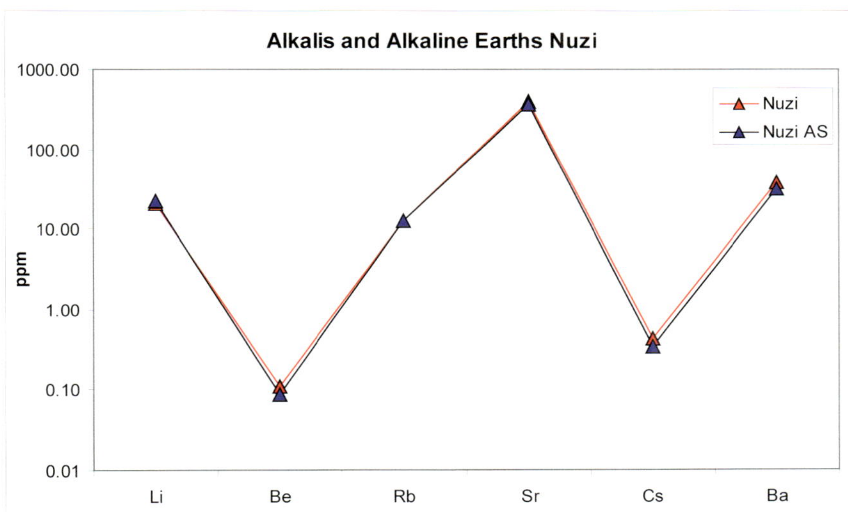

Figure 8-10 Line and dot plot of alkali and alkaline earth element data for glasses from Nuzi (Nuzi AS) analysed by Shortland *et al.* (2007: 784) and for glasses (Nuzi) analysed in the present study.

(Figure 8-11), the concentrations of these elements are closely similar. The Egyptian glasses have higher concentrations of Be and Sr and have lower levels of Li, Rb and Cs than the Near Eastern glasses. The observed differences in concentrations of the alkali and alkaline earth elements between the two regions indicate the use of different raw material sources.

Overall, the differences in base glass and trace element compositions of glasses found in the two regions show that the glasses were made to a similar recipe, but that different raw materials were used. This supports the view that the Egyptian and Near Eastern glasses were not made in the same place, but show regional compositional distinctions that separate them. This does not of course preclude the possibility that some of the glasses analysed might have been transported by trade or gift to the region in which they were found.

Elements which have been shown to represent the key differences between Near Eastern and Egyptian glasses will be examined in greater detail in the following sections.

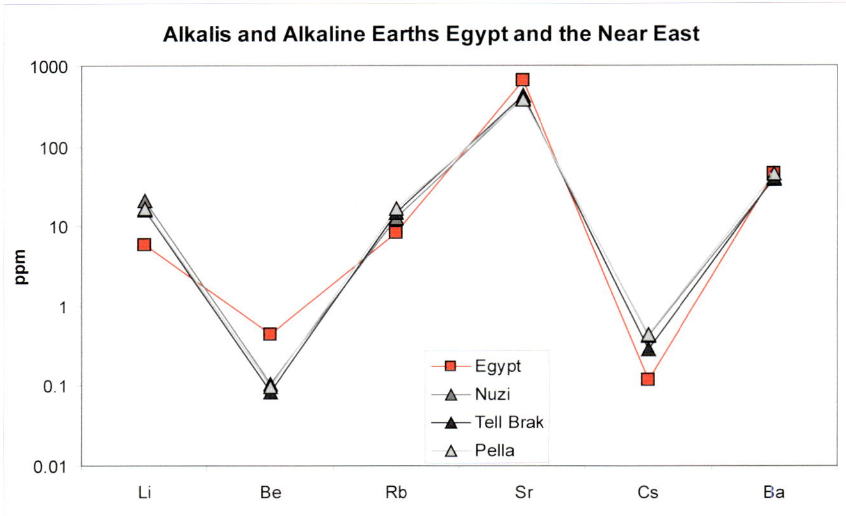

Figure 8-11 Line and dot plot of alkali and alkaline earth element data for glasses from Egypt and the Near East. Data for Egypt and Tell Brak from Shortland *et al.* (2007: 784-785) and for Nuzi and Pella from the present study.

8.3.3.1 Cr, La, Zr, Ti and Iron Oxide

The work of Shortland (2005: 2-3) explored the potential to provenance glasses using the ratios of elements found in clays that were introduced to the glass batch as contaminants during the manufacturing process. The ratios of Fe and Ti in the glasses were plotted together and it was found that the Egyptian glasses lay on a trend line that corresponded with Nile River silt values, whereas the Mesopotamian[59] glasses from Tell Brak followed a distinctly different trend closer to, but not exactly the same as, that of the local Tell Brak (Tell Beydar) clays. Shortland concluded that further data were required to confirm these preliminary findings. Iron oxide and Ti concentrations for glasses from Pella and Nuzi analysed in the present study have been plotted with the Nuzi and Tell Brak data from Shortland and Eremin (2006: 595) and from Shortland *et al.* (2007: 784) against the Egyptian data produced by Shortland and co-workers (Figure 8-12). The Egyptian glasses have higher Ti concentrations, and the different trend lines of the Near Eastern and Egyptian data are clear and support Shortland's finding (2005: 2-3).

That many trace elements are present in colourless and coloured glasses at similar levels was proposed in the study of Egyptian and Mesopotamian glasses by Shortland *et al.* (2007: 786). Although the colours analysed in the present study are limited, the results indicate that opacified glasses have lower REE and SRE concentrations (section 8.2.1). Shortland *et al.* (2007: 786) used the uniformity of trace element data to test the possibility of distinguishing glass compositions further by examining the ratios of Cr to La, and Zr to Ti more closely. These elements vary most between Egyptian and Near Eastern glasses, and as they are refractory and non-volatile they are commonly used as compositional discriminants in igneous geochemistry. Their differences are likely to reflect those of the raw materials, and their concentrations would not have been altered during the manufacturing process. The examination of the ratios of these elements is believed to emphasise the distinctiveness of glasses from the two major regional groups in the LBA. The combined data from Shortland and co-workers and from the present study are tabulated in Table 8-1 and plotted in Figures 8-13 and 8-14.

From the data presented in Table 8-1, it is evident that the concentrations of Zr, Ti, Cr and La are similar in the glasses from Nuzi, Tell Brak and Pella, although the glasses from Pella have on average slightly higher concentrations of Zr, Ti and Cr. The ratios of Zr to Ti fall within the range of 0.04:1 to 0.05:1 for all three groups of Near Eastern glass, and the ratios of Cr to La range on average between 7.18:1 and 11.57:1, with Pella glasses having the highest Cr to La ratio.

The Egyptian glasses have a low average Cr concentration (6.07 ppm) compared with the Near Eastern glasses (11.2 ppm to 19.3 ppm). The ratio of Zr to Ti in Egyptian glasses averages 0.08:1, which is approximately twice that found in the Near Eastern glasses, but the average ratio of Cr to La is 1.84:1, which is considerably lower than in the Near Eastern glasses.

[59] Although the 'Mesopotamian' glasses are indistinguishable from glasses from the wider Near Eastern region, Shortland and colleagues worked specifically on glasses from the Mesopotamian region, and hence this term is used advisedly here.

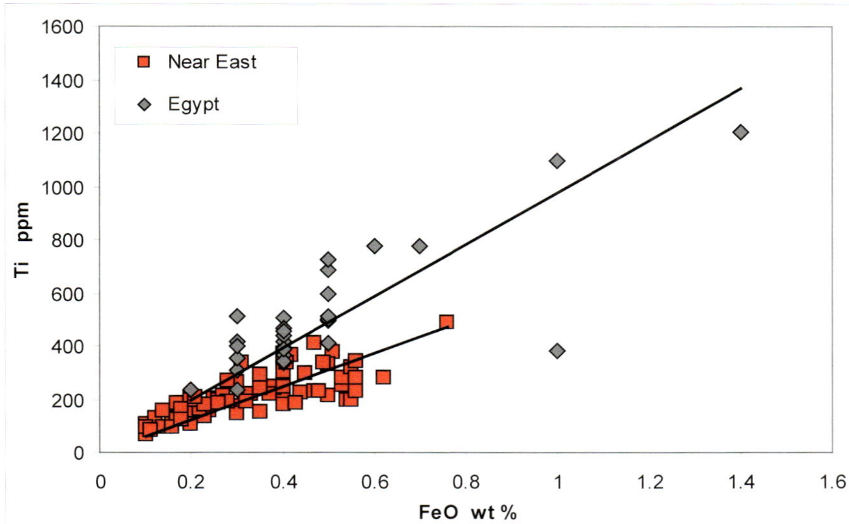

Figure 8-12 Scatter plot of iron oxide *vs* Ti for Near Eastern and Egyptian glasses. Data sources cited in the text.

Table 8-1 Concentrations in ppm of Zr, Ti, Cr and La and ratios of Zr/Ti and Cr/La for glasses from the Near East and Egypt.*

Averages	Zr ppm	Ti ppm	Cr ppm	La ppm	Zr/Ti	Cr/La
Nuzi						
n= 29	8.20	183	15.0	1.76	0.04	8.55
Pella						
n= 43	10.8	247	19.3	1.73	0.04	11.6
Tell Brak						
n= 11	7.27	165	11.2	1.49	0.05	7.18
Egypt						
n= 36	41.7	519	6.07	3.42	0.08	1.84

*Data for Egypt, Nuzi and Tell Brak from Shortland et al. (2007: 784-785) and for Nuzi and Pella from the present study.

These observations form the basis of an elemental distinction between LBA glasses from Egypt and the Near East, with a particular difference arising in the concentration of Cr. The Cr contents in the glasses from Pella and Nuzi were observed to be elevated relative to the content of the UCC (sections 3.3.2, 4.3.3), a feature which is not observed in glasses from Egypt.

Using Shortland and co-worker's method of comparing Zr:Ti *versus* Cr:La (Figure 8-13) reveals that the so-called Near Eastern glasses plot together with the exception of three Pella glasses (samples 77, 136, 194), and at least two glasses from Tell Brak, which plot closer to the majority of Egyptian glasses. This compositional variation perhaps indicates that the Near Eastern outliers are of Egyptian origin, although their fit with the Egyptian glasses is not close, which leaves this question of provenance open. The Egyptian glasses are shown to have a limited range of Cr to La ratios with higher and more widely ranging ratios of Zr to Ti than the average Near Eastern glasses. The Near Eastern glasses have, by contrast, greater and more widely ranging ratios of Cr to La, with lower concentrations of Zr, Ti and La, but higher concentrations of Cr compared with the Egyptian glasses. Similar observations were made by Shortland *et al.* (2007: 788) who suggested that the glasses from each region were made with raw materials sufficiently similar to suggest that there

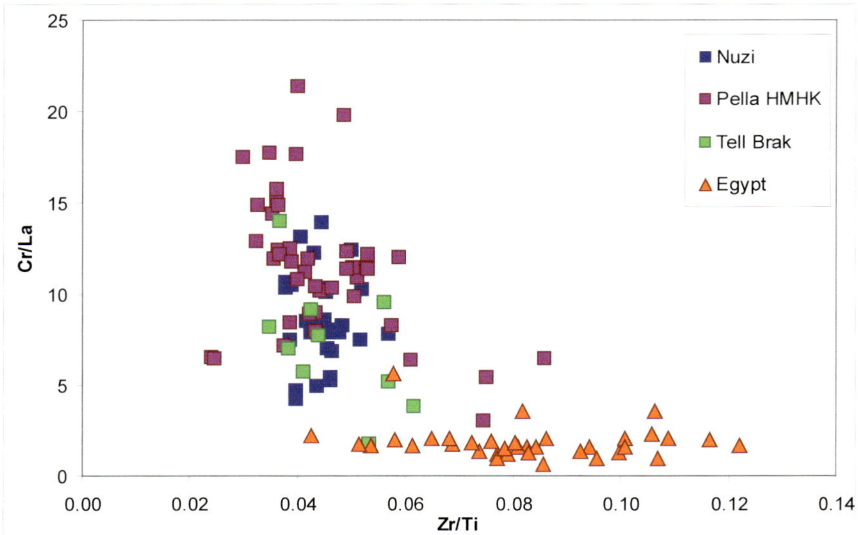

Figure 8-13 Scatter plot of Zr/Ti *vs* Cr/La for LBA Near Eastern (squares) and Egyptian (triangles) glasses. Data for Egypt, Nuzi and Tell Brak from Shortland *et al.* (2007: 784-785) and for Nuzi and Pella from the present study. This scatter plot is of the type introduced by Shortland *et al.* (2007).

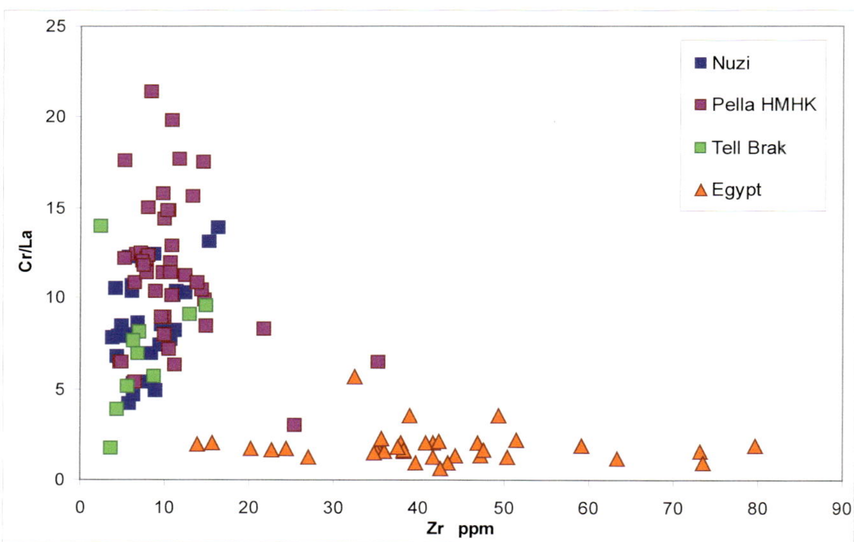

Figure 8-14 Scatter plot of Zr *vs* Cr/La for LBA Near Eastern (squares) and Egyptian (triangles) glasses. Data for Egypt, Nuzi and Tell Brak from Shortland *et al.* (2007: 784-785) and for Nuzi and Pella from the present study.

might have been only one major glass-making centre in each region. This is unlikely in view of the research by Rehren and co-workers, and Nicholson and co-workers in Egypt (section 1.3). The presence of high- and low-lime groups of glass at Nuzi, Pella and Tell Brak could indicate that they were the products of different workshops in the Near Eastern region (section 8.4).

When the data for the same sites are represented in a plot of Zr *versus* Cr:La (Figure 8-14), the distinction between Egyptian and Near Eastern glasses is clearer, and only two of the glasses from Pella plot more closely with the Egyptian glasses.

The data from the present study confirm the model developed by Shortland *et al.* (2007) for discriminating LBA Near Eastern from Egyptian glasses using ratios of Cr to La and Zr to Ti, although the comparison of Zr with Cr:La appears to give a clearer division. The results indicate that the Near Eastern glasses, with two possible exceptions, are not of Egyptian origin, and that none of the Egyptian glasses are of Near Eastern origin.

8.3.3.2 Li, Be, Sr and Cs

The average concentrations of Be and Li are distinctly different in Egyptian glasses by comparison with Near Eastern glasses. A more subtle difference may be observed in the concentrations of Sr and Cs in the glasses between the two regions (Figure 8-11). In addition to the discriminatory value of Cr, La, Zr and Ti explored previously in section 8.3.3.1, the possible value of Li, Be, Sr and Cs as regional discriminators of glass compositions is worthy of investigation. Averaged data for Be, Li, the ratio of Li to Be, Sr, Cs and the ratio of Sr to Cs are given in Table 8-2.[60]

Table 8-2 Concentrations in ppm of Li, Be, Sr and Cs and the ratios of Li/Be and Sr/Cs for glasses from the Near East and Egypt.*

Averages	Li ppm	Be ppm	Li/Be	Sr ppm	Cs ppm	Sr/Cs
Nuzi						
n= 29	21.6	0.096	265	393	0.383	1190
Pella						
n= 43	16.4	0.099	342	382	0.443	936
Tell Brak						
n= 10	17.3	0.087	207	414	0.323	1491
Egypt						
n= 36	5.91	0.438	28.3	646	0.119	9164

*Data for Egypt, Nuzi and Tell Brak from Shortland et al. (2007: 784-785) and for Nuzi and Pella from the present study. For Tell Brak, sample HH 545.30 has an anomalously high Li/Be ratio, and sample HH224 has anomalously low concentrations of these elements.[60] Both were omitted from calculations of the relevant averages. Pella samples 22, 119 and 136 have anomalously high Be concentrations are were omitted from calculations of averages for Be and Li/Be, but are included in the plots of Pella data below.

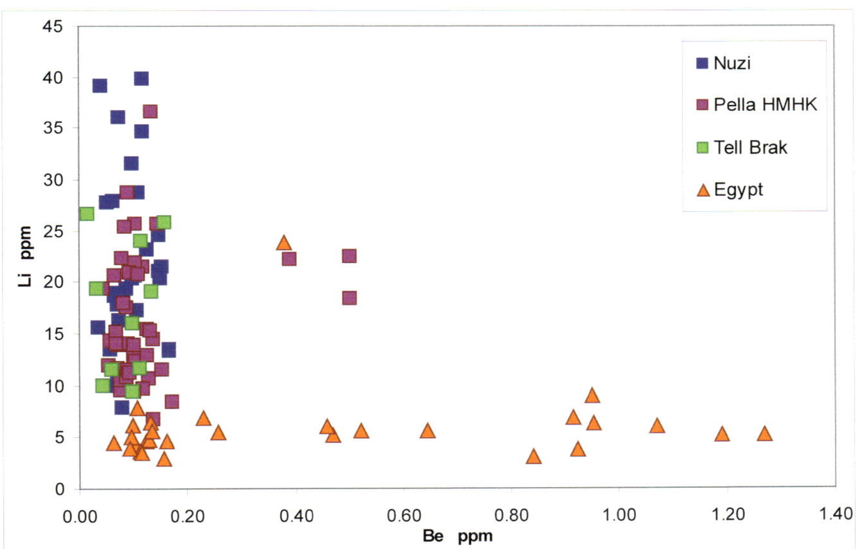

Figure 8-15 Scatter plot of Be *vs* Li for LBA Near Eastern (squares) and Egyptian (triangles) glasses. Data for Egypt, Nuzi and Tell Brak from Shortland *et al.* (2007: 784-785) and for Nuzi and Pella from the present study.

[60] Presumably sample HH224 was considered by Shortland *et al.* (2007) to have been unweathered but the anomalously low alkali and alkaline earth element contents could be indicative of depletion due to weathering (unpublished analyses of weathered glasses from Nuzi by the author).

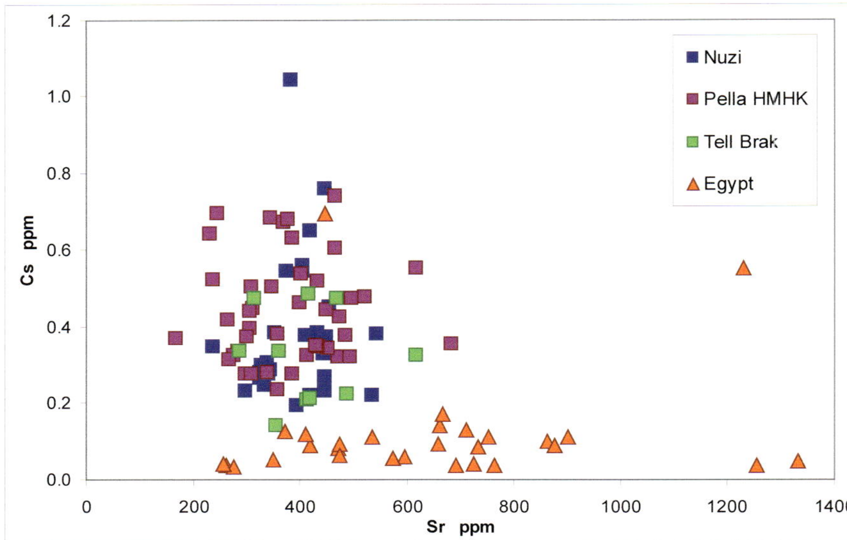

Figure 8-16 Scatter plot of Sr *vs* Cs for LBA Near Eastern (squares) and Egyptian (triangles) glasses. Data for Egypt, Nuzi and Tell Brak from Shortland *et al.* (2007: 784-785) and for Nuzi and Pella from the present study.

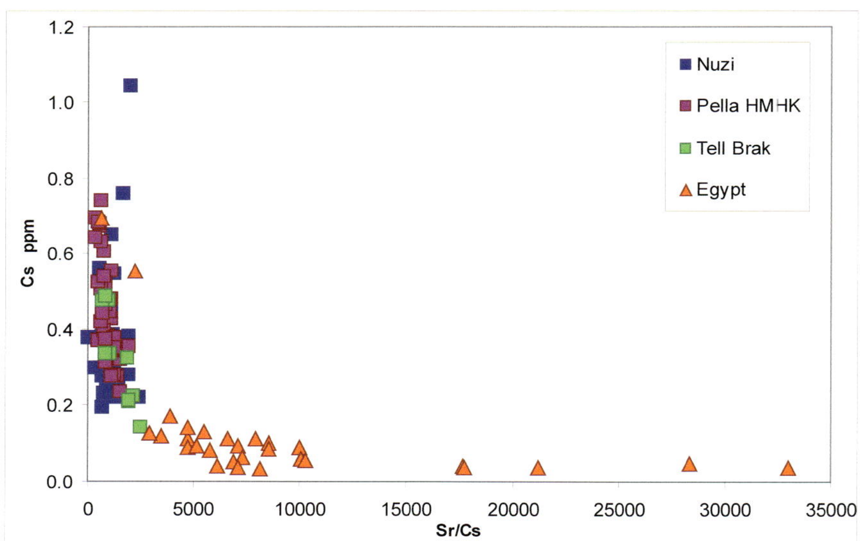

Figure 8-17 Scatter plot of Sr/Cs *vs* Cs for LBA Near Eastern (squares) and Egyptian (triangles) glasses. Data for Egypt, Nuzi and Tell Brak from Shortland *et al.* (2007: 784-785) and for Nuzi and Pella from the present study.

Egyptian glasses have an average content of 0.44 ppm Be, with a widely varying range from 0.06 ppm to 0.95 ppm. The Li content averages 5.9 ppm, with a low and narrow range from 3 ppm to 7.8 ppm. One sample contains a comparatively high level of 23.9 ppm (sample UPP22 from Malkata). The Li content of the Near Eastern glasses is clearly higher with averages between 16 ppm and 21.6 ppm, and a range from 6.7 ppm to 39.8 ppm. The Be content of the Near Eastern glasses averages between 0.08 ppm to 0.10 ppm and is at a considerably lower level than is found in most of the Egyptian glasses (Figure 8-15). The averaged ratios of Li to Be are higher, and the ratios of Sr to Cs are lower in Near Eastern glasses by comparison with those ratios determined for Egyptian glasses.

Although there are several Egyptian glasses with Be concentrations as low as those of the Near Eastern glasses, the low Li contents of the Egyptian glasses separate the glasses from the two regions. The single black HMHK glass from Pella (sample 74) overlaps with the Egyptian glasses in Figure 8-15, which did not

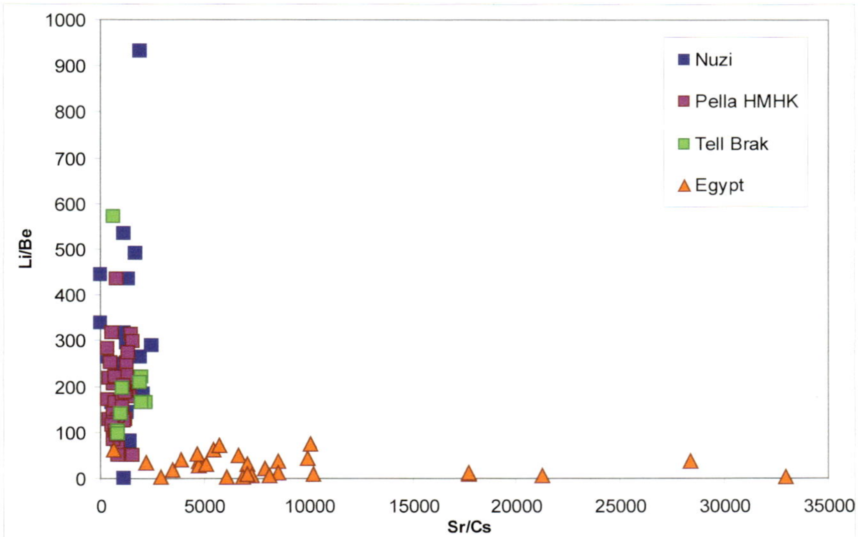

Figure 8-18 Scatter plot of Sr/Cs *vs* Li/Be for LBA Near Eastern (squares) and Egyptian (triangles) glasses. Data for Egypt, Nuzi and Tell Brak from Shortland *et al.* (2007: 784-785) and for Nuzi and Pella from the present study.

occur when the concentrations of Cr, La, Zr and Ti were compared. The blue glasses (samples 77, 136 and 194) which plotted near to Egyptian glasses in Figure 8-13 are shown by this comparison to fit unquestionably with the Near Eastern glasses. The three Pella glasses with anomalously high Be concentrations plot with one Egyptian glass as outliers.

A comparison of the concentrations of Sr and Cs in glasses from both regions (Figure 8-16) manifests a similar division, with Near Eastern glasses having lower average Sr concentrations (382 ppm to 414 ppm) and higher average Cs concentrations (0.32 ppm to 0.44 ppm) than are found in the Egyptian glasses (646 ppm Sr, 0.12 ppm Cs). The range of Sr concentrations for Near Eastern glasses is 168 ppm to 617 ppm, whereas the range is higher in Egyptian glasses at 255 ppm to 1332 ppm. Cs concentrations have a higher range, from 0.14 ppm to 1.04 ppm, in Near Eastern glasses, and have a low range from 0.03 ppm to 0.69 ppm in Egyptian glasses.

The differences between the Near Eastern glasses and those from Egypt is made clearer when the ratios of Sr to Cs are plotted against Cs (Figure 8-17). The Near Eastern glasses have a low ratio of Sr to Cs, with a relatively large range of Cs concentrations. The Egyptian glasses, on the other hand, have low Cs concentrations with a wider range of Sr to Cs ratios, a similar picture to that seen in the plot of Zr *versus* the ratios of Cr to La (Figure 8-14).

Other combinations of elements and ratios were plotted to find alternative and perhaps clearer discriminators between the glass compositions of the two regions. Figure 8-18 illustrates another view, this time plotting the ratios of Sr to Cs against the ratios of Li to Be, but without achieving greater clarification of the regional distinction.

The distinctive ratio of Cr to La was plotted against the equally distinctive ratio of Li to Be as an additional test of these potentially powerful regional discriminators (Figure 8-19). This plot again provides significant comparative results where the Egyptian glasses form a small tight group with low values for both the ratios of Cr to La and of Li to Be, whereas the ratios of these elements in Near Eastern glasses have both higher and more wide-ranging values.

It is evident that the use of trace elements as regional discriminators is certainly informative, and can be based on comparisons of a wider range of elements than has previously been attempted to ensure that anomalous data are not misinterpreted. The method of dividing the trace elements into meaningful groups of REE, SRE and alkali and alkaline earth elements, adopted in the current study, is used to make the initial comparisons between glass compositions. From these comparisons, the elemental concentrations which signify regional differences may first be identified and then selected elements compared in a range of bivariate plots. The elements Be, Li, Sr and Cs, and the ratios of Li to Be, and of Sr to Cs, are found to discriminate clearly between the LBA glasses made in the Near East compared with those made in Egypt. The use of these elements as regional determinants adds an important dimension to the trace elements already identified as indicators of

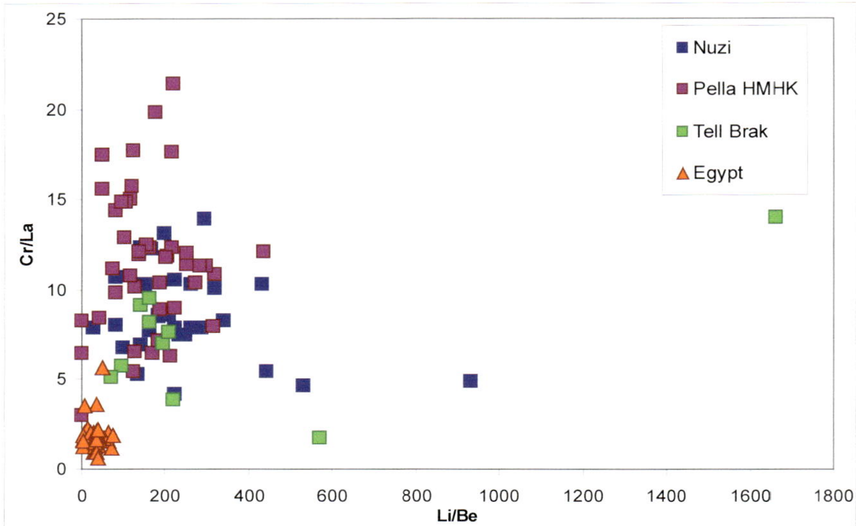

Figure 8-19 Scatter plot of Li/Be *vs* Cr/La for LBA Near Eastern (squares) and Egyptian (triangles) glasses. Data for Egypt, Nuzi and Tell Brak from Shortland *et al.* (2007: 784-785) and for Nuzi and Pella from the present study.

regional groups of LBA glass by Shortland *et al.* (2007). It is suggested that the use of additional trace elemental parameters is extremely helpful when assessing regional differences in LBA glass compositions.

8.4 High- and Low-Lime Groups in LBA Glasses

8.4.1 Introduction

Current knowledge and debate about plant ash compositions and their relationship to both the final glass compositions and the lime content of early glasses were reviewed in sections 1.7 and 1.8. In the following sections, the enigma of the lime content of the HMHK LBA glasses and its implications for understanding ancient glass-making are discussed.

8.4.2 The Source of Lime in Ancient HMHK Glasses

The manufacture of the earliest glasses from the Near East by the addition of a high-lime plant ash to a clean sand or quartz pebble silica source is the basis of the two-part theory of glass-making put forward by Turner (1956c), and discussed by many researchers since (sections 1.5, 1.6, 1.8).

According to this model of glass-making, a colourless glass is likely to have a minimum of two raw materials – silica and a soda-rich flux, which could be derived from either a mineral soda or a plant ash soda source. Plant ash can be rich in lime, but the concentrations can vary considerably, as is shown in the growing number of analyses of plant ash compositions (section 1.8).[61] If the silica was derived from quartz pebbles, or a clean sand, which would carry few contaminants, the source of the lime must have been plant ash. If the silica was derived from sand and the flux from mineral soda, it is assumed that the lime was incorporated as a constituent of the sand in the form of aragonite, calcite or limestone, for example. The only known ancient mineral soda source in this region is natron, or trona, from the Western Desert oases of Egypt. It is relatively free from contaminating trace elements and contains little lime (section 1.9). The LBA Near Eastern glasses analysed in the present study are believed to be fluxed with plant ash, and not mineral soda, because of their elevated magnesia and potash contents. Even with sensitive trace element data, it has not been possible to decipher the glass compositions in order to identify whether a glass has been made from a mix of ashes, or from a single plant ash. Hence the interpretation of the glass compositions is biased towards one with the minimum number of components.

The question of whether or not ancient glass-makers knew to add lime as a stabiliser has been discussed, but not resolved. The prevailing view has been that the ancient glass-makers did not recognise the value of lime, and that lime was added unintentionally as a component of the silica or plant ash (section 1.7). While lime might not have been added as a separate ingredient, the results of the present study indicate the possibility that its value, or the value of the raw materials which

[61] See Canti (2003: 341-347) for information on the chemistry of calcium compounds in plants, and pp. 347-355 for the chemistry of calcium and other components during ash formation.

introduced it, could well have been recognised by the Near Eastern glass-makers, who might have understood their raw materials more intimately than is generally perceived (sections 3.2, 4.2). It is possible that the addition of lime-bearing plant ashes was a well understood and carefully controlled part of a complex art.

While Shortland and Eremin (2006: 590) found that the lime content of plant ash did not correlate with any other base glass oxide introduced with it in the plant ash, Barkoudah and Henderson (2006: 317, 319-320), in their study of the composition of plant ashes from Syria, found correlations between lime, magnesia and potash in *Salsola* plants and noted that such correlations have been identified in some plant ash glasses. They observed a correlation between soda and lime in *S. vermiculata* in particular, and found that lime levels are high enough in 'some' ashes to provide enough lime to make the LBA glasses. In addition, they noted that correlations might be genus specific and dependent on soil geology where the plants were grown. Results of analyses of glasses from Nuzi, both from the present study, and from the study of Shortland and Eremin (2006: 595), indicate that there are correlations between lime and magnesia for instance, when the high- and low-lime glass groups are separated. To Shortland and Eremin (2006: 590-591), who did not recognise the lime groups, the lack of correlation when the glasses were examined as one group suggested, perhaps incorrectly, the deliberate addition of lime to the glass batch.

Shortland and Eremin (2006: 596) found that Lisht glasses were distinct from early Egyptian, Malkata and Amarna glasses because they have significantly lower lime concentrations, indicating that either a different plant ash was used to make these glasses, or that different amounts of lime were added as a controlled separate component. As the levels of potash and magnesia were similar between the glasses, Shortland and Eremin (2006: 591, 596-597) concluded that the lack of correlation of lime with other plant ash oxides and the difference in lime contents between Lisht and other Egyptian LBA glasses, indicated that lime, perhaps in the form of limestone, was deliberately added as a separate component in different amounts to the LBA Egyptian glasses. Shortland and Eremin considered it unlikely that the lime content of the plant ash could be internally consistent, and yet the lime content of the glasses could vary with no change in potash or magnesia. Soda levels in HMHK Malkata and Amarna glasses analysed by Shortland *et al.* (2007: 584-588) vary in average soda concentration from 16.7% to 18.9%. The examination of glasses from pre-Malkata Egypt, Malkata, Amarna and Nuzi by Lilyquist *et al.* (1993: 41) identified variations in the average concentrations of lime and potash between the glass groups, and soda varied between 17.4% and 19%. The variations in the content of soda, lime and potash in Egyptian glasses are difficult to explain by the addition of lime alone.

8.4.3 High- and Low-Lime Glasses from LBA Pella and Nuzi

Lime concentrations of the low-lime HMHK glasses from Nuzi range from 4.6% to 7.2%, and the lime concentrations of the high-lime glasses from Nuzi range from 8.5% to 11.5%. The Pella low-lime concentrations range more widely from 2.9% to 7.8%, and the high-lime concentrations range from 8.2% to 10.4%, similar to the findings for the glasses from Nuzi (sections 3.2, 4.4).

To show with confidence that the division into high- and low-lime groups of the glasses from Nuzi analysed in the present study is not the result of analytical variables, such as sample preparation or the chemical analysis itself, the compositional data from the analyses of a total of 44 glasses from Nuzi performed separately by Brill (1999a: 41-2; 1999b: 40-41) (n = 11), Shortland and Eremin (2006: 595) (n = 17) and the data from the present study (n = 16) are presented in the histogram in Figure 8-20. The data from these three studies concur to indicate clearly that there are both a high- and a low-lime group of glasses from Nuzi, with only one glass, from Brill's study, having a lime concentration between 7.5% and 8.5%.

The bimodal distribution of lime concentration is evident, although less striking, in the glasses from Pella (Figure 8-21) than in the glasses from Nuzi. The histogram of lime contents of Pella glasses shows a cluster of glasses with low-lime contents between 4.5% and 6.49%, and another cluster of glasses with high-lime contents between 9% and 10.49%.

On the basis of previous argument concerning the sources of lime in LBA glasses (section 1.7), the question arises as to whether the high-lime glasses are essentially low-lime glasses which had extra lime added to them, or whether a different plant-ash or combination of ashes was used to make the glasses in the high-lime group. Do the variations in composition between groups indicate glass manufacture by two workshops using different ashes? Does the extra lime in the high-lime glasses serve a purpose?

It has been established (section 3.2) that the two lime groups analysed in the present study do not appear to be directly related to opacification in the glasses from Nuzi, although the two opaque glasses analysed from Nuzi fall into the high-lime group. The remaining five out of seven high-lime glasses are translucent with low Sb contents, as are all of the glasses in the low-lime group. The high-lime glasses from Nuzi analysed

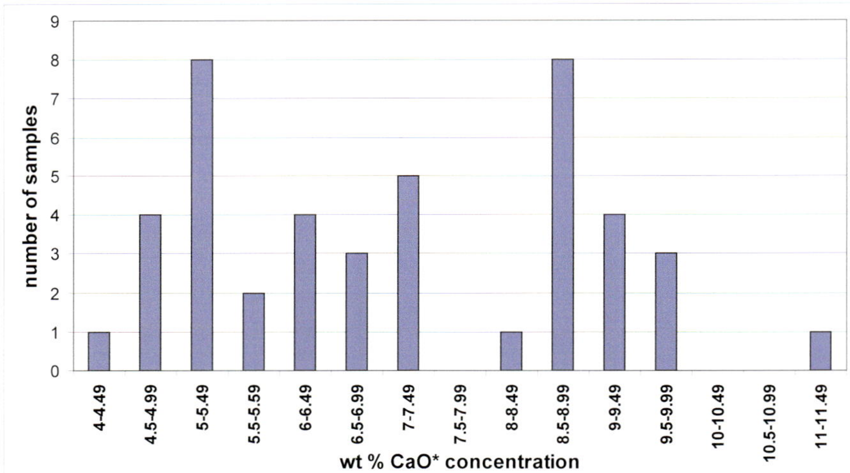

Figure 8-20 Histogram illustrating the bimodal distribution of samples in half per cent increments over the range of lime (CaO*) concentrations measured in wt % of 44 Nuzi glasses analysed by Shortland (2006: 595), Brill (1999b: 40-41) and samples analysed in the present study.

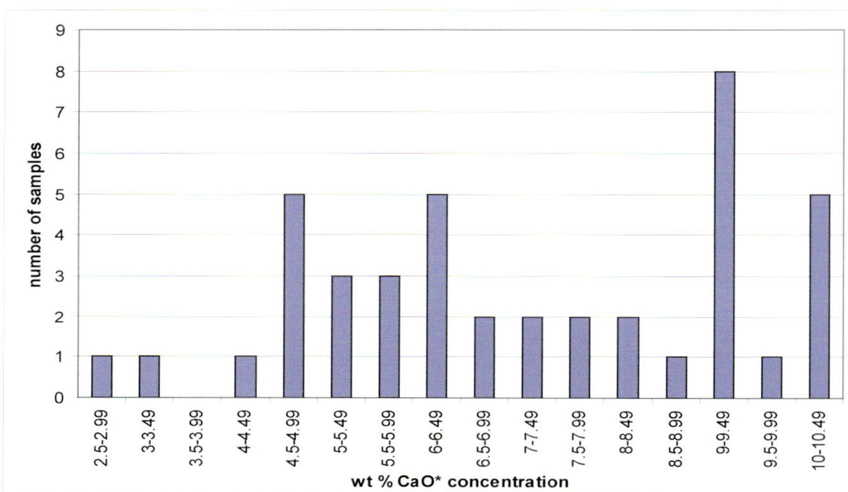

Figure 8-21 Histogram illustrating the bimodal distribution of samples in half per cent increments over the range of lime (CaO*) concentrations measured in wt% of 43 Pella HMHK glasses analysed in the present study.

by Brill[62] (1999b: 40) and Shortland and Eremin (2006: 595) similarly include both translucent, low Sb and opaque, high Sb examples, while the low-lime group is populated only by translucent, low Sb samples. In this regard, the data from all three studies are in agreement.

The high- and low-lime groups of glasses from Pella (Figure 8-22) appear to be more closely connected with opacification (high-lime) and translucency (low-lime). Intriguingly, one opaque glass with a high Sb content (sample 2-22) was made with a low-lime concentration, albeit at the upper limits of the low-lime group, with a lime content of 7.24%, and one translucent glass was made with a high-lime concentration (sample 194). Curiously, sample 209 has a high-lime content with a relatively low Sb content (4,237 ppb), but is nevertheless opacified.[63] These anomalous compositions show that glasses could have high-lime levels and be translucent, or could equally be opacified with a low-lime content, although it appears that the glass-makers might have chosen not to make opaque glasses with lower lime, even though they could (sample 222). From Figure 8-22 it is clear that there is no correlation between the

62 Sb concentrations were given for the opaque glasses only.

63 Other characteristics can cause opacity, such as bubbles and unmelted batch material (section 1.10.2), but the sample is unavailable to check.

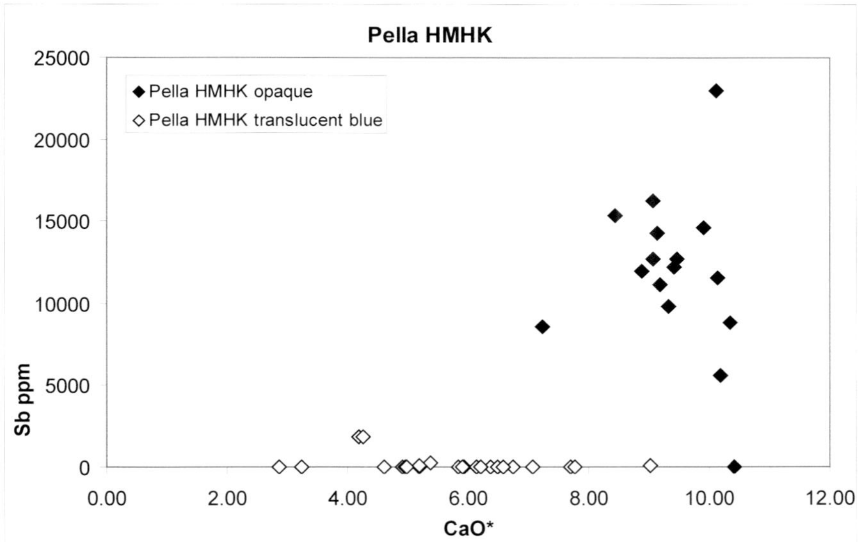

Figure 8-22 Scatter plot of lime *vs* Sb for Pella HMHK blue glasses. The low Sb opaque glass is sample 209, the low-lime opaque glass is sample 222, and the high-lime translucent glass is sample 194.

lime and Sb concentrations in opaque glasses, which supports the suggestion that these components were not added together during the manufacture of the opaque glasses (see sections 1.10.2, 3.4, 4.5.4).

Thirty-five glasses from Tell Brak were analysed by Brill (1999b: 39), Henderson and Velde (in Henderson 1997: 96-98) and Shortland and Eremin (2006: 595). All but one of the glasses analysed by Brill were opaque, and were of both high- and low-lime content. All glasses analysed by Shortland and Eremin were opaque, with the majority being of low-lime concentration. Translucent glasses analysed by Henderson and by Velde were of low-lime composition.

Although high- and low-lime glass groups are clearly present at Pella, Nuzi and Tell Brak, their correlation with the opacity or translucency of the glasses is clouded by the high- and low-lime contents of the opaque glasses from Tell Brak. Whether this suggests that lime was used differently by the makers of glasses from Tell Brak than by the makers of glasses from Nuzi and Pella is an area requiring further study.

Analyses of 283 Egyptian HMHK glasses from Amarna, Malkata, Lisht, Wadi Qirud and Timna, and pre-Malkata glasses from the reigns of Hatshepsut to Thutmosis IV (Lilyquist *et al.* 1993: 36-39; Brill 1999b: 27-35; Nicholson and Henderson 2000: 215; Shortland and Eremin 2006: 584-588; Jackson and Nicholson 2007: 181-182; Freestone unpublished data) include many high-lime translucent glasses and many low-lime opaque glasses, using the parameters defined by the lime contents of the Near Eastern glasses (sections 3.2, 4.4). It is tempting to suggest that some of the Near Eastern glass-makers, whose products were found at Pella and Nuzi, were consciously using, and hence showing a preference for, low-lime contents to make translucent glasses and high-lime contents to achieve opacity, even though they could have made opaque glasses with low-lime and translucent glasses with high-lime concentrations, as is shown by the contemporary Egyptian glass compositions, and by the opaque glasses from Tell Brak. The reason for these apparently different practices by the Near Eastern glass producers, within the parameters of high- and low-lime groups is unclear. Perhaps a higher lime content was preferred by some glass-makers to ensure crystallisation of the opacifying compound calcium antimonate ($Ca_2Sb_2O_7$). It is impossible to know whether the reason was technological, or was related to other factors, such as the dictates of ritual practice. Ritual procedures involved in ancient glass-making are mentioned in the Assyrian glass-making texts, but there is no specific reference to lime content (Oppenheim *et al.* 1970: 31-32, 44-45, 47, 52; section 1.3).

It is curious that these choices, if technological, are more clearly illustrated at Pella than at Nuzi or at Tell Brak, where the connection between lime content and opacity/ translucency is less obvious. No explanation for this can be offered at present. The data for lime concentrations for the glasses from Pella, Tell Brak and Nuzi are combined in Figure 8-23 to show clearly, with a more substantial data-set, the two lime-based groups of Near Eastern glasses.

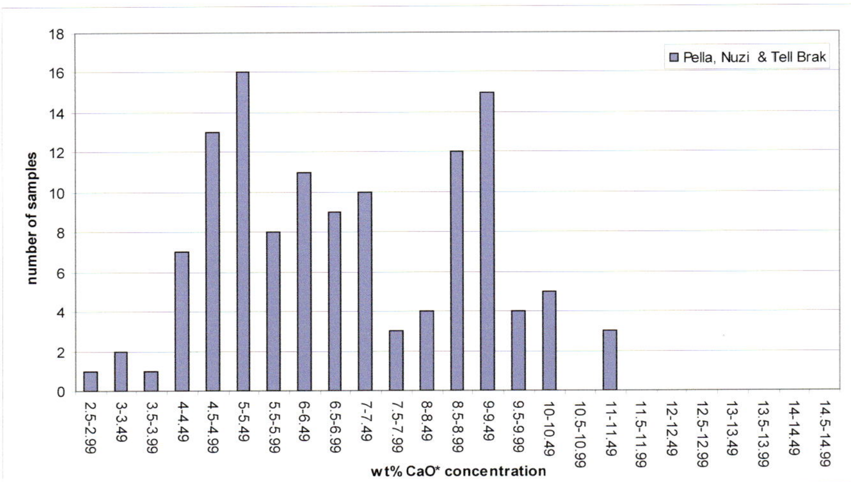

Figure 8-23 Histogram illustrating the bimodal distribution of samples in half per cent increments over the range of lime (CaO*) concentrations measured in wt% of 124 Near Eastern glasses. Nuzi glasses n=46, Pella glasses n=43, Tell Brak glasses n=35. Sources cited in the text.

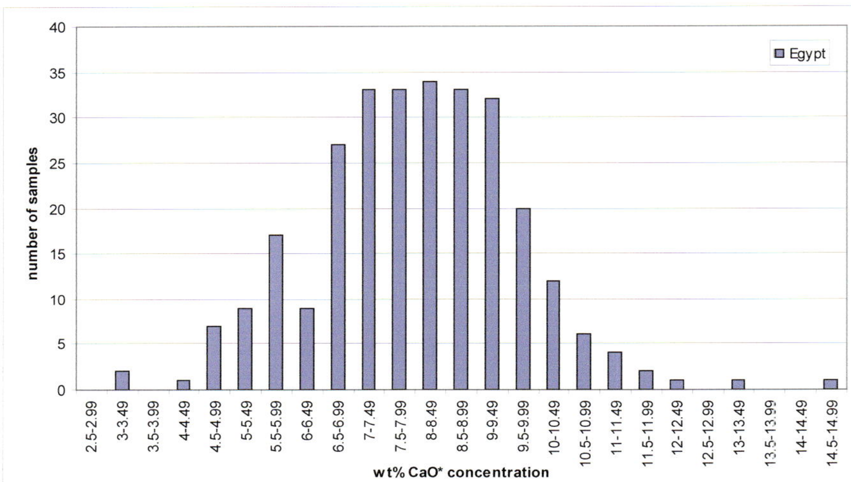

Figure 8-24 Histogram illustrating the distribution of samples in half per cent increments over the range of lime (CaO*) concentrations measured in wt% of 283 Egyptian HMHK glasses. Sources cited in the text. Although Lisht glasses and some pre-Malkata glasses have lower average lime contents than other Egyptian glasses, there is some overlap between the ranges of concentrations for these and other Egyptian glasses, which forms the continuum of lime concentrations plotted in this figure.

Figure 8-24 presents the lime concentration data from analyses of LBA Egyptian HMHK glasses compiled from several analysts, cited above, as a comparison with the Near Eastern data. It is clear from this analysis that the Egyptian glass-makers used a similar range of lime concentration to make their glasses, but the distribution of lime concentrations is not bimodal, which was suggested by this plot of lime against magnesia. The Egyptian glass-makers opacified glasses with both high- and low- concentrations of lime within the range they used (generally above 5% lime). This is by direct contrast with the practice of their Near Eastern counterparts at Pella, Nuzi and Tell Brak who used high- and low-lime concentrations to make their glasses, and made opaque glasses predominantly with high-lime concentrations at Pella and Nuzi.

The reasons for the different approaches to the use of lime in glasses from Egypt by comparison with those from the Near East are difficult to determine. It appears that the Near Eastern glass-makers were deliberate in their choice of lime-bearing raw material, whereas the Egyptian practice was non-specific. An examination of available glass compositional data reveals that opaque and translucent glasses could be made with concentrations of lime above approximately 5%, within the range of lime concentrations used to make LBA glasses.[64] Does this suggest that the Near Eastern craftsmen were aware of the different lime contents of their raw materials? If glass could be opacified with both high- and low-lime concentrations, what was the significance of the two lime groups?

The two lime-based groups of glass from the Near Eastern region could represent the products of more than one glass manufacturing site, although the glasses are otherwise chemically nearly identical. Such uniformity of production from more than one workshop would perhaps be difficult to reconcile unless it is accepted that the production of glass, including raw material sources and processing, and glass recipes, were strictly controlled by state and/or temple authorities (section 1.3). Each lime group contains both translucent and opaque glasses making a separation of workshops based on glass opacification seem unlikely.

The small difference in the concentrations of the SRE and the REE of the opaque and translucent glasses from Pella and Nuzi suggests that there is a difference in raw material proportions, or types, for these two kinds of glass, regardless of lime content. Does this represent the use of different recipes for making opacified and translucent glasses? No trace element data have been published for translucent glasses from Tell Brak, making it impossible to compare their REE and SRE contents with those of the opaque glasses. That the high- and low-lime groups from Pella were contemporary, and not the products of different chronological periods has been established in section 4.4. It seems reasonable to assume that the same is true for the glasses from Nuzi and Tell Brak.

Does the lack of bimodality in the lime content of Egyptian glass suggest that all LBA Egyptian HMHK glasses were made in one workshop? In the light of work by Rehren and Pusch (2005: 1757) and Nicholson (2007), this is unlikely (section 8.3.3.1). Another significant point which arises from the comparison of Egyptian and Near Eastern glasses is that, apart from subtle differences identified above in their base glass compositions (sections 8.3.3 and 8.4), the Egyptian and Near Eastern use of lime was not the same. The different use of lime between the two regions implies that not all of the Egyptian glasses could have been imported from the Near East, a finding which supports the growing evidence for local Egyptian production of HMHK glass in the New Kingdom (section 1.3). The lime data are not sufficient on their own to classify individual glasses as Egyptian or Near Eastern, because both regions use the same range of lime concentrations, although, based on current evidence, opaque glasses with low-lime content are not likely to have been products of the workshops which produced the glasses from Pella and Nuzi, but they could have come from the workshop which made glasses found at Tell Brak.

8.4.4 The Relationship of Lime to Other Oxides and to Sr

Having established that there are two lime-based groups in LBA HMHK Near Eastern glasses, it is of interest to examine the relationship of lime to the other base glass oxide concentrations to ascertain whether lime was being added as an individual ingredient, or as part of the plant ash component of the glasses. While there is little apparent difference in the absolute concentrations of most oxides between the low- and high-lime groups, both magnesia and potash concentrations are elevated on average in the low-lime glasses, and soda levels are slightly elevated, while phosphorus, silica, alumina and iron oxide levels remain relatively constant between the two groups (sections 3.2, 4.4).

The narrower range of magnesia concentrations in the high-lime groups from both Pella and Nuzi (2.5% to 8% in the low-lime, compared with 2% to 5% in the high-lime) supports the proposition that the alkali source of the two groups is somehow different, and that a high-lime concentration was not simply achieved by the addition of lime to a low-lime glass batch. If limestone or seashells were added to the low-lime glasses to achieve the concentrations of lime found in the high-lime glasses, it could be expected that the range of magnesia content would remain the same. If dolomitic limestone were added, the concentration of magnesia would increase with the lime, but it does not. If extra lime were added, only the lime content would change relative to the content of the other oxides, but this also is not the case.

Sr appears to be elevated in LBA HMHK glasses analysed in the present study by comparison with the SBS level established for these glasses (sections 3.3.1, 4.5.1). The concentration of Sr in high-lime glasses falls in a similarly restricted range to that of magnesia (Figure 8-4, Figure 8-25, Figure 8-26), with a greater range of Sr concentrations in the low-lime glasses (Nuzi 238 ppm to 447 ppm, av. 355 ppm, and Pella 168 ppm to 682 ppm, av. 380 ppm) compared with that in the high-lime glasses (Nuzi 384 - 544 ppm, av. 441 ppm, and Pella 266 ppm to 617 ppm, av. 385 ppm). The Sr to lime ratio is lower in the high-lime glasses.

To explore further the relationships of those oxides considered to be associated with plant ashes, the concentrations of the oxides of magnesium, potassium,

[64] Glasses from both Egypt and Tell Brak were opacified with as little as 5% lime content (Henderson 1997; Shortland and Eremin 2006). Glasses with a lime concentration below this amount are translucent in all groups examined in the present study.

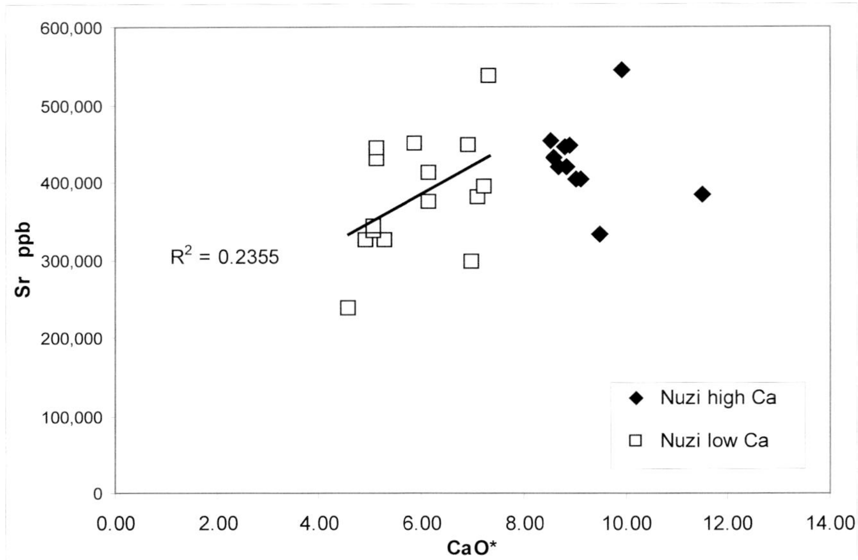

Figure 8-25 Scatter plot of lime *vs* Sr for high- and low-lime glasses from Nuzi, represented by the combined data for 28 samples from Shortland and Eremin (2006: 595) and the present study.

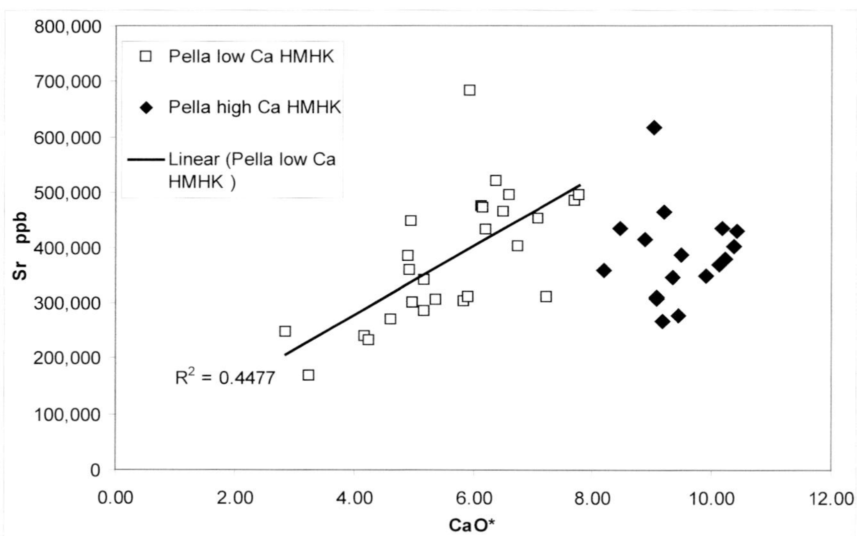

Figure 8-26 Scatter plot of lime *vs* Sr for LBA high- and low-lime glasses from Pella, from the present study.

calcium and sodium were re-summed to 100%. The data for the high- and low-lime groups are compared in Figure 8-27 (Nuzi) and Figure 8-28 (Pella), and show that where lime levels have increased, the concentrations of magnesia, potash and soda have decreased in the glasses from both sites.

Was lime added as a separate ingredient or as part of a plant ash to make the high-lime glasses? An examination of the ratios of the averaged concentrations of the base glass oxides from the high- and low-lime groups, presented in Table 8-3, attempts to clarify their relationships to each other, and hence to establish whether lime in the high-lime group is related to other base glass oxides and to the relative decreases in the concentrations of magnesia, potash and soda.

The ratios of base glass oxides in glasses from Pella and Nuzi are similar, although the differences in the ratios of soda to both magnesia and to potash are smaller

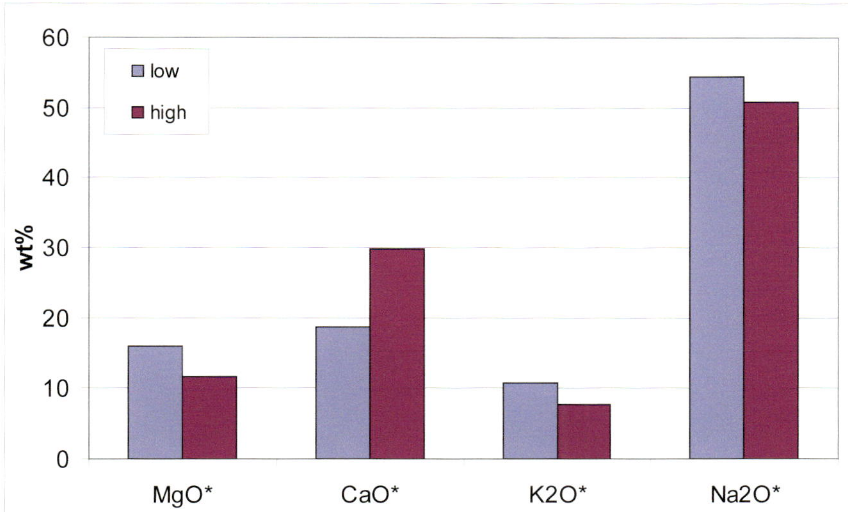

Figure 8-27 Histogram comparing the concentrations in wt% oxides of plant ash associated elements in the high and low lime groups at Nuzi. Data for these oxides are re-summed to 100%.

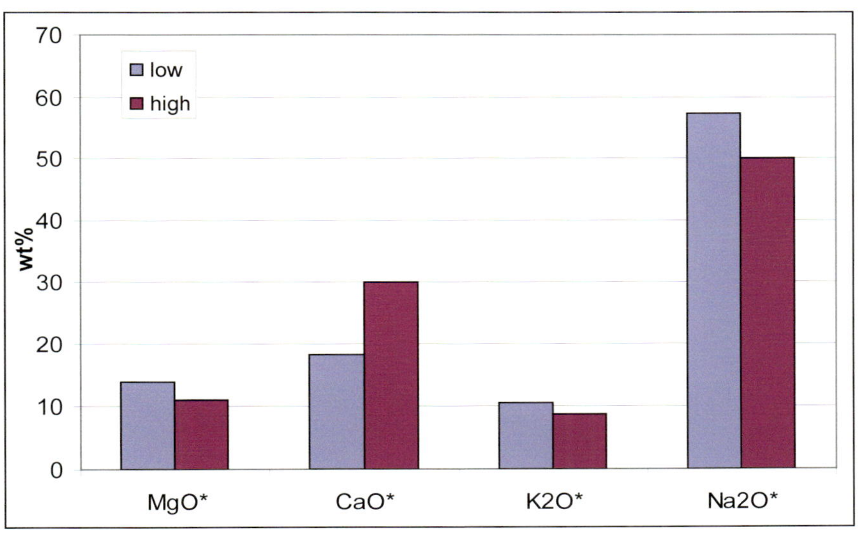

Figure 8-28 Histogram comparing the concentrations in wt% oxides of plant ash associated elements in the high and low lime groups of Pella HMHK glasses. Data for these oxides are re-summed to 100%.

in the glasses from Pella, than in the glasses from Nuzi. The differences in the ratios of lime content to other oxides for both glasses from Pella and Nuzi are obvious (Table 8-3). The ratios of the concentrations of lime to the other oxides in the high-lime glasses are approximately twice those found in the low-lime group, and the ratios of soda to other oxides differ between high- and low-lime groups. The ratios of the non-plant ash related oxides, alumina and iron, remain closely similar in the Pella and Nuzi glasses, as do their absolute concentrations. The ratios of magnesia to potash are similar between the high- and low-lime groups at both sites despite the differences in absolute concentration between the groups.

It is difficult to determine whether the differences in the ratios of oxides other than lime are statistically significant such that they indicate the use of another plant ash which increased the lime levels, or whether the increase in lime content was due to the addition

Table 8-3 Average ratios of oxides in low- and high-lime glasses from Nuzi and Pella glass data from the present study

	MgO^*/K_2O^*	CaO^*/MgO^*	CaO^*/K_2O^*	CaO^*/Na_2O^*	CaO^*/SiO_2^*	Na_2O^*/MgO^*	Na_2O^*/K_2O^*	$Al_2O_3^*/FeO^*$
Nuzi Low lime *n=9*	1.51	1.18	1.78	0.35	0.08	3.46	5.20	2.21
Nuzi High lime *n=7*	1.57	2.55	4.00	0.59	0.14	4.37	6.83	2.09
Pella Low lime *n=26*	1.35	1.31	1.77	0.32	0.08	4.06	5.48	1.99
Pella High lime *n=17*	1.28	2.68	3.43	0.60	0.14	4.48	5.73	1.81

of lime alone. If the high-lime group is the result of adding lime to the low-lime group, then the changes in the ratios of the concentrations of soda to each of lime, magnesia and potash would be difficult to explain.

The ratios of lime to magnesia and of soda to potash are deemed to be those that dictate the suitability of a plant ash for glass-making (Tite *et al.* 2006: 1286-1291). The ratios of soda to potash of 5.2:1 (Nuzi) and 5.5:1 (Pella) in the low-lime group, and 6.8:1 (Nuzi) and 5.7:1 (Pella) in the high-lime group, fall within the main range determined by Tite and Shortland (2006: 1289) for the plant ashes they analysed from the Near East. Assays of the ashes from this region show that the soda to potash ratios, and percentages of alkali present as carbonates, are variable (see also Barkoudah and Henderson 2006), and hence might be responsible for the variations observed in the HMHK glasses.

8.4.5 Statistical Significance of Oxide Contents of the High- and Low-Lime Glasses

To ascertain the statistical significance of the differences between the oxide concentrations of the two lime groups from both Pella and Nuzi, a two-tailed unpaired t-test (see section 2.5.5) applied to this data gave a value of $p = 0.0001$ when the mean lime contents were compared. The difference observed in lime concentrations between the two groups was highly statistically significant, both when the Nuzi data of individual researchers were analysed separately, and when all the data were combined and analysed together. The difference in the average lime data for the high- and low-lime groups of Pella glasses was equally significant. Results of the t-tests applied to the means of the base glass oxides from both Pella and Nuzi glasses are presented in Table 8-4. The t-test was applied to the data for glasses from Nuzi from each individual researcher (Shortland and Eremin 2006; Brill 1999b; and the present study) and then to the combined data for glasses from Nuzi.

When averaged data for oxide concentrations from the present study of glasses from Nuzi were subjected to a two-tailed unpaired t-test, differences between the concentrations in the high- and low-lime groups of lime, magnesia and potash were shown to be statistically significant (Table 8-4). Differences in the average concentrations of soda and phosphorus between the high- and low-lime groups were significant when Shortland and Eremin's data were tested, but not when data from the present study were tested. Differences in the concentrations of soda, potash and magnesia were not significant when Brill's data were analysed.

In addition, the combined data, which provide a larger number of samples for Nuzi glasses from all analysts, were subjected to the t-test. Differences between the results from individual researchers might be due to analytical error, or sensitivity of the analytical technique used, and could be influenced by the smaller number of samples analysed, which increased the significance when the results were combined. The differences in the mean values of lime, potash and magnesia content were all found to be statistically significant, but soda and phosphorus concentrations were not significantly different.

When mean data for Pella glasses from the present study were subjected to the t-test, the differences in

Table 8-4 Results of two-tailed unpaired t-tests for Nuzi and Pella high- and low-lime group oxide data*.

	n total (high/ low lime)	Confidence Interval	p CaO*	p Na_2O*	p MgO*	p K_2O*	p P_2O_5*	p Al_2O_3*	p FeO*	p SiO_2*
Nuzi (present study)	16 (7 / 9)	95%	<0.0001	*=0.4172*	=0.0064	=0.0147	*=0.5197*	*=0.5684*	*=0.7639*	*=0.4359*
Nuzi (Shortland)	17 (5 / 12)	95%	<0.0001	=0.0171	=0.0042	*=0.2123*	=0.0238	*=0.7024*	*=0.9753*	*=0.6073*
Nuzi (Brill)	11 (5 / 6)	95%	=0.0003	*=0.9120*	*=0.1468*	*=0.6656*	no data	*=0.3882*	*=0.5391* (Fe_2O_3)	*=0.2639*
Nuzi (Combined)	44 (17 / 27)	95%	<0.0001	*=0.2022* (present study and AS data only p=*0.4558*)	=0.0097	=0.0382	*=0.5984*	*=0.3915*	*=0.7703*	*=0.2616*
Pella (present study)	43 (17 / 26)	95%	<0.0001	=0.0033	=0.0067	=0.0153	*=0.5227*	*=0.1245*	*=0.1303*	*=0.6027*
Nuzi summary			√	X	√	√	X	X	X	X
Pella summary			√	√	√	√	X	X	X	X

* $p < 0.05$ is considered to be statistically significant (√). *$p > 0.05$ is considered to be not statistically significant (X).*

concentrations, between high- and low-lime groups, of lime, magnesia, potash and soda were found to be statistically significant. Phosphorus was measured with SEM-EDS at low concentrations and was shown not to be significantly different.

In both Nuzi and Pella glasses, silica, alumina and iron oxide concentrations were not significantly different between the high- and low-lime groups, reflecting that silica concentrations do not change, while those of plant ash related constituents do. Iron oxide and alumina are sediment-related contaminants in the glass, and their concentrations remained unaffected by the changes in alkali and alkaline earth oxides in the two lime groups. This accords with the observation that the REE and SRE contents of high- and low-lime glasses are the same in glasses from both Nuzi and Pella (section 8.2.1).

The significant differences in the concentrations of plant ash related oxides, besides lime, suggest the use of different ashes, or a mix of ashes, between each of the high- and low- lime groups. These differences cannot be due to the addition of lime to the low-lime group to achieve a high-lime composition.

If the observed differences in the oxide concentrations are due to the use of a different plant ash, or ash mix, it would have to be a particularly high-lime ash with relatively less soda, magnesia and potash than that used to make the low-lime glasses. It is not possible, however, to distinguish the chemical signature of an individual plant ash from that of a mixture of ashed plants, especially after the plants have undergone the modifications and possible contaminations encountered during the ashing and glass-making processes (see Canti 2003; section 1.8).

The difference in the soda content of the two lime groups from Pella were significant, but they were not significant in the glasses from Nuzi. It is interesting to consider that if the glass-makers were trying to maintain particular concentrations of silica and soda, as the two prime base glass ingredients, the concentrations of these oxides might remain relatively similar between the glass groups. Is it possible that craftsmen found two ashes, or ash mixes, which delivered similar soda content with the required concentration of lime? The relative constancy of the soda content of Near Eastern LBA glasses suggests that the glass-makers could have known which ashes would provide the required amount of soda to make the glass in the first place. Does it follow that they knew which plants produced high- or low-lime ashes? The weight of the evidence presented here lends support to the proposition that they could have understood the subtleties of different ash compositions.

When the data for the glasses from Nuzi and Pella are examined, the addition of lime as a separate ingredient to the low-lime group to create high-lime glasses seems unlikely. This conclusion is based on the changes observed in the ratios not only of lime, but of soda to the other oxides between the high- and low-lime groups, on the significantly different concentrations of lime, magnesia and potash in the high- and low-lime glasses from both Nuzi and Pella, and on the significantly different concentrations of soda in the glasses from Pella. The concentrations of Sr, though not subjected to the t-test, appear also to be different between the groups, as would be expected because Sr behaves chemically in the same way as calcium (Figures 8-25, 8-26). Altogether, the observed variations between high- and low-lime glasses indicate that the lime-bearing material in both groups was plant ash.

8.5 Conclusions

If the high- and low-lime groups were the products of two or more different workshops, which used local ashes of different alkali content, several questions arise. Why were the colours of glass so limited? Why were the opacified glasses from Nuzi and Pella made almost exclusively with high-lime contents, while at Tell Brak there was a greater range of colours and there was no connection between opacity and high-lime content? Why, regardless of opacity, do the glasses from Tell Brak fall into the same two lime groups? These questions are, as yet, difficult to answer.

If both lime groups were the products of one workshop, it could be that their compositional differences were due to seasonal variations in the composition of the plant ashes used. If this were so, a range of compositions from low- to high-lime might be expected, and not the two distinct groups found in the present study (section 1.8).[65] Hence the existence of two lime groups based on seasonal plant ash variation is improbable. As a sophisticated occupation of skill and experience, glass-making was probably continuous rather than seasonal. Moorey (1999: 195), amongst others, discussed the likelihood that LBA glass workshops were based in courts and/or temples where a variety of artisans would have worked, most probably full-time (section 1.3). The traditional glass-makers of Herat in Afghanistan, as recorded by Brill (1979, *The Glass Makers of Herat* video), practice their craft continuously, and not seasonally.[66]

[65] Studies by Smedley and Jackson (2006) and by Jackson and Smedley (2008) have revealed the graduated increase and decline in alkali and alkaline earth oxide concentrations in bracken, during its growth cycle. Although bracken was used to make plant ash for glass production in Europe, it was not the plant type used to make plant ash for ancient Near Eastern glass production. Such studies have not been published on Near Eastern plants proposed for ancient glass-making, but the patterns of seasonal growth are important to consider.

[66] Unfortunately, few contemporary traditional glass workshops have been recorded in the Middle East, and extrapolations from comparing them with ancient glass workshops should be undertaken with extreme caution. The modern artisans who fashion objects from

Although this is not direct evidence for patterns of ancient glass production, it is tempting to consider that the practices of the ancient glass-makers might have been similar to those of the Herat craftsmen, but within the regulated environment of the palace and temple.

It is important to note that the LBA glass object forms are limited and universal (section 1.4), and are not related to the opacity or to the lime content of the glasses. If two or more workshops were operating simultaneously, it is tempting to suggest that the overall similarity of the glass object shapes produced and of the composition of the two lime-based glass groups indicates that LBA glass-making workshops used similar recipes within a relatively homogeneous and tightly controlled technological framework. It is proposed here that LBA glass-making, in general, followed such strict protocols that it left little to the choice or to the individual creativity of the craftsmen for many generations. The glass-makers appear to have been bound by an industry that was governed closely by state and temple because essentially its purpose was to produce highly valued and commercially competitive imitations of precious stones. These artificial stones were valuable also because they were imbued with ritual and healing powers, and were requested by rulers (for example in the Amarna Letters; Robson 2001; section 1.3) and buried with kings, for example Thutmosis III (Tatton-Brown and Andrews 1991: 26-27).

Glass production was consequently confined by prescribed rituals, some of which are recorded in the neo-Assyrian glass-making texts, which, although they date to the 7th C BC, are thought to have originated in the earlier LBA (Oppenheim 1973a: 261; section 1.3). Oppenheim (Oppenheim *et al.* 1970: 6) noted that the glass texts themselves were, like all cuneiform documents, subject to stylistic requirements and 'their wording and their literary form were historically conditioned', perhaps paralleling the nature of the glass-making practices they described.

Whereas Oppenheim (Oppenheim *et al.* 1970: 7) viewed the glass craftsmen as independent and itinerant artisans, the highly regulated and homogeneous physical and chemical characteristics of LBA Egyptian and Near Eastern glass products do not reflect the work of individuals. Oppenheim (Oppenheim *et al.* 1970: 6-7) formed this view because the cuneiform administrative texts do not mention glass-makers. The texts refer to other occupations, such as the lapidary and metal smith, leading him to conclude that the glass-maker was independent of valuable materials and did not require substantial infrastructure. Evaluation of the available data indicates that some raw materials, such as Co alum used to colour Egyptian glasses during this time, could have been considered to be valuable because they do not occur commonly. Moreover, even if all the glass-makers' raw materials were not precious, the glass-makers' products certainly were, and for this reason the apparently regulated art of glass-making was unlikely to have been the preserve of independent individuals.

If Near Eastern glasses were produced by one workshop, it was one which inexplicably made both high- and low-lime, translucent and opaque glasses and is an aspect of glass-making that requires further research. On the basis of both the compositional and the archaeological evidence presented in this study, it is considered highly probable that there were at least two glass-making workshops, which operated simultaneously, and which used different plant ashes to produce both high- and low-lime glasses. It appears that in the Near East, the glasses from the sites of Nuzi and Pella were characterised by low-lime translucent, or high-lime, both translucent and opaque compositions. At Tell Brak, however, the picture is different. While the compositional evidence shows that there are the same two groups based on lime content, the low-lime glasses included both translucent and opaque examples, whereas the high-lime glasses were opacified. The reason for this variation might be sought in the produce of more than one workshop, which used different sources of similar raw materials, but according to a specific recipe. It seems reasonable, given the nature of the glass industry as currently understood, that the involvement of practices not related to technology, but perhaps to ritual dictates have influenced the choice of raw materials and glass-making processes.

The closely similar trace element compositions of both high- and low-lime glasses from Nuzi and Pella suggest that raw materials from a similar geological region were used in standard, well-controlled recipes. For reasons which are difficult to establish, however, the glass-makers used two different sources of plant ashes, which were both used to make opaque and translucent glasses. The compositions of opaque and translucent glasses exhibit small variations in REE and SRE signatures.

The nature of LBA glass-making has been explored and many significant insights have been gained, but the enigma of the varying lime concentrations in LBA Near Eastern glasses remains just that. It is hoped that further research will eventually clarify the questions raised here about LBA glass-making.

recycled glass at Da'our in Cairo work all year round, but those who, until recently, made glass in Hebron in Jordan only worked for six months of the year (Fischer 2008: 97, 103).

Chapter 9

Iron Age Glasses – Comparisons and Interpretations

9.1 Introduction

There is little archaeological or textual evidence for the production of glass during the IA in either the Near East or in Egypt. The chemical analyses of the IA glasses from Pella and Nimrud, undertaken in this study, have provided important information to fill gaps in current knowledge of 1st millennium BC glass production. The insights gained into this early era of glass-making are a valuable addition to our understanding of the development of this industry.

To this end, in this chapter the compositional characteristics of the IA glasses from Pella and Nimrud, identified in this study, are discussed. The analytical results are placed in the broader context of glass compositions from the Mediterranean region, as far as they are currently known. Data from the few published analyses of IA glasses from the Near East[67] (Hasanlu), from France and from Egypt (Nesikhons) are compared with compositional data for the glasses from Pella and Nimrud. The implications of the glass compositions are discussed in terms of continuity and innovation following the period of LBA glass-making. In particular, the procurement and use of raw materials for glass-making, and the wide distribution of glasses at this time, are addressed. The resulting synthesis of information from comparable studies is intended to provide a more comprehensive understanding of glass production in the Near East during the IA.

9.2 IA Base Glass Compositions

The regular production of glass was re-established in the Near East early in the IA, after the period of decline at the end of the LBA. Production of IA glasses involved the use of fluxes low in magnesia, potash and lime compared with those used in LBA HMHK glasses. Mineral soda, such as natron, was introduced, while plant ashes continued to be used. From the compositional data obtained from the IA HMHK glasses, it appears that the plant ashes were of different types to those used previously in the LBA. The use of calcareous sand-based silica sources, as an alternative to pure quartz pebbles, would have been necessary to provide lime for the mineral soda glasses. Near Eastern glass-makers of the IA had to modify the technological practices of the LBA to achieve suitable compositions when using new combinations of raw materials that were perhaps less well-known to them. That this adaptation took time to implement is illustrated by the unstable, low-lime compositions of the LMLK glasses of Nesikhons (Schlick-Nolte and Werthmann 2003), and new glass formulations were developed in a more diversified industry.

Four HMHK glasses, closely resembling the LBA compositional type, were found in IA contexts at both Pella and Nimrud (Group 2). It is tempting to conclude from this finding that the HMHK technology of the LBA Near East continued into the IA, although the number of HMHK glass beads unearthed from IA contexts is much reduced by comparison with the numbers found in the LBA (section 4.2, Table 4-2). Typical LBA plant ash glass continued to be made into the IA at 1st millennium BC Hasanlu (north western Iran) and at many sites in France where HMHK glasses are well represented (see especially Brill 1999a, 43-4; Gratuze 2002; Stapleton 2003).

Three of the four IA HMHK glasses from Pella are from questionable stratigraphical contexts, suggesting that they were in fact LBA glasses, and that at Pella, the LBA HMHK glass-making tradition did not continue into the IA (section 4.2). The only other glasses unearthed from secure IA contexts at Pella were the LMLK black glasses.

It seems reasonable to speculate that the four typical LBA HMHK glasses from Nimrud, which do not have detailed stratigraphical information, were likely also to be LBA artefacts. These glasses (Group 2) are distinctly different from the other groups of Nimrud glass, including the Group 1 glasses which are not, despite being HMHK glasses, of the typical LBA base glass composition (section 6.4.1).

The Nimrud Group 1 glasses are more similar to the LBA plant ash glass tradition, but their composition indicates that a different recipe, or raw materials, such as different plant ashes, were used. The Group 1 glasses contain lower average levels of soda and higher average levels of lime than the LBA glasses, or the glasses of Nimrud Group 2, LMLK Group 3, Co blue Group 4, the Pella LMLK glasses and the glasses from the tomb of Nesikhons (Schlick-Nolte and Werthmann 2003) (Figures 9-1, 9-3).

[67] Glasses from IA Marlik Tepe in Iran, analysed by Brill (1999a: 45), were weathered and not included as comparative material in this study.

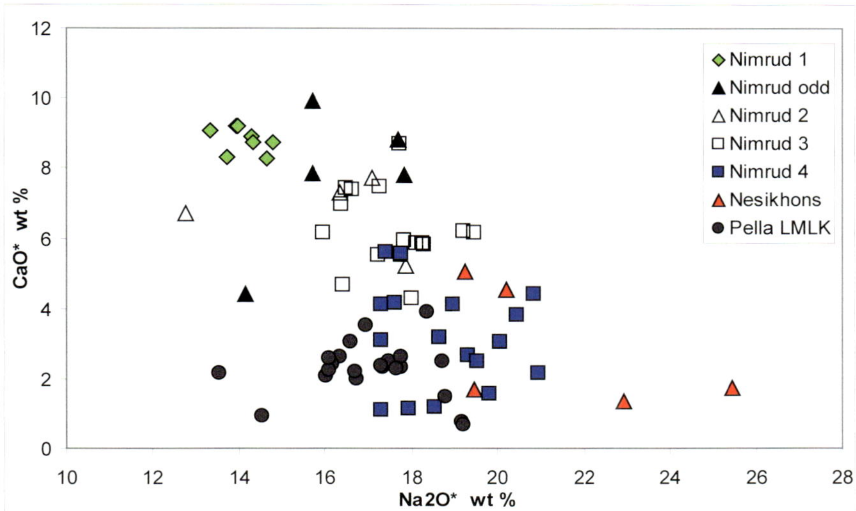

Figure 9-1 Scatter plot of soda *vs* lime for IA glasses from the Near East and Egypt. Data for Nesikhons' glass from Schlick-Nolte and Werthmann (2003: 26). Pella and Nimrud data are from the present study.

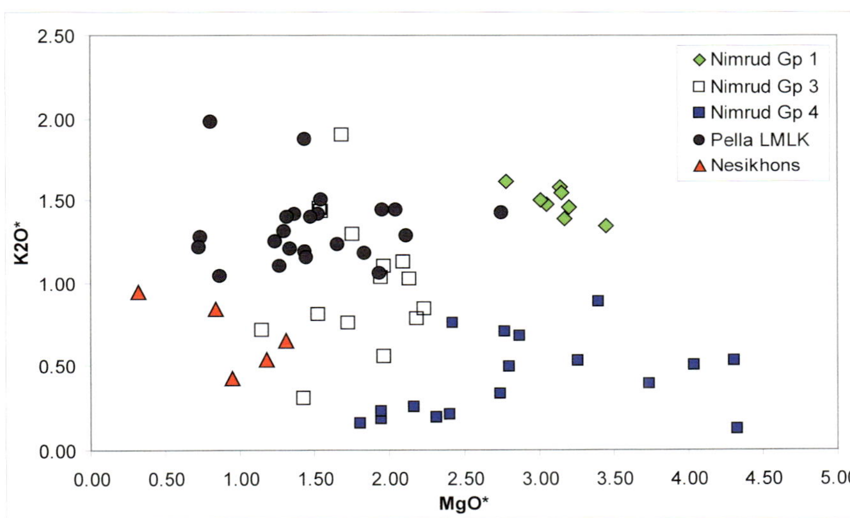

Figure 9-2 Scatter plot of magnesia *vs* potash for low potash IA glasses from the Near East and Egypt. Data for Nesikhons' glass from Schlick-Nolte and Werthmann (2003: 26). Pella and Nimrud data are from the present study.

The plot (Figure 9-1) of soda and lime contents for the IA HMHK glasses shows that Groups 1 and 2 have relatively high lime contents by comparison with the other IA glasses, and lower soda contents, while Group 3 has higher lime contents than the remaining LMLK glasses of Nesikhons, Nimrud Group 4 and Pella LMLK.

While the magnesia and potash levels of the Group 1 glasses are lower than those of Group 2 HMHK glasses, the Group 1 magnesia levels are higher than those of the LMLK groups (excepting Group 4 which has magnesia levels boosted by the alum colourant) (Figure 9-2). Group 1 glasses have a relatively low average potash level by comparison with glasses of the LBA HMHK tradition (section 6.4).

Although the Group 3 glasses have low levels of potash and phosphorus, their magnesia contents are, by definition, high for LMLK glasses. Compared with LBA plant ash glasses, they are categorised as being of a marginal LMLK type (sections 1.5, 1.9, 6.4.3). It appears that the Group 3 glasses were made with a soda flux from a natron-type evaporitic source because their chemical compositions imply that they were made with

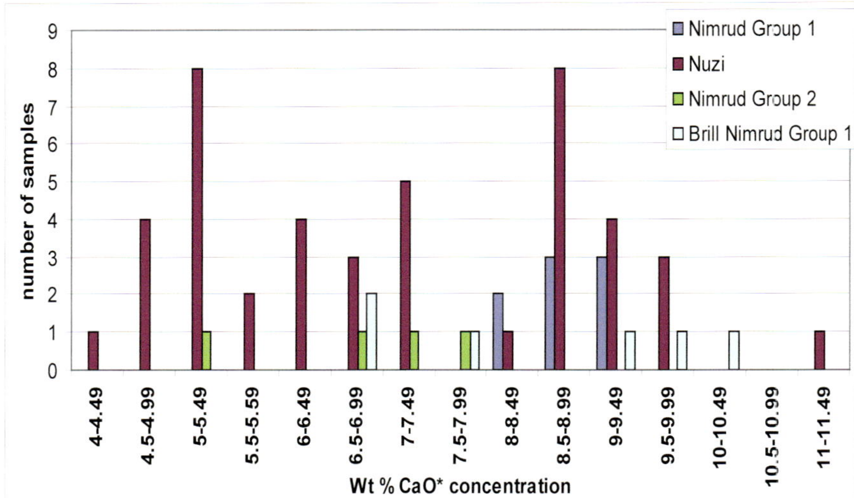

Figure 9-3 Histogram comparing the bimodal distribution of lime in Nuzi HMHK glasses with the lime contents of Group 1 and Group 2 HMHK glasses from Nimrud using data from Brill 1999b (Nuzi and Nimrud), Shortland and Eremin 2006 (Nuzi) and the present study (Nuzi and Nimrud).

relatively pure alkali and pure silica raw materials. It is possible that the Group 3 average magnesia level of 1.8%, with a lime content of 6.2%, suggests that these glasses could alternatively have been made with a plant ash containing low contents of magnesia, potash and phosphorus combined with a pure quartz silica source. The average levels of 0.6% alumina and 0.3% iron oxide for these glasses are low, and hence are suggestive of a pure silica source. If, on the other hand, the Group 3 glasses were made with a mineral soda, the silica source was presumably a calcareous sand with a very low level of contamination.

The Pella LMLK glasses have a relatively low magnesia content on average, but have higher potash levels (but below 1.5% on average). The Pella LMLK average potash level is only slightly less than that of the Group 1 HMHK glasses, which are distinguished by their relatively higher magnesia and lime contents. The plant ash used to make them seems to have been a low potash variety.

In Figure 9-3 the lime contents of typical 2nd millennium BC HMHK Nuzi glasses are compared with those of Nimrud HMHK Groups 1 and 2 of the 1st millennium BC. This comparison shows that Nimrud Group 1 HMHK glasses have lime contents similar to those of the 2nd millennium high-lime HMHK group from Nuzi, while the more typically 2nd millennium HMHK glasses of Nimrud Group 2 may be classed as low-lime. Brill's (1999b: 47-49) Nimrud Group 1 glasses adhere to this grouping, although two low-lime glasses are included also, but as noted previously, these glasses do not all fit into Group 1 when elements other than magnesia and potash are considered (section 6.4). Overall, the IA Nimrud HMHK glasses have higher average lime concentrations in a narrower range than do the 2nd millennium BC HMHK glasses.

The LMLK Group 3 glasses have lime and phosphorus levels that are relatively higher than those of the Nimrud Group 4 glasses, and which are not dissimilar to those of the IA HMHK glasses. The low potash, lime and phosphorus levels of the Group 4 glasses, with moderately elevated magnesia and elevated alumina, are reminiscent of the composition of LBA Egyptian Co blue glasses.

The difference between 1st and 2nd millennium BC HMHK glasses is further illustrated in Figure 9-4 where the concentrations of magnesia, potash, soda and lime are re-summed to 100% for Nimrud HMHK Groups 1 and 2 and are compared with the concentrations of those oxides found in Nuzi high- and low-lime HMHK glasses. Each group is distinct, having different average contents of these oxides, which suggests that they were made with different plant ashes. The Nimrud HMHK glass compositions are different to those of the LBA plant ash glasses in other respects, and hence might not represent a direct continuation of the LBA plant ash glass tradition.

9.3 Trace Element Compositions

An examination of trace element compositions of Nimrud Groups 1 to 4 glasses (sections 6.5.2, 7.3) shows that the SRE are enriched, to a greater or lesser degree by comparison with the SRE content of the UCC, in the concentrations of Sc, Cr, Zr, Hf and U, whereas Cr, W, Tl

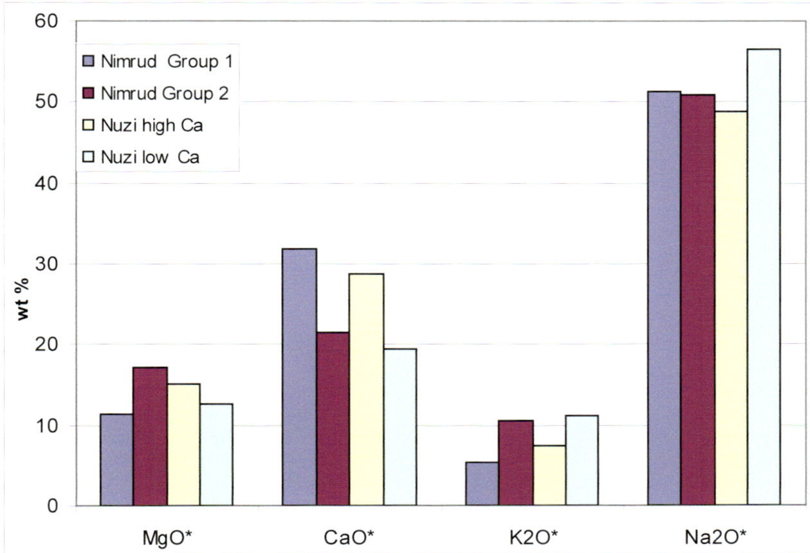

Figure 9-4 Histogram comparing the concentrations of plant ash associated elements in typical 2nd millennium glasses from Nuzi and Nimrud 1st millennium plant ash glasses. Data for these oxides are re-summed to 100%.

and U are enriched in Pella HMHK glasses, and only Cr is enriched in the Nuzi glasses from the 2nd millennium BC. If Group 2 glasses are the remnants of earlier LBA production, they do not appear, on this evidence, to have come from the same source as the LBA glasses from Pella and Nuzi. The Group 4 glasses are enriched in Y, which is related to the general enrichment of the REE in these glasses, probably as a result of the use of Co alum as a colourant (section 7.2). The concentrations of REE in Groups 1 to 3 glasses are depleted relative to the UCC, and are at comparable levels to those of the LBA glasses (sections 6.5.1).

The Pella LMLK glasses are, by contrast, enriched in REE and in SRE compared with the LBA HMHK glasses as a result of contamination from the iron oxide and/or sand sources used to make them[68] (sections 5.4.1, 5.4.2). The elements Cr, Zr, Hf, W and U are enriched to different levels in the Pella LMLK glasses. Sc is not enriched as it is in the Nimrud glasses, and Tl, which was enriched in the Pella HMHK glasses, is considerably depleted in the Pella LMLK glasses, and in the Nimrud Groups 1, 3 and 4 glasses.

Essentially, the positive Cr anomaly is common to all of the Near Eastern glasses from both the LBA and the IA, serving to distinguish them from LBA Egyptian glasses and showing that trace elemental values can be used as regional discriminants. The IA Near Eastern glasses all have elevated concentrations of Zr, Hf and U. The relative enrichment of Sc is seen only in Nimrud glasses, and the enrichment of W is seen only in the Pella glasses of both the LBA and IA (sections 5.4, 6.4).

The low REE and SRE concentrations reflect low levels of sedimentary contamination in the Nimrud IA glasses, especially those of Group 3. When the concentrations of the alkali and alkaline earth elements of Nimrud Groups 1 to 3 and LBA Nuzi glasses are compared, it is apparent that the Nimrud Group 2 glasses are closest in composition to the glasses from Nuzi, lending support to the proposition that the Group 2 glasses might be remnants of LBA production (section 6.5). Otherwise, there is little to distinguish the glasses based on this group of elements.

9.4 Regional Discriminants

The averaged data for the concentrations of Zr, Ti, Cr and La, and their ratios, for IA glasses analysed in the present study, are presented in Table 9-1. Comparative LBA Egyptian material was analysed by Shortland *et al.* (2007: 784-785) and included 19 colourless and Cu blue HMHK glasses, five mixed Cu-Co glasses with low potash, and six low potash Co glasses (n=30). Six Co blue low potash Egyptian glasses are included from unpublished analyses of the author.[69]

[68] Sayre and Smith (1974: 56), who analysed early black glass, commented that, 'Perhaps the enhanced amounts of rare earths found in the "black" glass will eventually provide an indication of the precise material added to darken this glass', thereby making the link between the increased levels of REE contamination and the iron oxide colourant, which is also responsible for colouring the black glasses examined in the present study.

[69] Egyptian Co glass samples were generously provided by I.C. Freestone, while at The British Museum, and include four unprovenanced vessel fragments donated by Howard Carter, and two cane fragments from Amarna, all dated by Dr Dan Barag to Dynasty

Table 9-1 Average concentrations in ppm of Zr, Ti, Cr and La and ratios of Zr/Ti and Cr/La for glasses from the IA Near East and LBA Egypt

Averages	Zr ppm	Ti ppm	Cr ppm	La ppm	Zr/Ti	Cr/La
Pella LMLK						
n=23	70.6	1069	98.7	8.82	0.07	8.64
Nimrud Gp 1						
n=8	86.6	746	13.6	3.61	0.12	3.81
Nimrud Gp 2						
n=4	23.1	537	6.54	2.66	0.05	19.7
Nimrud Gp 3						
n=13	17.3	240	6.83	1.51	0.07	4.58
Nimrud Gp 4						
n=15	38.1	354	13.1	3.22	0.10	4.16
Egypt						
n=35	41.7	519	6.07	3.42	0.08	1.84

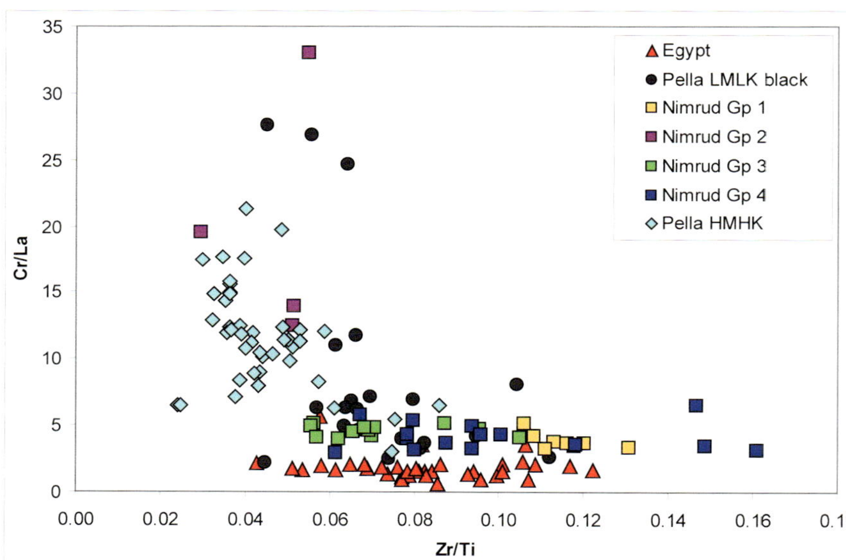

Figure 9-5 Scatter plot of Zr/Ti *vs* Cr/La for IA Near Eastern glasses from Nimrud and Pella compared with LBA HMHK glasses from Pella and LBA Egyptian glasses of both HMHK and Co blue low potash glasses. Data sources are cited in the text.

The narrow and low range of ratios of Zr to Ti, with a wide range of Cr to La ratios, typifies the LBA HMHK Near Eastern glasses (section 8.3.3.1), (Figure 9-5). These ratios are not, however, identifying features of the IA Near Eastern glasses, with the exception of Nimrud Group 2 glasses (HMHK), which exhibit the characteristics of LBA HMHK glasses. The separation of the different groups of glasses is more clearly illustrated by the plot of Zr *versus* Cr/La (Figure 9-6), which distinguishes Pella HMHK from Nimrud Group 2 glasses by their Zr contents.

The Pella LMLK glasses have higher Zr contents than the LBA HMHK glasses. Whereas the Pella IA glasses have relatively low average Cr to La ratios, there are several glasses with higher Cr to La ratios reminiscent of the LBA glasses (samples 59, 61 and 225), but which are also separated from the Pella HMHK and Group 2

18 (Freestone, personal communication). For data see Appendix VI.

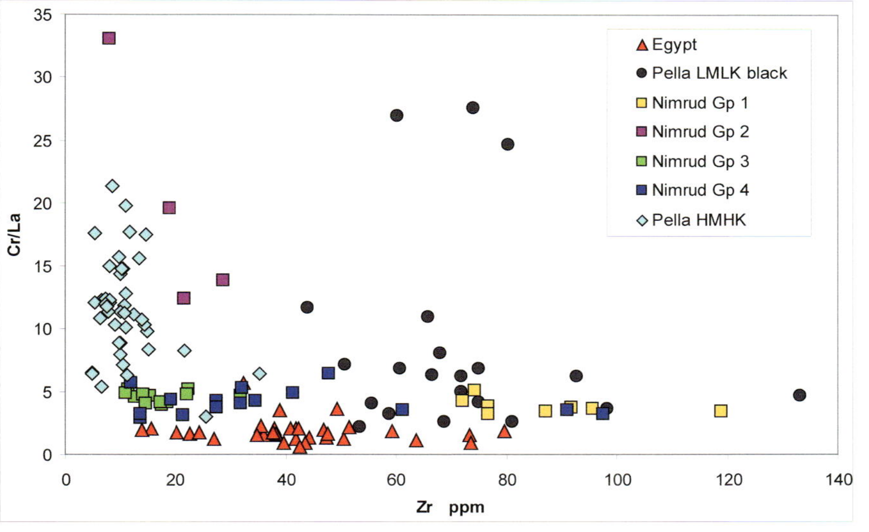

Figure 9-6 Scatter plot of Zr *vs* Cr/La for IA Near Eastern glasses from Nimrud and Pella compared with LBA HMHK glasses from Pella and LBA Egyptian glasses of both HMHK and Co natron-type glasses. Data sources are cited in the text.

glasses by Zr content in Figure 9-5. Pella samples 59 and 61 were found together in the temple Area XXXIIB and belong to a group of glass beads some of which have lower ratios of the concentrations of Cr to La. The variation in their composition probably derives from natural variation in the composition of the batches used to make the glasses.

The Pella LMLK glasses have particularly high levels of Ti, Cr and La, and a relatively high average Zr content, as does Nimrud Group 1 glass compared with other IA and LBA glasses. Nimrud Groups 2 and 3 have lower Cr levels, close to those in the LBA Egyptian glasses, with only the Pella LMLK and Nimrud Group 1 glasses having high Cr values. Nimrud Groups 1, 3 and 4 glasses do not have Cr values as high as those that distinguish the LBA glasses from the Near East, but they do exhibit elevated Cr levels relative to their SBS (sections 6.5.2, 7.3). Nimrud Groups 3 and 4 glasses have low Zr contents close to those of the Egyptian glasses, while Nimrud Group 1 glasses have higher levels of Zr. None of the glass groups has ratios of Cr to La as low as those of the Egyptian glasses.

The Egyptian glasses all plot together regardless of whether they are HMHK, or Co blue low potash glasses, while Nimrud Groups 1, 3 and 4 plot just above the Egyptian glasses, with a similarly narrow range of low Cr to La ratios, and wide ranging Zr concentrations. Could this represent a change in the IA Near East to the use of a raw material that gives a chemical signature similar to that of the LBA Egyptian glasses? Jackson and Nicholson (2007: 98; section 1.6) have shown that LBA Egyptian glasses could have been made with sand and plant ash, and hence it is interesting to consider whether a sand source of silica is responsible for aspects of the composition of the IA Near Eastern glasses. Even Group 3, made with comparatively pure raw materials (sections 6.4.1.3, 6.4.1.4, 6.5), exhibits the same tendency as the other IA glasses, with a low Cr to La ratio and more wide-ranging Zr concentrations.

With the exception of Group 2 glasses, the Nimrud glasses cannot be distinguished from the Egyptian glasses using Cr/La and Zr/Ti ratios in the way that Shortland *et al.* (2007) established for the LBA Near Eastern and Egyptian glasses. The Pella IA glasses plot closely with the Nimrud Groups 1, 3 and 4 glasses, with the exception of several glasses, with wide-ranging ratios of Cr to La, which plot with the Group 2 and LBA Near Eastern glasses, when ratios of Cr to La and Zr to Ti are compared.

The plot of Cr to La against Zr (Figure 9-6) separates the Nimrud Group 2 and Pella LMLK glasses by their Zr content. The Pella LMLK glasses appear to be distinct from the Nimrud glasses, and both Nimrud and Pella IA glasses appear to be distinct from the Egyptian glasses, although they are closely similar by comparison with the compositions of the LBA glasses.

The elements Li, Be, Sr and Cs and the ratios of Li to Be and Sr to Cs were shown in section 8.3.3.2 to be useful regional discriminants between LBA Egyptian and Near Eastern glasses. Averaged data for these elements and their ratios are presented in Table 9-2. In the IA glasses from Nimrud and Pella, Be levels are generally low, as they were in the LBA Near Eastern glasses. Only the Nimrud Group 4 glasses are distinguished by their high Be content, which is probably added with the Co alum

Table 9-2 Averaged concentrations in ppm of Li, Be, Sr and Cs and ratios of Li/Be and Sr/Cs for glasses from the IA Near East and LBA Pella and Egypt.*

Averages	Li ppm	Be ppm	Sr ppm	Cs ppm	Li/Be	Sr/Cs
Pella HMHK						
n= 43	16.4	0.099	382	0.443	342	936
Pella LMLK						
n= 23	45.7	0.351	109	0.437	130	248
Nimrud Group 1						
n= 8	5.27	0.169	438	0.142	31.2	3091
Nimrud Group 2						
n= 4	15.3	0.164	399	0.203	95.4	1962
Nimrud Group 3						
n= 13	4.44	0.103	210	0.147	43.1	1430
Nimrud Group 4						
n= 15	11.3	2.50	106	0.126	4.51	837
Egypt						
n= 35	6.42	0.462	622	0.113	27.0	8902

*Data sources are cited in the text.

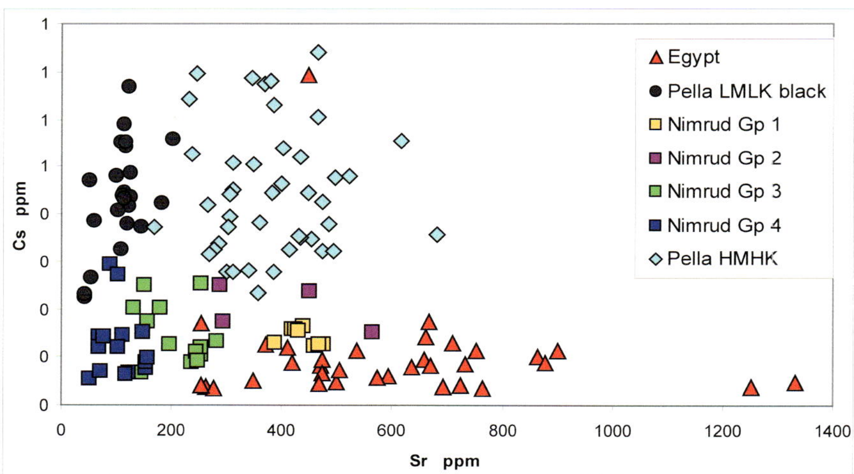

Figure 9-7 Scatter plot of Sr *vs* Cs for IA glasses from Pella and Nimrud compared with LBA glasses from Pella and Egypt. All non-Egyptian data are from the present study. Egyptian data from Shortland *et al.* (2007) and analyses of the author (Appendix VI).

colourant (sections 7.3, 7.5). The levels of Be are not as high in Egyptian Co blue glasses, which might be a result of using alum from a different source in the same region (section 7.5).

When the elements Sr and Cs are plotted (Figure 9-7), the groups of IA glasses from Nimrud and Pella and the groups of LBA glasses from Pella and Egypt are clearly separated.

The ratios of Sr to Cs plotted against the ratios of Li to Be for the IA and LBA glasses (Figure 9-8) show that the Near Eastern glasses have low ratios of Sr to Cs, regardless of whether they were made in the IA or in the LBA, whereas the Egyptian LBA glasses have larger and more wide ranging ratios of Sr to Cs. The Pella LBA and IA glasses and the Nimrud Group 2 glasses have wide ranging ratios of Li to Be, which distinguish them from the Egyptian glasses. Nimrud Groups 1, 3 and 4 have low

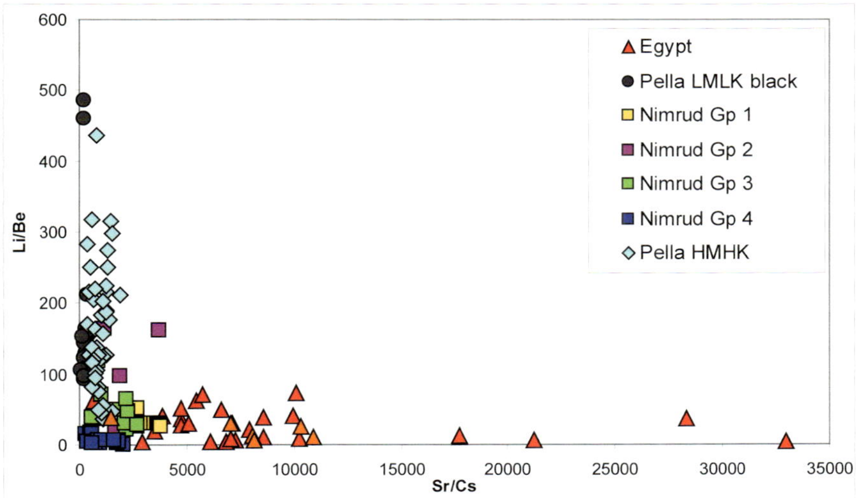

Figure 9-8 Scatter plot of Sr/Cs *vs* Li/Be for IA glasses from Pella and Nimrud compared with LBA glasses from Pella and Egypt. All non-Egyptian data are from the present study. Egyptian data are from Shortland *et al.* (2007) and analyses of the author (Appendix VI).

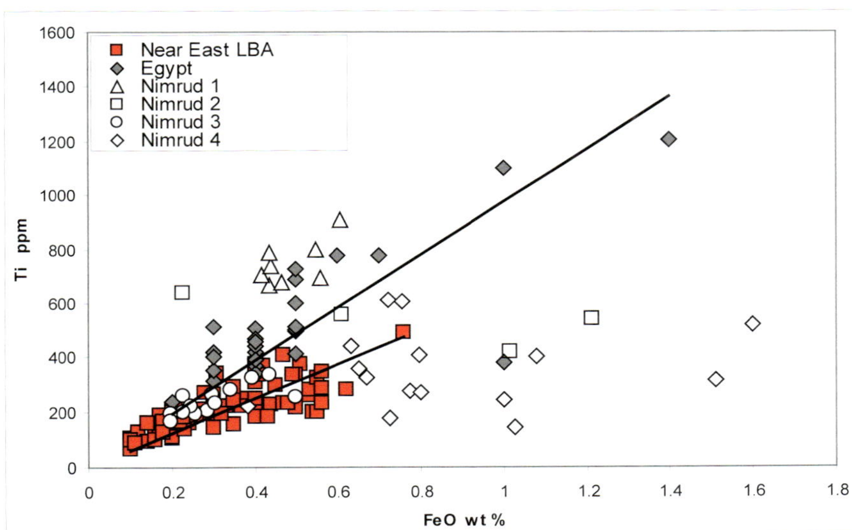

Figure 9-9 Scatter plot of FeO *vs* Ti for Nimrud Groups 1 to 4 compared with LBA glasses from the Near East and Egypt. Data sources cited in the text.

Li to Be ratios and low Sr to Cs ratios, which are lower than those of most of the Egyptian glasses. These ratios do not provide the clear distinctions illustrated by Cs and Sr in Figure 9-7, and therefore should be used with caution as regional discriminators for IA glasses.

The examination of iron oxide and Ti contents for Near Eastern and Egyptian glasses as regional discriminators (section 8.3.3.1) produces interesting results (Figure 9-9).

Nimrud Group 3 glasses plot with the LBA Near Eastern glasses when iron oxide is plotted against Ti, but Group 1 glasses plot close to Egyptian LBA glasses, although with a higher average Ti content. Nimrud Group 2 glasses do not form a group, but have widely diverging iron oxide concentrations. Nimrud Group 4 glasses plot closer to the LBA Near Eastern glass trend line, but have greater concentrations of iron oxide. The regional classification of IA glasses using iron oxide and Ti contents provides a different view of the juxtaposition of the Near Eastern and Egyptian glass compositions, but does not necessarily clarify the origins of the IA glass. The nature of the Groups 1 and 2 raw materials, and their relationship to those of the Egyptian glasses is unclear.

Base glass and trace element compositions of the Nimrud glasses, and the lack of comparative glass found in Egypt at this time indicate, however, that an Egyptian origin for the Nimrud glasses is unlikely. Pella LMLK IA glasses are not plotted because their iron oxide content is disproportionately large. The Ti content of the Pella LMLK glasses is higher on average than in the other glasses, and their chromite content suggests a link with more northerly regions where chromite is found (section 5.4.2).

9.5 An Emerging Technology

The proliferation of LMLK glass compositions in the IA Near East and Egypt represents a change in glass-making technology that involved variation in compositions and raw materials during the re-emergence of glass-making at this time. The experimentation and diversification in glass compositions are evident in the Nimrud, Pella and Egyptian glasses, and probably represent the production of more than one workshop.

The variety of compositions exhibited by the Nimrud glasses informs us that these are not all typical Near Eastern plant ash glasses as defined in the LBA. Plant ash glass is present, but the compositions of two, and possibly three, of the four main groups have low potash contents, closer to glasses made with mineral soda, this being particularly so for the Co blue glasses. As at Pella, the importance of this finding is that LMLK, non-Co glasses occur somewhat earlier than has generally been documented. The only other group of glasses to have been analysed come from the 10th century BC tomb of Nesikhons in Egypt (Schlick-Nolte and Werthmann 2003).

The IA black beads from Pella represent a significant addition to the increasing number of LMLK glasses with low lime, recognised in the Near East from the late 2nd to early 1st millennia BC. At least 5% lime is considered essential to render a glass resistant to corrosion (section 1.7), but in the Pella LMLK glasses, high concentrations of iron oxide colourant have served to stabilise and preserve the glasses. Low lime, natron glass from the tomb of Nesikhons in Egypt owes its survival to very dry burial conditions (Schlick-Nolte and Werthmann 2003; section 1.5), while the Co coloured natron glass from Nimrud appears to have survived only because of the high alumina content imparted by its Co alum colourant (Reade *et al.* 2005; section 9.7.3).

The technology of the Pella LMLK glass appears to be based upon mineral soda, similar to the 9th C BC Co blue glasses from Nimrud. While the favoured interpretation is that the soda source for the Pella glass was natron, the possibility that a plant ash low in potash, magnesia and lime was utilised cannot be eliminated, in the light of the recent compilation of data by Tite *et al.* (2006) and Barkoudah and Henderson (2006) (section 1.8).

The Pella glass appears to be the earliest in this series of innovative LMLK glasses identified so far. The Pella material dates from 1050 to 950/900 BC, or the mid 11th to 10th C BC, while Nesikhons' glass is dated to the early 10th C (974 BC) and Nimrud glass to the 9th to 8th C BC. LMLK glass appears in both Egypt and the Levant at similar times. Where it developed and who developed it first is unknown, although as Egypt was the source of natron, it would not be surprising if this technical advance occurred there first, and the glass made from these earlier unsuccessful attempts has not survived. As natron might have been known and transported to the Near East for other purposes, besides glass-making, the origin of natron fluxed glass technology remains unresolved.

The question remains as to whether the Near Eastern IA glasses represent the products of a single technology based on natron, or a combination of technologies which utilised either natron or an LMLK plant ash particularly rich in soda. Given that in the early IA in the Near East, there is a resurgence of the glass industry after a period of low activity at the end of the LBA, the possibility of widespread experimentation with glass formulations is quite possible. Moorey (1985: xiii) commented aptly that craftsmen acquired their knowledge 'by trial and error over long periods of time, by careful selection from amongst the raw materials available ... and by the native ingenuity of generations of craftsmen'. The Nimrud ungrouped glasses, which appear as compositional outliers compared with other Nimrud IA glass compositions, might be examples of compositional variation due to experimentation in glass-making, especially as such compositional outliers are rare in chemical groupings of LBA glasses.

Experimentation with glass compositions, or recipes, could be seen in terms of the 'cascade' model for invention processes, described by Schiffer (2005). In relation to the development of complex technological systems, where invention takes place in a series of steps or 'cascades', craftsmen progress from striving to achieve their desired product by a process of creation, manufacture, replication and use, until an accepted and successful product is manufactured and the process can be replicated. This model is used to explain the variants which are often treated as inexplicable and idiosyncratic in archaeological assemblages, such as the Nimrud ungrouped glasses, and the glasses from the tomb of Nesikhons, which as early and serendipitously preserved examples of the LMLK technology, had not achieved complete technical success in terms of their resistance to decay.

The variety of compositions in evidence at Nimrud is not only suggestive of experimentation and diversity in technology, but is perhaps a result of a greater freedom of choice in the IA regarding raw materials, and it probably implies that there were a number of glass production centres. If it is assumed that the different chemical groups of glass were the products of different workshops, it suggests that both blue and colourless glasses were being made by different workshops to slightly different base glass recipes. Nimrud Group 1 consists largely of colourless (and two pink) glasses and might represent the products of an individual workshop. Groups 2 and 3 were likely to be the products of two different workshops, because the same range of colours was made, but with separate compositions, which must represent the use of different raw materials, combined in particular recipes. Group 4 glass composition is different again. The ungrouped glasses may be all that is left of yet more variations, which could have been the result of experimentation, or were the products of multiple workshops, perhaps from further afield and hence not so widely represented at Nimrud.

The glass of Nesikhons provides evidence that a variety of different colours was made even in the earliest IA glasses. The survival of the Nesikhons glasses, and the Nimrud Co blue and Pella black glasses may be chance amongst other low lime glasses whose colourants or burial conditions did not preserve them. That these three groups, found in different parts of the ancient Near East and Egypt, date to within approximately 100 to 200 years of each other seems to be more than a coincidence. It is likely that these glasses represent the remnants of a nascent LMLK glass technology. It seems reasonable to suggest, on the current evidence, that much of this glass must have decayed and disappeared and only the 'tip of the iceberg' is now visible.

At the time of writing, there are no published IA parallels for the Nimrud Co blue compositions from the Near East or Egypt. It is unfortunate that the evidence for glass-making in the 11th to 7th centuries BC is so limited, but this is not surprising considering the results of the compositional analysis of glass from Nimrud, Pella and the tomb of Nesikhons (Schlick-Nolte and Werthmann 2003). As might be expected, the earliest glass-makers using natron did not have good control of their raw materials. They might not have been aware of the importance of lime as a stabiliser (section 1.7) and commonly used sands or pebbles, low in lime, as sources of silica (section 1.6). Natron contains essentially no lime (section 1.9) and, as a consequence of the low lime content, the Egyptian 10th C BC glasses are very unstable and appear to have been preserved only because of the extremely dry conditions in the tomb.

Similarly, the Nimrud Co blue glasses are low in lime, but their resistance to weathering was enhanced by the addition of alumina and magnesia as part of the Co colourant. The low-lime Pella LMLK glasses have been preserved by their high iron oxide content. Brill's (1999b: 48) analyses show that opaque yellow and opaque red glasses from Nimrud are also comparatively low in lime. These coloured glasses are likely to have been preserved due to the presence of divalent Pb and Cu added to colour them. Thus it is quite likely that a substantial number of glasses from this period, which were not coloured with large amounts of stabilising compounds, or preserved in the extremely dry conditions such as those encountered in the tomb of Nesikhons, have simply weathered away, and that much of the evidence for LBA to early IA use of natron-based glass has been lost.

The Group 3 LMLK glasses, on the other hand, have sufficiently high lime to have preserved them without elevated levels of iron oxide or alumina. This may represent an advance in the technological understanding of glass stabilisers from the time of Nesikhons over approximately 200 years to the date of the Nimrud non-Co LMLK glasses. The Nimrud Co glasses, on the other hand, have much lower lime concentrations on average than the other glasses from Nimrud. It is possible that this Co glass-making tradition, or recipe, is linked to the Egyptian technology of the 2nd millennium BC, and has therefore developed separately (section 9.7.1).

9.6 Iron Black Glasses

Iron oxide was identified as a colourant to make 'very dark or black glasses' from the 2nd to 1st C BC, and later (Turner 1956b: 178T-179T).[70] Glasses of a very dark green-black colour were noted by Sayre and Smith (1974: 56) to contain elevated iron oxide and alumina, a feature found in 1st millennium BC plant ash glasses from India by Brill (1987: 5).[71] Henderson (1985a: 283) documented the occurrence of dark 'black' glasses from IA Europe intentionally coloured with iron oxide, and

[70] Iron is known to have co-existed in glass in both its oxidised and reduced forms producing the colour green (Biek and Bayley 1979: 14-15; see also Weyl 1953: 102-3, 108; Bamford 1982: 6). Iron is usually considered to be a natural impurity in raw materials, but was sometimes added intentionally. A concentration of over approximately 3.5% is considered intentional by Gratuze (2001c) who noted that intentional iron colouration is frequently found in glasses from the IA.

[71] The black Indian glasses analysed by Brill (1987) were high in iron (and sometimes alumina which greatly improved their chemical durability). Iron oxide levels were usually greater than 5% and included early dark olive glasses from Rupar (sample 2900 dated to 1000-700 BC, sample 2901 dated to 600-200 BC) and Hastinapur (sample 2903 dated to 600-300 BC) (1987: 14). They are soda-lime-silica glasses with unusually low lime contents, although they are plant ash glasses with iron oxide from 2.8% to 8.5%, and alumina from 4.7% to 8.8%, and therefore are different from the Pella LMLK black glasses

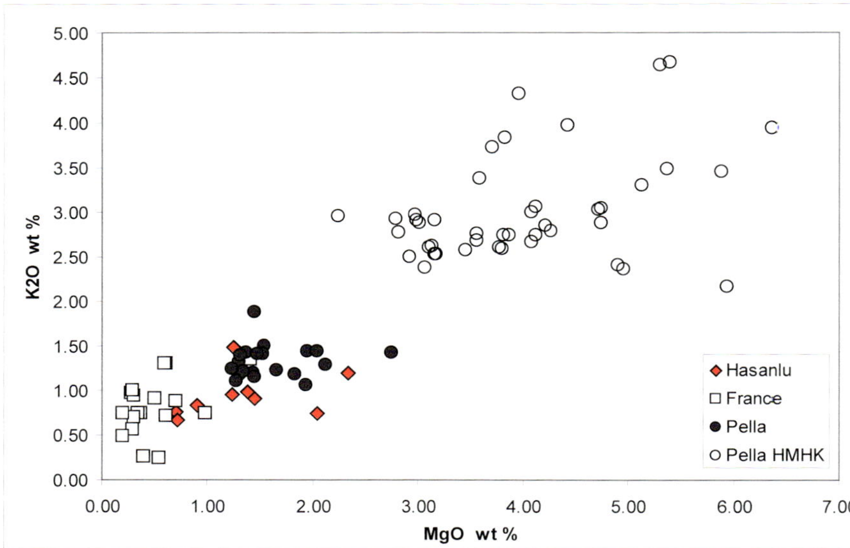

Figure 9-10 Scatter plot of magnesia *vs* potash for iron black glasses from Hasanlu, France and Pella of the early 1st millennium BC compared with LBA Pella HMHK glasses. Data sources cited in the text.

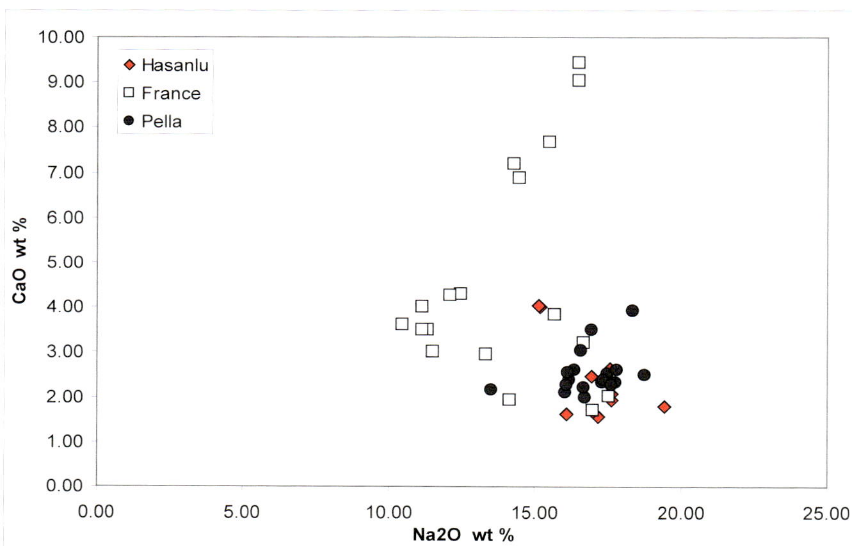

Figure 9-11 Scatter plot for soda *vs* lime for iron black glasses from Hasanlu, France and Pella of the early 1st millennium BC. Data sources cited in the text.

one iron black glass was analysed by Santopadre and Verità (2000: 30) from MBA Poviglio in Italy (1600 BC to 1500 BC). Brill (1999b: 43-44) analysed HMHK black glasses from Hasanlu in Iran, which have high magnesia, potash and lime contents that suggested these glasses were made with plant ash.

There are few high iron oxide, black glasses, of LMLK composition which are comparable to the Pella LMLK glasses included in the present study. Eighteen such glasses were analysed by Gratuze from various sites in France, dating to the early 1st millennium BC (11th to 9th C BC). Gratuze analysed samples from Champ-de-Mont, Quitigny (n.d.(b)), L'Alveyron, including Bouche Rolland, Bringairet and Lac du Bourget (n.d.(d)), Mons (2001b), Mez-Notariou (2002), Puech de Mus (2005), and Gratuze and Cosyns analysed glasses from Neufchâteau-Le Sart (Gratuze and Cosyns 2007).[72] Most French

[72] The presence of chromite providing iron oxide and Cr in glass-making sands has also been observed in Britain by Henderson (1985: 271).

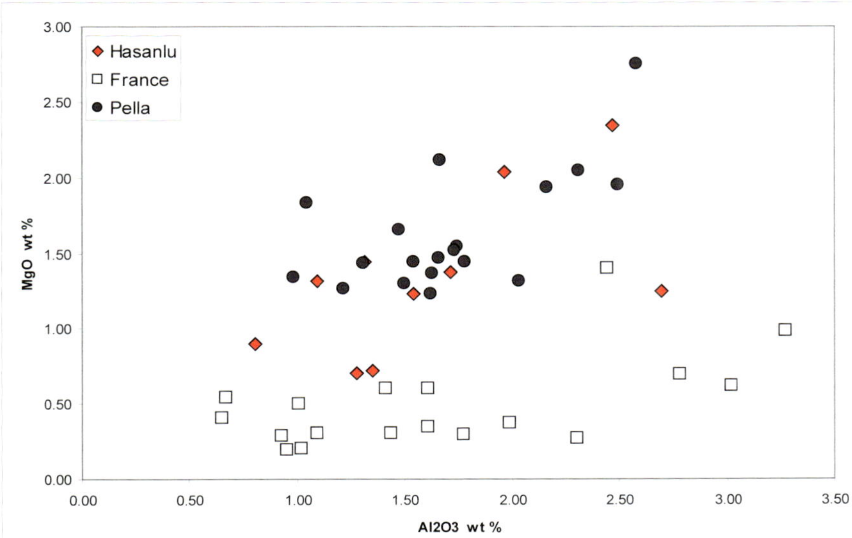

Figure 9-12 Scatter plot of alumina *vs* magnesia for iron black glasses from Hasanlu, France and Pella of the early 1st millennium BC. Data sources cited in text.

glasses analysed come from the Languedoc region, near the Mediterranean littoral, and provide valuable comparative material for the Pella LMLK natron glasses.

Comparative IA LMLK natron black glasses were also analysed by Stapleton and Swanson (2002a; 2002b), and Stapleton (2003: 107-108) from the 11th to 9th C BC site of Hasanlu in Iran. The composition of high-iron, black LMLK glass from Hasanlu is identical to the Pella glass and is closely similar to the French glass compositions. The iron oxide (Fe_2O_3) levels range from approximately 7.5% to 15%. The compositions of the French LMLK glasses are closely similar to those of the LMLK glasses from Pella and Hasanlu (Figures 9-10, 9-11, 9-12), with similar low lime levels, and iron oxide (Fe_2O_3) levels between 9% to 20%. Iron oxide (FeO) levels in Pella glasses range from 5.5% to 16% (section 5.3).

Stapleton (2003: 153) preferred to classify the Hasanlu black glasses as plant ash fluxed because they made up a small proportion of a largely plant ash glass assemblage at a time when mineral soda was not recognised as having been used to flux Near Eastern glass. Given the apparent widespread occurrence of this kind of black glass (Hasanlu, Pella, France), the low concentrations of magnesia, potash and lime in the glasses, and the identification of mineral soda fluxed glasses from as early as the 10th C BC in Egypt and the Near East, the possibility that the Hasanlu black glasses were made using natron should not be discounted. The identical base glass compositions and close dates of the Pella and Hasanlu black glasses argue for the same origin, which Gratuze and Cosyns (2007: 6) advocated was Syro-Palestinian for the French black glasses.

While the French glasses are closely similar in composition to the Near Eastern glasses, they have lower magnesia, potash and soda levels with higher lime contents. Their concentrations of alumina and silica are close to those of the Pella and Hasanlu glasses, while their iron oxide content is a little higher on average.

The IA black glasses from Pella are very early examples of LMLK glasses with low lime, and which may be natron-based. They survived only because the high iron oxide content, which was used to colour them, and incidentally served to stabilise the glass. Their base glass compositions are identical to those of the glasses from Hasanlu and closely similar to those of the contemporary beads from France. This unusual and clearly diagnostic glass was traded, either as beads or possibly in ingot form, across a wide region.

9.6.1 Chromite

Chromite was identified in the Pella LMLK glasses (section 5.4.2). It is not known whether the Hasanlu or French black glasses also contained chromite. In their study of the LBA black faience from the site of Tell Deir 'Alla, several kilometres south of Pella in the Jordan valley, Groot *et al.* (2006: 235) found chromite, which they stated had not been found in vitreous materials from the LBA of the Near East, or at any time before the Roman period. It is interesting to note that vitreous materials containing chromite are attested from two sites in the Jordan valley, suggesting that a supply of chromite-containing raw material, such as sand, must have been accessible.

Chromite sand grains are typically derived from ophiolites, fragments of the earth's mantle pushed up into the crust, or from other mafic or ultramafic rocks. They are resistant to weathering, and are natural contaminants in glass raw materials. They could have been derived from a high iron, silica sand.[73] The total Cr contents of these glasses range up to 400 ppm, which is high relative to most ancient glass (Freestone, personal communication). The presence of chromite and high Cr levels tantalisingly suggests the possibility that the glass-making sand or the colourant was derived from the ophiolites of northern Syria and Anatolia, north of Pella (Robertson 2002: Fig. 1). Ophiolite deposits are scattered from Croatia, across the Balkans, Cyprus, Anatolia and coastal Syria, but are not found in the southern Levant or Mesopotamia. Chromite deposits do occur in north eastern and south eastern Iran, near Tepe Yahya (Shaystehar *et al.* 2001; Potts, personal communication). The geographical occurrences support a northern origin for the black, high iron glasses, as it is highly unlikely that such a low-quality sand would be transported deliberately to make black glass somewhere else. It is not known to what extent chromite might have been naturally transported from Anatolia along the Tigris and Euphrates Rivers into Mesopotamia.

A range of evidence, although none conclusive on its own, makes an increasingly strong case for a Syrian glass-making industry, which may be the origin for the Pella and Nimrud LMLK glasses, which in turn could have developed at about the same time as experimentation in the production of LMLK glass was occurring in Egypt. Much of the evidence for this nascent glass tradition is likely to have disappeared due to chemical instability of the resultant products (section 9.5).

9.7 Co Blue Glasses

The 9th to 8th C BC Co blue glasses from Nimrud represent a unique group of IA glasses that were classified as LMLK because of their low potash and lime base glass composition, with magnesia elevated by the addition of the Co alum colourant rather than by the alkali source (section 6.4.4). This composition suggests that they could have been made with natron, but a low potash, low magnesia plant ash source of alkali might also be possible in view of the work done by Rehren (2001) on the composition of Egyptian Co blue glasses, and the work done by Barkoudah and Henderson (2006), for example, on plant ash compositions (sections 1.8, 1.10.1).

[73] The presence of chromite providing iron oxide and Cr in glass-making sands has also been observed in Britain by Henderson (1985: 271).

The Group 4 glasses were found to be coloured with a Co alum probably from an Egyptian source in the Western Desert oases (sections 1.10.1, 7.5; map Figure 9-13), but the glasses themselves were not of Egyptian origin as is shown below. These blue glasses were in the form of inlays believed to have been attached to Phoenician style ivories from Syria. Although Egyptian glass production declined during the Third Intermediate Period when no Co coloured glasses were produced (section 1.2), Shortland *et al.* (2006: 163) noted that Egyptian alum sources were exploited until at least the 9th C BC. The apparently Egyptian origin of the Co alum colourant used in the glasses from Nimrud makes the observations of Kaczmarczyk and Hedges (1983) (section 1.10.1) on the nature and periods of use of Co in Egypt significant in the discussion of the Co blue coloured glasses from Nimrud analysed in the present study.

The use of Co alum as a colourant in IA Nimrud was unexpected. Co alum began to be used as an intentional colourant for dark blue glasses and faience in Egypt from the time of Thutmosis III in the mid-15th C BC (Lilyquist *et al.* 1993). The investigation of Egyptian faience glazes by Kaczmarczyk and Hedges (1983) indicated that no Co was used in Egypt after the New Kingdom, from the 11th to the 7th C BC, during the unsettled Third Intermediate Period. Co was again used as a colourant in Egypt from the beginning of the Late Period, in the mid-7th C BC, but Co-coloured glasses from this time do not exhibit the elevated levels of Mn, Ni or Zn observed in Co glasses of the earlier period. Kaczmarczyk (1986) inferred that this indicated the use of a Co source outside of Egypt (section 1.10.1).

While most Near Eastern blue glasses appear to be coloured with Cu from the 3rd to 1st millennia BC, a handful of Co-coloured glasses, faience and glazes have been reported from Eridu, Nippur and Ugarit (Garner 1956b; Neumann 1927). These were found to be coloured with a non-Egyptian, non-alum Co source (Kaczmarczyk 1986: 374). Kaczmarczyk concluded that while the Great Western Oases were the most likely source of Co alum to the 2nd millennium BC Egyptian craftsmen, the Co source for 'Mesopotamian' glasses of all periods, and Egyptian glasses and glazes of the 1st millennium BC, was Iran. This picture, however, was clearly an over-simplification.

It can be concluded that the Nimrud Co-coloured glasses, were produced in the middle of a period of over four hundred years during which Co has not been reported to be a colourant in Egyptian glass; indeed the Near East is not a region generally associated with the production of Co blue glass (Figure 9-14). Were it not for these chronological and geographical gaps, it would be logical to assume that the Co alum in the Nimrud glass represented a continuation of the use of the Egyptian

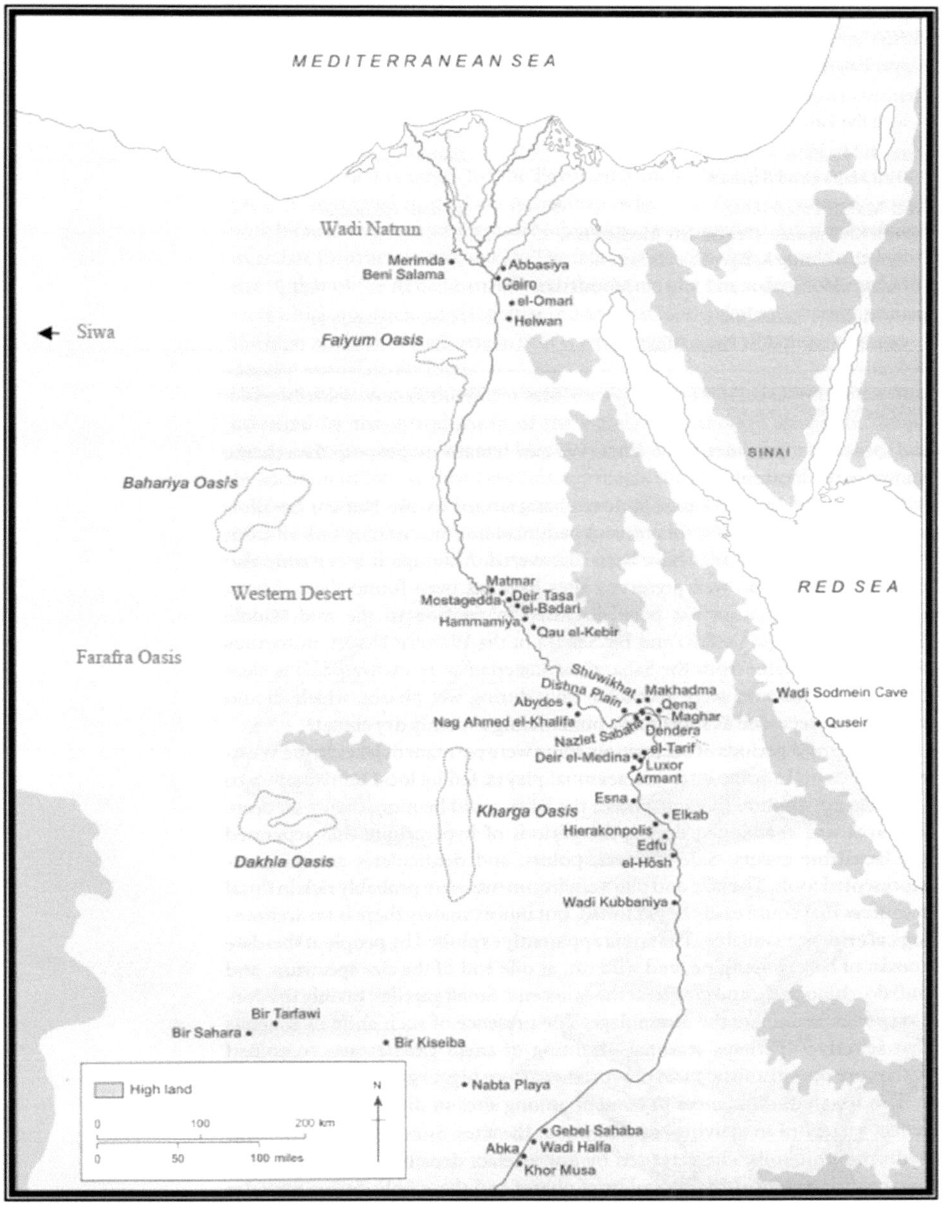

Figure 9-13 Map of Egypt showing the locations of the Western Desert oases in relation to the Nile Valley and the Mediterranean coast. Adapted from Hendrickx and Vermeersch (2000: 22).

source from the LBA, but such a proposal clearly needs careful consideration.

There is little evidence for the production of the Co blue glass in the Near East itself. It should be noted that the use of red alum (Co-bearing alum is pink in colour) as an additive in the manufacture of 'lapis lazuli' coloured glass is mentioned in the famous cuneiform glass-making texts from Ashurbanipal's (668-627 BC) library at Nineveh (Oppenheim et al. 1970: 41) (section 1.3). This indicates that the properties of alum as a colourant were known to Near Eastern glass-makers. The Nimrud Co blue glass inlays seem most likely to have originated from Syro-Palestine, and were probably not made locally by expatriate craftsmen imported by Ashurnasirpal II (section 6.3.1). If the glass arrived in Nimrud from the west, there is the possibility that Egyptian Co alum was used by Levantine glass-makers, or by glass-makers in Egypt itself, although the failure to detect alum-derived Co in Egyptian faience glazes of this period (Kaczmarczyk and Hedges 1983) renders the latter possibility unlikely.

9.7.1 *Compositional Comparisons of Co blue Glasses*

The analyses by Gratuze (n.d.(a); n.d.(b); n.d.(c); 1999; 2000a; 2001a; 2001b; 2002), Gratuze and Picon (2005) and Gratuze and Cosyns (2007) of glasses from contemporary

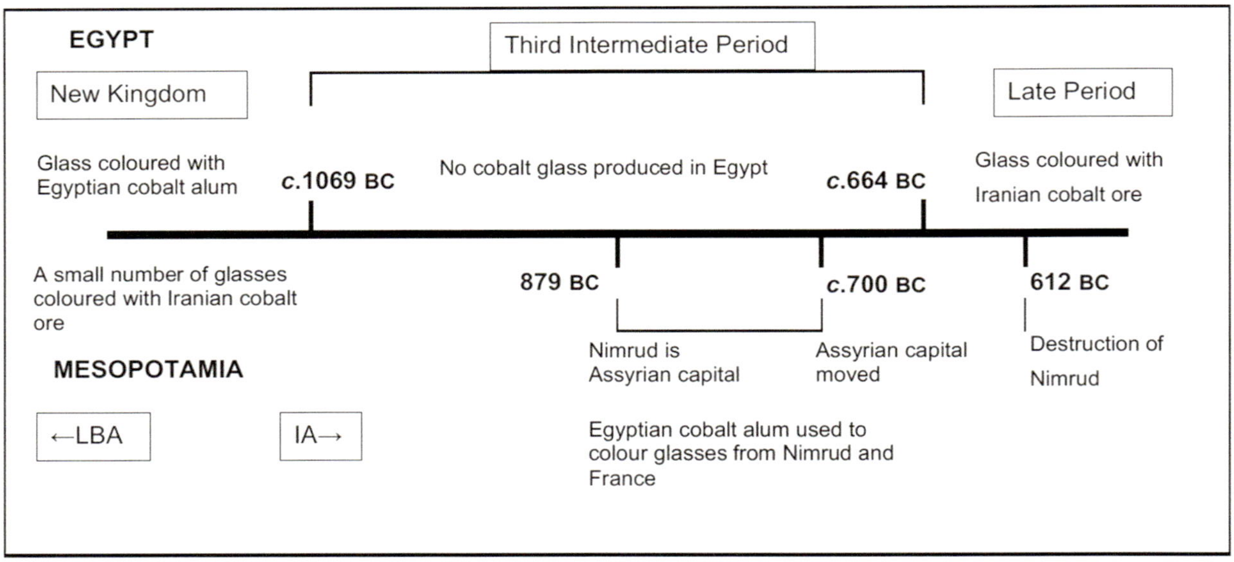

Figure 9-14 Diagram of the usage of Co to colour Egyptian and Mesopotamian glasses between the LBA and IA.

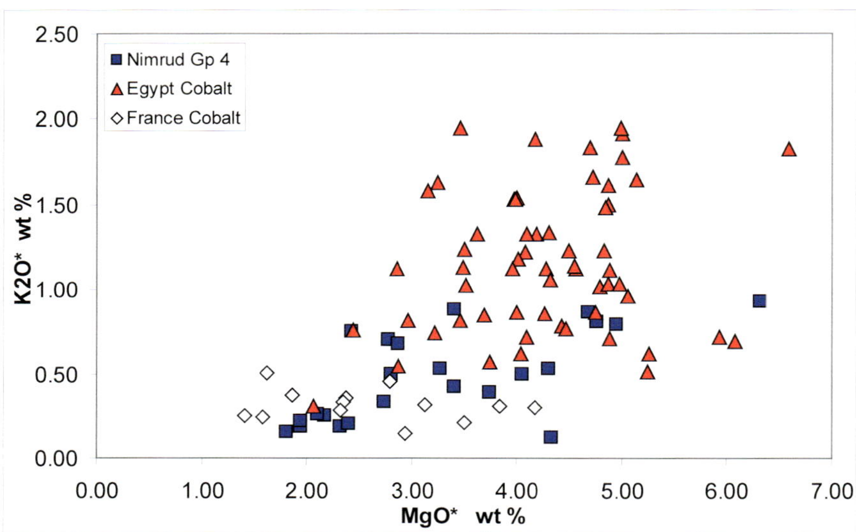

Figure 9-15 Scatter plot of magnesia vs potash for Co blue glasses from Nimrud, Egypt and France. Data sources are cited in the text.

France[74] suggested that the Co blue glasses from Nimrud were of closely similar composition. These similarities in base and colourant composition of the Co blue glasses from Nimrud to those analysed from early 1st millennium BC France led Gratuze and Picon (2005) to attribute the French glasses to a Syro-Palestinian origin.[75]

The base compositions of 23 Co blue glasses from IA Nimrud and 13 contemporary LBA/IA (8th C BC) Co blue glasses from France[76] have been compared with 62 earlier Co blue glasses from LBA Egypt to determine similarities and differences in their compositions. The Nimrud data are from the present study, *n*=17, and from Brill, *n*=6 (1999b: 47-48). The data for the Egyptian glasses are from Shortland and Eremin's study

[74] Glasses from BA Mariconda and Frattesina in Italy, analysed by Towle et al. (2001: 23), were found to be coloured with Co which was not derived from alum and consequently these glasses are not considered in the present study

[75] See also Gratuze et al. (1992) for a study of the origins of Co blue pigments used to colour glasses from the LBA to the AD 18th C in France.

[76] Note that material dated to the 8th C BC from France is classified as LBA to IA transition of the 9th to 7th C BC (Gratuze and Picon 2005: 270), but material of this date from the Near East is classified as IA.

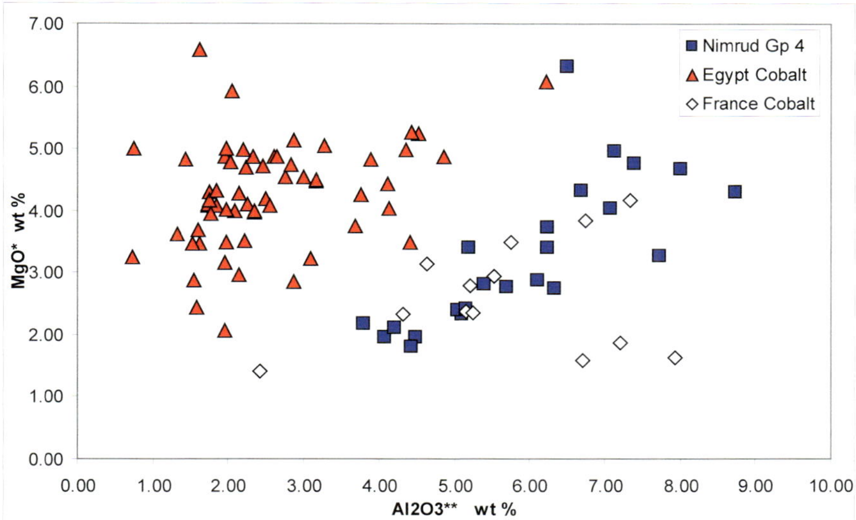

Figure 9-16 Scatter plot of alumina vs magnesia for Co blue glasses from Nimrud, Egypt and France. Data sources are cited in the text.

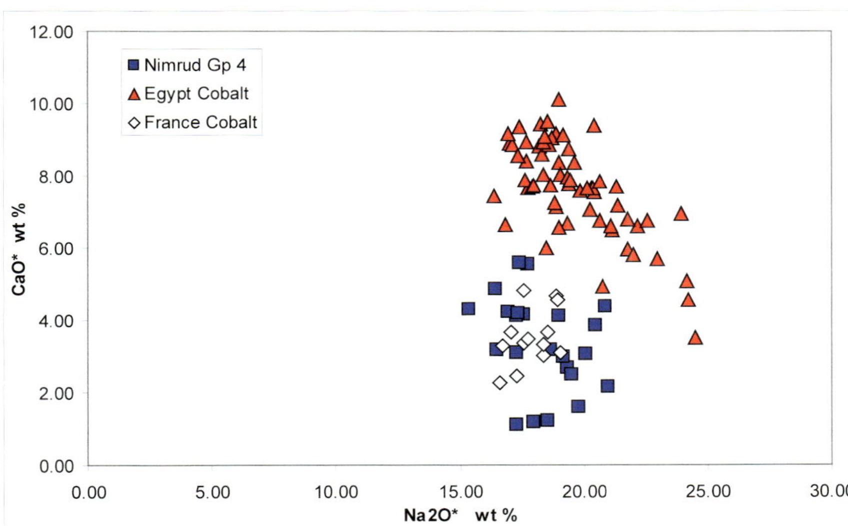

Figure 9-17 Scatter plot of soda vs lime for Co blue glasses from Nimrud, Egypt and France. Data sources are cited in the text.

of glasses from Amarna and Malkata, *n*=34 (2006: 585-586), Brill's analyses of glasses from Amarna, Malkata and Timna, *n*=14 (1999b: 27-29, 31, 34), the pre-Malkata glasses analysed by Lilyquist *et al.*, *n*=8 (1993: 36-39) and from 18th Dynasty glasses analysed by Freestone, *n*=6 (unpublished data); (see also section 8.3.2). The French glasses analysed by Gratuze come from the sites of Planaise de Sampzon (1999), Ventavon (n.d (c)) and Marmilhat (1999).

The concentrations of magnesia and potash in Co blue glasses from LBA Egypt, IA Nimrud and France are compared in Figure 9-15. The French and Nimrud glasses have closely similar compositions, while the Egyptian glasses have higher average potash and slightly higher average magnesia levels.

The compositions of New Kingdom Egyptian and Nimrud dark blue glasses are compared in Figure 9-16 as a plot of alumina *versus* magnesia. There is a clear positive correlation between these components in the Nimrud glasses (section 6.3.4), which are separated from the Egyptian Co blues by their higher alumina contents. The ratio of magnesia to alumina is approximately 0.5:1 in Nimrud glass, but is three to four times higher in the Egyptian glasses. Gratuze and Picon (2005: 273)

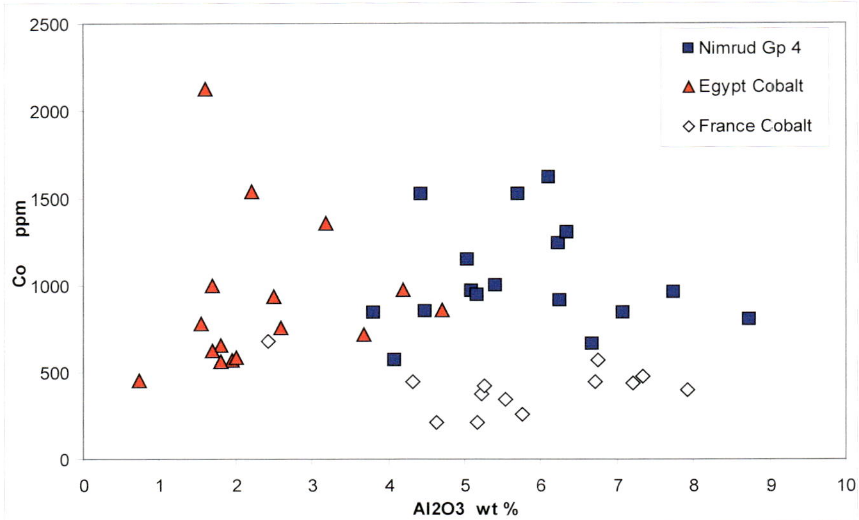

Figure 9-18 Scatter plot of alumina vs Co for Co blue glasses from Nimrud, Egypt and France. Data sources are cited in the text.

noted the lower alumina contents of the French glasses by comparison with the Egyptian ones. They agreed with Rehren (2001) that the excess of magnesia in the Egyptian glasses is explained by the use of plant ash, whereas the lower potash and magnesia content of the French glasses indicates the use of natron, with the magnesia and the alumina coming from the alum or from sand containing dolomite and/or alumina (Gratuze and Barrandon 1990).

There is no correlation between alumina and iron oxide in any of the three groups of glasses due to the elevated alumina contents that result from using Co alum as a colourant. In both the French and Nimrud glasses, silica and alumina are negatively correlated (r^2= -0.8), but this correlation is only weakly exhibited in the glasses from Egypt.

It is evident from Figure 9-17 that while the soda contents are similar for the three groups of Co blue glass, the average lime content of the Egyptian glasses is distinctly higher than that of the Nimrud and French glasses, which again plot together. This provides support for the view that the Egyptian glasses were made with plant ash, whereas the Nimrud and French glass compositions are indicative of the use of a natron flux.

The observed differences in composition between the French and Nimrud glasses on the one hand, and the Egyptian glasses on the other, could indicate the use of a different alum source, either within or beyond Egypt. Alternatively, the differences suggest that, while there was continuity in the use of the LBA alum source, there was a change in the technology of separation of the Co pigment from the alum. Gratuze and Picon (2005: 274) argued that the observed differences indicated that the IA preparation practice was different from the LBA practice, where alums were probably purified and concentrated before being incorporated into the glass. They suggested that the French Co blue glasses of the IA seem to have used the alum without further processing. They argued this with reference to the composition of the French black beads, which they determined were made with an iron-rich mixed alum salt, containing some Co, from the Dakhla and Kharga oases. The Co concentration in the Pella LMLK black glasses does not contain elevated levels of Co by comparison with other non-Co coloured glasses analysed in the present study.

There are suggestions of continuity between the LBA and IA approaches to glass-making from other compositional similarities. The use of Egyptian Co alum, for example, and the low potash contents of both the LBA and IA glasses, might be expected if the Egyptian craftsmen, or at least their recipes, had moved north to the Near East and were responsible for the manufacture of Co blue glass in this region in the IA. The change from a plant ash flux, which was probably used in LBA Egypt, appears to have been substituted by a mineral soda flux in the IA Co glasses of Nimrud, representing a technological innovation that developed from an LBA tradition, possibly inspired by Early IA experimentation with natron as a flux.

The similarly low potash contents of the New Kingdom and the Nimrud Co blue glasses was discussed in section 1.10.1 and will be reviewed briefly here. Shortland and Tite (2000: 146-147) argued that low potash content in the Egyptian glasses indicated that the soda in the

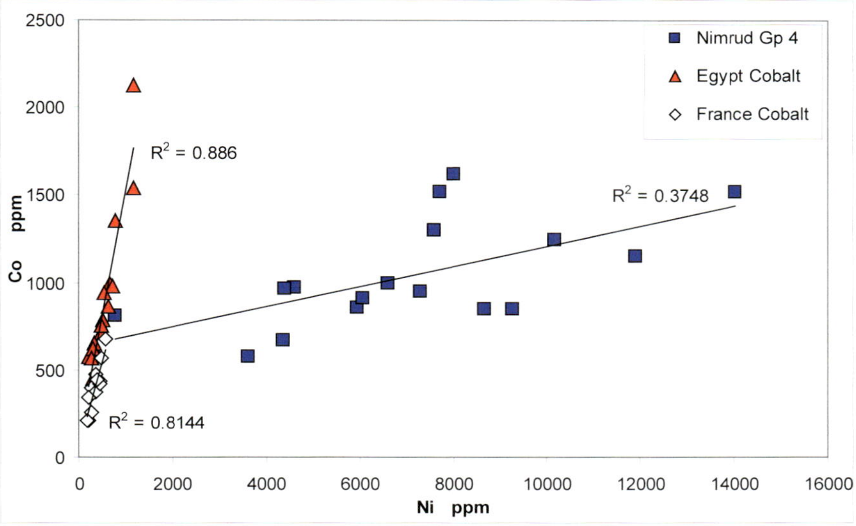

Figure 9-19 Scatter plot of Ni vs Co for Co blue glasses from Nimrud, Egypt and France. Data sources are cited in the text.

blue glasses was derived from natron, because plant ash generally provides a potash concentration well in excess of 1.5%. They considered the high magnesia content of the glasses to be the result of the presence of magnesia in Co alum. This view was challenged, however, by Rehren (2001; 2008: 1350) who preferred the proposition of a plant ash source for the soda in the New Kingdom glasses. Nicholson and Henderson (2000: 216-217) had already suggested that three low potash glasses they analysed might have been made with a different ash (analyses 6, 8 and 15), and analysis by Brill (1970a: table 2) of the ash of the chinan plant showed that it had high concentrations of soda and lime, and low concentrations of magnesia and potash (section 1.8). Tite and Shortland (2003) modified their position, suggesting that Egyptian Co blue glass contained both natron and plant ash. In the case of the Nimrud and the French glasses, it seems probable that the soda is derived largely from natron, as their lime contents are generally much lower than those expected in glasses where the alkali is added in the form of plant ash. This leaves open the question of the soda source in the New Kingdom Co blue glasses, because they have high lime contents (Figure 9-16).

The Co alum imparted high magnesia to the glass, and consequently a low magnesia glass was preferred in order to avoid devitrification. The resulting glass was low in potash and lime, but its stability was not compromised due to the relatively high level of alumina imparted by the Co alum.

There are no correlations between alumina and Co in either the French, Nimrud or Egyptian glasses (Figure 9-18) and, interestingly in the plot of the components of these glasses, the groups remain largely distinct, whereas the French and Nimrud glasses have closely similar base glass compositions. When the Co and other related colourant elements are compared, it is the French and Egyptian glasses which have the closest compositions.

Nimrud Co blue glasses have a considerably higher Ni content than the French or Egyptian glasses (Figure 9-19) and the correlations between Co and Ni are weak for the Nimrud glasses (r^2= 0.4), but are strong for the French and Egyptian glasses (r^2= 0.8 to 0.9). The correlations of the concentrations of Zn and Co are similarly weak for Nimrud glasses and strong for Egyptian and French Co glasses. The concentrations of Ni and Zn are well correlated for all groups between approximately r^2 = 0.7 and 0.9, with similar concentrations of Zn in the Egyptian and Nimrud glasses, with lower average Zn content in the French group.

This mixture of compositional similarities and variations in the colourant components of the French, Nimrud and Egyptian Co blue glasses is suggestive of the exploitation of several different alum sources within the Western Desert oasis region, and/or of different alum processing methods. The lower average Co content of the French glasses led Gratuze and Picon (2005: 274) to conclude that a different source of Western Desert alum was used for these glasses.

The only high Co (and high Cu) plant ash glass analysed in the present study was sample 5BM from Nimrud which is classed with Group 2 glasses according to base glass analysis (section 6.4.2). When trace element data are examined, its REE pattern is also found to match those of Nimrud Group 2 glasses. The REE pattern of sample 5BM is identical to the averaged REE data

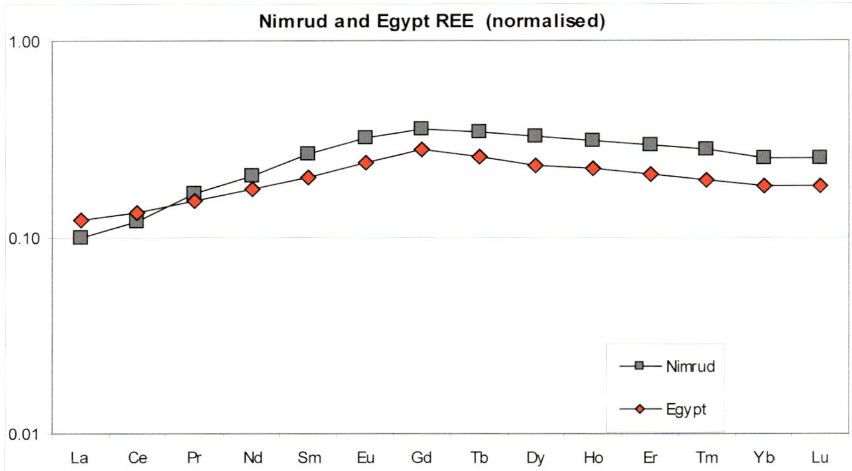

Figure 9-20 Line and dot plot comparing the average REE concentrations for Nimrud Group 4 and Egyptian Co blue glasses. Data are normalised to MUQ.

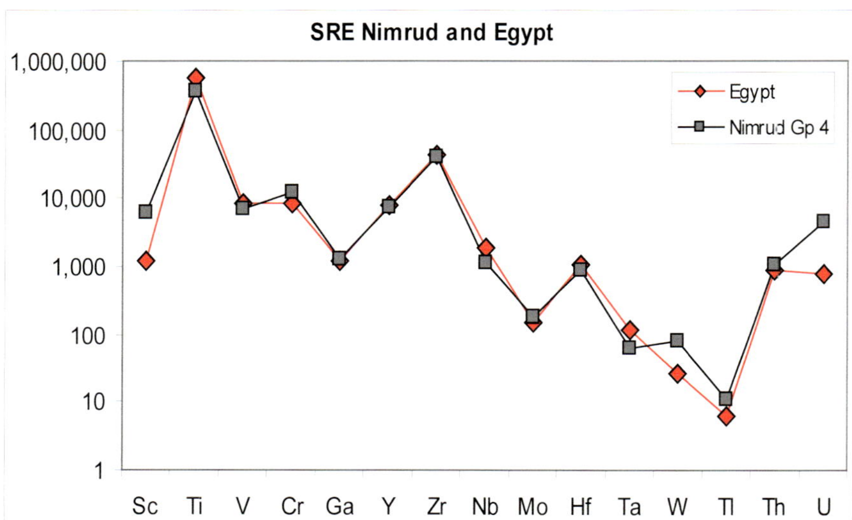

Figure 9-21 Line and dot plot comparing the averaged raw concentrations for SRE for Nimrud Group 4 and Egyptian Co blue glasses.

for Pella HMHK glasses, representing typical 2nd millennium BC plant ash glasses, although sample 5BM has a higher concentration of REE.

The averaged REE data from the analyses of Nimrud Group 4 glasses are compared with the averaged REE data from the analyses of six samples of Egyptian Co blue glasses from analyses of the author (Appendix VI) in Figure 9-20. The REE concentrations of the Nimrud Group 4 glasses were shown to have been influenced by the Co alum used to colour them (section 7.5). The similarity of the Nimrud Group 4 REE patterns with those of the Egyptian Co blue glasses suggests that the alum used to colour the Nimrud and Egyptian Co blue glasses originated from a similar source, presumably the Western Desert oases of Egypt.

Differences in the SRE content of the Nimrud Group 4 and Egyptian Co glasses occur most noticeably in the higher concentrations of Sc, W, Tl and U in the Nimrud Group 4 glasses, showing small variations in the sedimentary pattern between the two groups of glass (Figure 9-21).

Comparison of the alkali and alkaline earth elements from the Nimrud Group 4 and the Egyptian glasses in Figure 9-22 reveals a higher Be content in the Nimrud glasses and a higher Sr content in the Egyptian glasses.

Colourant related elements are plotted for Nimrud Group 4 in Figure 9-23 and for Egypt in Figure 9-24. Individual samples are plotted because of the variations in the content of elements such as Cu, Zn, Sn, Sb and

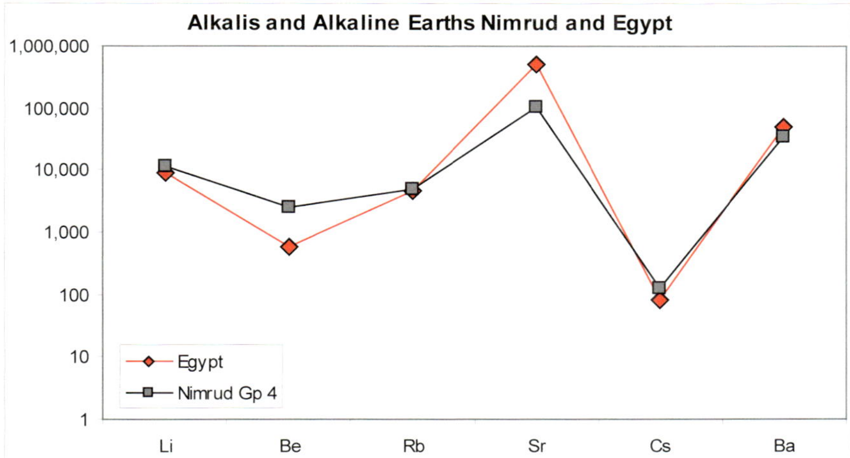

Figure 9-22 Line and dot plot comparing the averaged raw concentrations for alkali and alkaline earth elements for Nimrud Group 4 and Egyptian Co blue glasses.

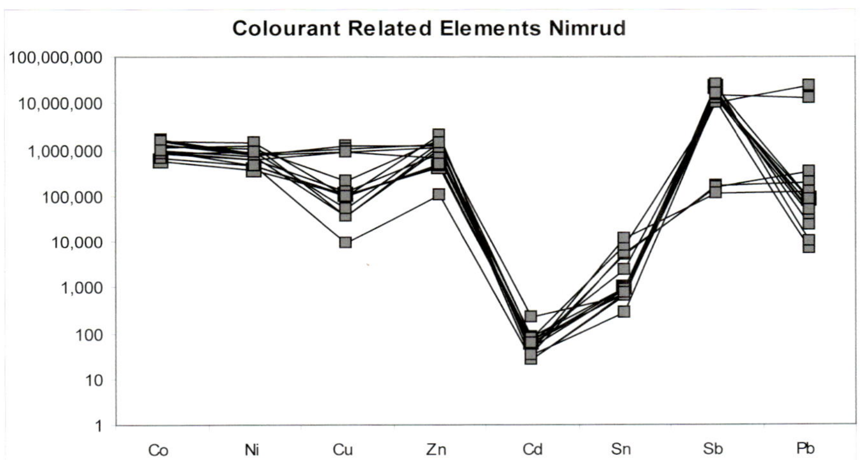

Figure 9-23 Line and dot plot of the raw concentrations for colourant related elements for Nimrud Group 4 Co blue glasses.

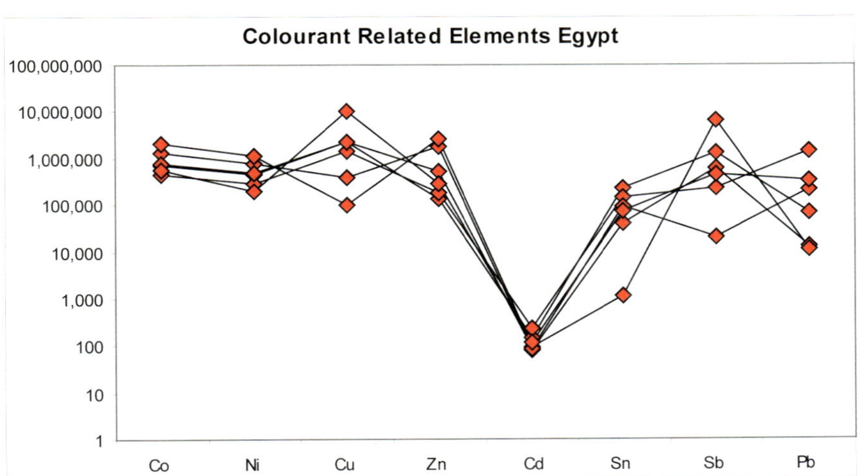

Figure 9-24 Line and dot plot of the raw concentrations for colourant related elements for Egyptian Co blue glasses.

Pb which are evident in both groups of glass. The concentrations of the colourant related elements are similar in both groups of glass.

The similarities in the REE concentrations, influenced by the Co alum colourant, and in the other colourant related elements, indicate a similar origin for the Co colourant and a similar colourant technology in both the Egyptian and Nimrud groups of glass. The variations in the concentrations of SRE and alkali and alkaline earth elements between the two glass groups suggest that these groups of Co blue glass were made with different base glass raw materials, originating in different regions. This finding is supported by the differences in the base glass compositional data observed between the LBA Egyptian glasses and the Nimrud IA glasses.

9.7.2 Raw Materials from Egypt in the Third Intermediate Period

Circumstantial evidence for an Egyptian source of Co lies in the apparently precocious use of natron as an alkali in these and other glasses from Nimrud (Group 3) and from IA Pella. The work of Schlick-Nolte and Werthmann (2003) suggested that, as might be expected, Egypt was already exploiting its own natron sources at this time for non-Co blue glass. Thus, it is quite possible that the natron used in the Nimrud and Pella glasses was of Egyptian origin. If so, it is also possible that the Near Eastern glass-makers obtained Egyptian Co.

In searching for a reason for the decline in Egyptian glass-making and the disappearance of Co as a colourant in vitreous materials, it is interesting to speculate as to whether the political upheaval of the Third Intermediate Period and its consequences might have been responsible (Kitchen 1986).

The Libyans were the chief political force that governed Egypt in the Third Intermediate Period from 1069-664 BC, encompassing Dynasties 21 to 25 (Leahy 1985: 51; Shaw 2000: 481). It was a period characterised by a fragmented and decentralised political structure (Leahy 1985: 51), in which significant changes to political organisation, society and culture were principally related to an influx of Libyans and Nubians. Egypt's interactions with regions beyond its boundaries were reduced, and this was among the factors which had an impact on the Egyptian economy, the structure of society, religious attitudes and funerary practices. At times, tensions arose over control of territory and resources (Taylor 2000: 330). The northern oasis of Bahariya had fallen to Libyans as early as the reign of Merenptah in the late 13th C BC New Kingdom. During the Third Intermediate Period, the oases of the Western Desert were certainly settled by Libyans who installed garrisons in the southern oases of Dakhla and Kharga. The Libyans were present also in the Delta (Leahy 1985: 53, 56, 60; Shaw 2000: 481), at Tanis, for example (Taylor 2000: 331).[77]

The makers of the Co blue glass inlays, found at Nimrud, must have obtained a supply of Co alum colourant, apparently from Egypt, but there is no evidence of how this might have been achieved. There is no evidence for the trade in Co for glass colouring, but Co alum is known to have been exploited until at least the time that these glasses were produced, i.e. the 9th C BC (Shortland *et al.* 2006).[78] If the Nimrud glasses were produced in the Syrian coastal region of the Near East, and not in Egypt, as the archaeological and chemical evidence suggests (sections 6.3.1, 9.4), the alum must have been gathered from the Western Desert oases and somehow transported northwards, possibly in the same goods trading network that transported natron.

Giddy (1987: 5-6, 9) described the string of Western Desert oases as geographically distinct from the Nile valley, being separated from it by 200km of desolate and treacherous terrain. Communication between the Nile valley and the oases had to be direct because no alternative existed to the 'unrelieved desert', where today there are no reliable water sources. The journey across the desert from the major oases to the Nile valley was possible via a number of direct routes which are attested by textual and archaeological evidence.

From textual evidence of the First Intermediate Period, Giddy (1987: 14) proposed a possible early date for the existence of a link between the Farafra and Siwa oases, which would have constituted the first stage in a land connection between the oases of Farafra and Siwa and the Wadi Natrun, to the west of the Nile delta. The other principal desert crossing from Siwa to the Nile valley went via Bahariya oasis to the region of Cairo. Siwa occupied a unique position because of its accessibility from the Mediterranean coast and from Libya to the west (Giddy 1987: 17-18).

That routes between the oases through the desert were already in use by the Third Intermediate period is also attested by the Kamose stela (mid 16th C BC, Shaw 2000: 481) which mentioned a north-south route running parallel to the Nile, past Farafra, which appears to have been a focal point for movements to the east and west as well as an important inter-oasis link in the 19th Dynasty. An inscription of Merenptah also mentions movements of Libyans to the west (Giddy 1987: 164) in the late 13th C BC (Shaw 2000: 481). The combined evidence is sparse, but indicates that materials could have been

[77] White (1994) reviewed current archaeological evidence showing a considerable pre-Greek Libyan presence in Egypt.

[78] Giddy (1987: 5) cited documentary evidence for the sourcing of alum from the Western Desert oases during the New Kingdom.

transported either across the desert between the oases, the Wadi Natrun and the coast, or alternatively, across the desert from the major oases to the Nile valley, and from there by river to their destination.

From the coast, traders such as the Phoenicians, could have carried the natron and Co alum cargo by ship to the coast of Syria, and hence to the glass-makers. Although there is no archaeological or textual evidence for the transportation of these raw materials overseas, the use of natron and of an alum source of Co is attested by the compositional analyses of the glasses from Nimrud, which are believed to be of Near Eastern origin because of their chemical composition and their association with Phoenician ivories (section 9.4).

The question remains as to why the Egyptians, themselves, made little glass and apparently stopped using the alum as a colourant during this time. The answer might lie in the possible migration of the Co blue glass-makers to more stable and profitable Near Eastern regions to the north. Egyptian craftsmen, and their successors, could have made Co blue glass in the Near East to a modified version of the traditional Egyptian recipe using a local silica source, fluxed with natron and coloured with Co alum acquired from Egypt. The expatriate production of Co alum coloured glass might, in part, explain the compositional differences observed in the otherwise closely similar IA Near Eastern and LBA Egyptian Co blue glasses. The glass-makers would have known that Egypt was the source of these raw materials, and would have had the technological knowledge to make low potash, Co coloured glasses, either from a low potash plant ash or with the new natron formulation.

That Co blue glass was not made in Egypt in the Third Intermediate Period is presumably not due to a disruption of the raw material supplies. The glass-makers in the Near East could obtain both natron and Co alum from Egypt, and it is assumed that there was no other source for these materials, because none has been discovered. The mineral soda composition of the 10th C BC glasses of Nesikhons show clearly that natron was being gathered from a Western Desert source, such as Wadi Natrun, to make glass during the Third Intermediate Period, although it is not certain where it was made. With access to the Western Desert oases, attested by the use of natron, it is not easy to explain why Co was not used in Egypt at this time, unless the specialist Co glass-makers had moved to Syria. Whether the glasses of Nesikhons were made by glass-makers who still worked in Egypt or not, these glass-makers appear to have been experimenting with a mineral soda fluxed glass that had too little lime to have survived outside the dry tomb where they were found.

Following the theory proposed by Rehren and Pusch (2007: 231-233; section 1.3) that there were different glass workshops to produce each of the specialist colours, it could be suggested that those who made Co coloured glass had, for some reason, relocated to Syria. No copper red glasses, which had been made at Qantir in the LBA, were made in Egypt at this time either. Other, non-specialist, glass-makers might have remained behind, working in changed circumstances that led to their experimentation with making mineral soda glasses (Nesikhons), rather than the traditional LBA plant ash glasses. Although much of the early unstable glass of this type would not have survived (section 9.5), it is tempting to speculate that the mineral soda tradition began in Egypt, near to the source of natron. Although the low-lime Pella LMLK glasses are earlier in date, they too would not have survived without their preservative high iron oxide content.

Experimentation with new glass raw materials might have been possible in an environment of political and cultural change that provided the impetus and the liberty to diverge from centuries old techniques. Under a decentralised administration, the glass-makers might not have adhered to religious or other imposed practices which governed LBA glass-making (section 8.5).

9.7.3 Phoenician Traders

Maritime trade was the most likely means by which Egyptian raw materials could have reached the Near East. Dominant amongst the maritime traders in the Mediterranean region during the LBA and IA, were the Phoenicians, who occupied the northern Syro-Levantine coast and colonised widely around the Mediterranean. They were recognised as the manufacturers and exporters of metal vessels, jewellery, ivory furniture components and glass objects (of particular significance when considering the Co blue inlays associated with the Phoenician-type ivories found at Nimrud).

Archaeological evidence is meagre in the Phoenician homeland, because much of it lies beneath modern cities (Akkermans and Schwarz 2003: 386). In *c.*1200 BC, technological innovations for boats enabled people from Syro-Palestinian harbours to sail long distances around the entire Mediterranean. The Phoenicians became the leading maritime commercial force in the Mediterranean after the decline of Hurrian populated Ugarit, which had monopolised maritime trade with the major Palestinian ports in the LBA (Markoe 2000: 29; see also Marfoe 1987) and had been linked by trade routes to the Euphrates (Kuhrt 1995: 303).

By the 11th C BC, according to the archaeological record, the Phoenicians were active traders who were expanding their network of commercial interests, particularly southwards along the coast of Israel. This was reflected in the urban growth of Phoenician cities at that time. The ports of Sidon, and to a lesser extent Tyre, were in a dominant position among the Phoenician coastal cities in the late 12th and 11th C BC (Markoe 2000: 31).

During the 10th C BC, in the absence of textual evidence, the Phoenician towns probably suffered an economic decline owing to the changed political environment in the surrounding Syrian region. The Aramaeans, an amalgamation of West Semitic pastoralist tribes, now occupied large tracts of Syria and central Mesopotamia. The eastward movement of these tribes severely affected neighbouring Assyria and Babylonia by obstructing internal communication and blocking traditional trade routes westward from the middle Euphrates to the Mediterranean. The disruption to commercial trade with the west, as well as the decline in Egyptian trade during the Third Intermediate Period, would have affected coastal Phoenician cities, for example Byblos, that acted as outlets on the Mediterranean coast for trans-Euphratian commerce (Markoe 2000: 36; Master 2003: 47, 59).

At the end of the 10th C BC, the economic and military fortunes of Egypt and Assyria changed as both sought to reassert themselves. This was the precursor to their aggressive military campaigns of the 10th to 9th C BC. Under the ominous threat of Assyria, the Phoenician cities allied themselves with Egypt after the accession of Sheshonq I in the mid-10th C BC. This was the beginning of the 22nd, or 'Libyan', Dynasty in Egypt (Markoe 2000: 37).

By the early 1st millennium BC, after the decline at the end of the LBA, the Phoenicians were able to revitalise Mediterranean maritime trade and establish colonies as far west as Spain and the Atlantic coast of North Africa (Markoe 2000: 29, 41, 170-188; Akkermans and Schwarz 2003: 8-9; Master 2003: 57). By the 9th C BC, there is evidence of a Phoenician presence in inland north Syria, which Markoe (2000: 38) associated with Phoenician involvement in the Taurus region and in the nascent Syrian trade with the Euphrates Valley.

In the 9th to 8th C BC, the Neo-Assyrians subjugated the Philistine states in the southern Levant and the more northerly Phoenician cities and states of inland Syria, requiring them to provide continual tribute to their overlords. In particular, the Phoenicians acted as major suppliers of raw materials for the Assyrian empire, until it was destroyed in the late 7th C BC by an alliance of the Babylonians of southern Mesopotamia and the Medes of western Iran (Akkermans and Schwartz 2003: 387-389; Master 2003: 48). The Phoenicians offered gifts, including ivory, to the Assyrians to secure trade concessions, and Ashurbanipal's exploratory campaign to the Mediterranean coast is thought to have been commercial in nature (Markoe 2000: 39), indicating the Assyrian concern with access to goods and trade routes.

By the end of the 9th and in the first half of the 8th C BC, the Phoenicians had greater political freedom and were presumably free to conduct their commerce (Markoe 2000: 40). According to Master (2003: 50), only the Phoenician city of Tyre, which was a highly successful Mediterranean merchant town, was allowed to remain independent. The Assyrians had an interest in extracting revenue from international trade, and trade using water transport was a relatively cheap means of moving goods in quantity.

The campaigns of Tiglath-Pileser III, in the mid 8th C BC, ended the political independence of the Phoenicians. The Assyrian demands for metals exerted financial pressure that spurred the Phoenicians to search more widely for Mediterranean ore sources, around the time of Sargon II, in the late 8th C BC (Markoe 2000: 41, 98; Boardman 2001: 36; Master 2003: 48). The south of France, for example, was a valued outlet for tin transported overland from Cornwall and Brittany, and which was then transported by Phoenician vessels around the Mediterranean (Markoe 2000: 187). The south of France was also the location where the French LMLK black and Co blue glasses were found, no doubt as a result of these trading links.

Essentially, the Phoenicians transported commodities that they neither produced nor consumed, and as such their economy was largely reliant on their role as intermediaries. They possessed no material resources of their own, although they produced and traded pottery, as was revealed by the Levantine and Phoenician ceramic contents of a Phoenician shipwreck off the coast of Spain. The Phoenicians also produced glass and traded it abroad. Bartoloni (2001: 97) commented that the high value of glass goods was enhanced by 'the virtues of minimum space occupied and maximum profits earned in shipping by sea'. Two 8th C BC Phoenician ships, which sank off the coast of Ashkelon in Israel, carried wine, which the Phoenicians did not produce. Their colonies and agreements with established powers around the Mediterranean enabled them to move goods anywhere in this region in an innovative system that was unsurpassed until the Roman period (Fulford 1989: 171-172; Bartoloni 2001; Master 2003: 57-60).

Ships traveled in a counter-clockwise direction around the Mediterranean, staying close to the coast as much as was possible. There were few good harbours between the Nile Delta and Lebanon and it was possible that trading ships entered the Nile to find anchorage (Van De Mieroop 2004: 133). On the basis of the wide-ranging trading network operated by the Phoenicians, it is likely that they were responsible not only for transporting the raw materials to make glass, but also the finished Co-blue and iron black glasses from the IA Near East to France. Their cargoes were described as highly heterogeneous by Horden and Purcell (2000: 369), and it can be surmised that small glass objects and ingots were easily distributed along with wine, metals and other goods. Such cargoes were known from the LBA to the Early IA, from shipwrecks such as that of Ulu Burun in *c.*1300 BC and Cape Gelidonya in *c.* 1200-1150 BC (Horden and Purcell 2000: 347; Gratuze and Billaud 2003: 13-14).

Many of the French glasses were found in the region of Languedoc, a short distance from the Mediterranean littoral, with access to shipping trade. That glass found its way to France from the Near East, from as early as the 14th C BC, is revealed by the identical composition of some of the beads found at French sites with the glass recovered from the Ulu Burun shipwreck (Gratuze and Billaud 2003: 13-14). Bass (1987: 716-718) proposed that either Canaanites or LBA Phoenicians played a major role in the maritime commerce of the eastern Mediterranean (see also Larsen 1987).

In his discussion of Phoenician ivories with their blue glass inlays, Barag (1985: 51-52) proposed that the Co blue inlays could have been made by Phoenician craftsmen, but that there is a general lack of knowledge of the role of Phoenician craftsmen in glass-making in the first millennium BC. Nevertheless, Barag (1985: 54) believed that it was clear that Phoenician craftsmen were embellishing ivories and glass vessels, such as those found at Nimrud, with glass inlays. Pieces of raw Co coloured glass and frit were reported at LBA Ugarit by Matoïan (2000b: 41; see also Caubet and Kaczmarczyk 1987), providing evidence that this material was worked, if not made, on the Syrian coast. Matoïan (2000a) considered that the association of the ivories with the glass inlays suggested that they were produced in the Syrian coastal region (see also Liebowitz 1987).

Plant ash glasses were exported from the Near East to France in the LBA, and the new mineral soda glasses were exported in the IA. Gratuze and Picon (2005: 269) maintained that the European plant ash glass of the MBA to LBA was probably introduced by travellers. At this time, there were practically no Co blue beads, but at the beginning of the French IA, mineral soda glasses first appear and supplant the earlier glasses, during the 9th to 7th C BC.

9.8 Colourless Glasses

A small number of colourless glasses are known from the LBA in Egypt and the Near East (section 1.10.2). They appear to have been either naturally colourless, due to the use of pure raw materials, or were decoloured by the careful control of the kiln environment (Nicholson 2007: 1; Jackson and Nicholson 2007: 106). The use of Sb or Mn as a decolourant was not introduced until after the LBA (section 1.10.2).

Colourless glass is rare in the Pella IA assemblage, which dates between one to three centuries earlier than the IA Nimrud glass. One colourless glass was analysed from an 11th to 10th C BC context from Pella (sample 119), but might in fact have been made in the LBA (section 4.2.1, 4.5.4). Its low Sb content of 0.64 ppm indicates that Sb was not used to decolour the Pella glass sample and it is, therefore, typical of LBA colourless glasses from both Egypt and the Near East.

From the IA at Nimrud come ten glasses, decoloured with levels of Sb averaging 3500 ppm. These glasses belong both to Group 1 HMHK and Group 3 LMLK. Figure 9-25 compares the averaged raw colourant element data for Nimrud colourless glasses with the Pella colourless glass. Whereas the Pella glass is older, its colourant and related trace element signature, apart from its low Sb content, is remarkably similar to that of the Nimrud colourless glasses, whether HMHK or LMLK.

Examination of the Sb contents of the translucent Cu blue glasses from LBA Pella reveals that there are two samples (210 and 216) which have highly elevated levels of Sb compared with the other translucent glasses. The average Sb content of the translucent blue glasses, excluding the two high values, is 16 ppm (a range of 1 ppm to 51 ppm). The two high Sb glasses both have approximately 1830 ppm of this element (section 4.5.4). This falls within the range of Sb values for the Nimrud glasses deliberately decoloured with Sb (section 6.5.4.1). Similarly, from Nimrud there is one translucent Cu blue glass (37P), which has in excess of 2000 ppm of Sb, equivalent to levels in the colourless glasses, and considerably less than in the intentionally opacified glasses. The significance of, and the reason for the high Sb content of these particular Cu blue glass compositions remain unclear.

Could these high levels of Sb have been added to the Cu blue glasses as a contaminant together with another ingredient, or is it possible that intentional decolouring with Sb had started earlier in the LBA, and that sometimes decolourised glass was coloured blue with Cu as a secondary process, even in the 2nd millennium BC? This is unlikely given that no colourless glasses intentionally decoloured with Sb are known from the LBA. Nuzi glasses of this period, analysed in the present study, do not appear to have been decoloured before being intentionally coloured, i.e. unless opacified, their Sb content is low, at approximately 10 ppm Sb on average (section 3.3.4). It would be unwise to draw conclusions on the basis of such slim evidence, and to wait until further information should come to

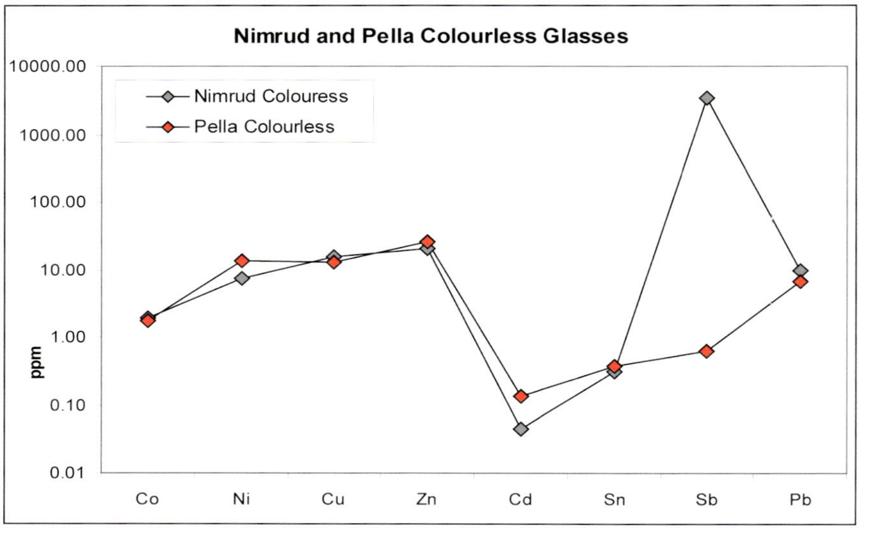

Figure 9-25 Line plot of averaged raw Nimrud colourless glass colourant and related trace element data compared with Pella colourless sample 119. A logarithmic scale is used.

light through new discoveries. Importantly, it is now clear that the earliest evidence for the use of Sb as a decolourant is found in the IA glasses from Nimrud, showing that intentional decolouring with this element began half a millennium earlier that has previously been thought (section 1.10.2).

It is significant that while the production of glass had declined in Egypt, the IA was a period of renewed vigour in glass-making in the Near East. Natron from Egypt was introduced as an alternative flux to plant ash, with the earliest examples of natron fluxed glass coming from Egypt and Pella. Plant ash continued to be used, but appears to have been derived from different plants to those used in the tightly controlled LBA glass industry. Co alum from Egypt continued to be used as a colourant in glasses produced somewhere in the Near Eastern region, probably along the coast of Syria, although Co alum was no longer used as a colourant in Egypt, the Near Eastern glasses appear from the results of this study to have been modelled on the Egyptian Co blue glasses of the LBA. Glasses in the Near East were now deliberately decoloured by the addition of Sb, whereas in the LBA decolouring was either natural or produced by appropriate kiln conditions.

Diversity in glass compositions from the IA Near East is evident. LBA traditions, which provided the basis for the IA industry, evolved and were modified to supply new demands. Extensive maritime trade in the IA transported natron and Co alum from Egypt to the Near East, and then carried Co blue and iron black glasses as far afield as France, to the west. Iron black glasses would have travelled by overland and river routes to Hasanlu in the east. The combined IA data presented in this study have provided important insights into the development, experimentation and innovation which occurred in glass production during this period in the Near East. Diversified glass compositions which introduced natron as a flux, Sb as a decolourant, new plant ash glass formulations, and the use of Co alum from Egypt as a colourant, characterise an industry that archaeological and chemical compositional evidence indicate as likely to have been centred in Syria.

Chapter 10

Conclusions and Future Research

10.1 Conclusions: LBA Glass-making

The present study has sought to answer archaeological and technological questions regarding the production of the earliest consistently manufactured glasses, during almost 1000 years of the LBA and IA of the Near East (Chapter 1). The technology of the production of both LBA and IA glasses was shown to have been far more complex than has previously been understood. The investigation of the characteristics and the development of this early pyrotechnical industry was made possible by the collection and interpretation of chemical compositional data obtained from analyses of the glasses included in the current study.

Data from compositional analyses of a selection of LBA and IA glasses from the key sites of Nuzi, Pella and Nimrud have provided important new information on the little studied glasses of the ancient Near East (Chapters 3 to 7). A large number of oxides and elements was analysed by highly sensitive methods to produce a detailed database. In order to study and interpret these data, an innovative method of organising and presenting the data was developed (Chapter 2). Trace element data were grouped as REE, SRE, alkalis and alkaline earths, and colourant and related elements. Concentrations of the elements in each group were compared with those occurring naturally in the UCC, and an SBS was determined from the REE and SRE contents of the glasses. The comparison of elemental concentrations in the glasses with the SBS of those glasses provided a means for assessing the degree of elevation or depletion of elemental concentrations compared with natural sedimentary levels. This process was central to the identification of key trace components in the glasses, which were used as discriminators between individual glasses and between groups of glasses.

The data for LBA glasses obtained from the present study have been examined together with the limited published analyses of other LBA Near Eastern and of Egyptian glasses to define the characteristics of their base glass compositions (Chapter 8). The relative homogeneity of the soda-lime-silica glass compositions of Egyptian and Near Eastern glasses of the 2nd millennium BC has been demonstrated, but regional distinctions have been identified also in the present study (section 8.3). Trace element data have proved particularly useful for this purpose, but, at the time of writing, a lack of comparable published trace element data has limited comparison with other glasses. Significant variations in trace element concentrations of glasses from the LBA Near East and Egypt were identified, with the chief indicators of regional distinctions being the REE, Cr, La, Zr, Ti, Be, Li, Cs and Sr.

Interesting variations occur in the lime contents of the LBA Near Eastern glasses, which fall into two groups based on high- and low-lime concentrations (section 8.4). These groups do not appear to be related to opacification of the glasses from the sites of Nuzi and Tell Brak, but more closely conform to a high-lime formula for opaque glasses and a low-lime formula for transparent glasses from Pella. These observed differences are associated with significant differences in the concentrations of the related alkalis magnesia, soda and potash, which suggests that different plant ashes have been used to make the glasses from the two lime groups. The same range of lime concentrations, but not the two lime groups, are found in contemporary Egyptian glasses, suggesting a difference in selection and/or use of raw materials between the regions.

The two lime groups in Near Eastern glasses are probably indicative of the existence of more than one glass production centre in the Near Eastern region, and do not appear to be related to the seasonal growth of plants ashed to make the flux, or to the separate and intentional addition of lime (section 8.5). This is not to say that the ancient glass-makers were unaware of the properties of the plant ashes they used. On the contrary, the apparently deliberate choices of standard raw materials, which characterised LBA glass-making, suggest that each of the major ingredients, namely silica, soda and lime, were consistently incorporated, even if the lime component was not individually recognised. Just as plants were recognised for their soda contents, so were they understood to produce an enduring glass, the result of having a sufficiently high lime content. Analyses of plant ashes from many regions have shown that plant ash composition is considerably varied for a number of reasons (section 1.7, 1.8). The glass-makers had to be able to assess which were suitable to make glass.

The consistent formulation of the LBA glasses, as evidenced by their base composition, use of colourants and an opacifier, and their trace element signature, suggests that glass production in the LBA was tightly controlled, probably by royal court and/or temple interests. Archaeological, chemical and textual evidence support the proposition that glass was made to imitate precious stones, and was in itself a highly valued and perhaps competitive material, made in a closely regulated industry that followed standard recipes (sections 1.3, 8.5).

10.2 Conclusions: IA Glass-making

The LBA ended with political and social upheavals that saw a decline in glass manufacture in a period traditionally known as a 'Dark Age'. Glass production all but ceased in Egypt and little glass has been found dating to the IA in the Near East (section 1.2). Never-the-less, this period proved to be one of change heralding a new era of technological experimentation that led to significant developments in glass production in the IA. The changes in political control and religious dictates in the IA of the Near East and the Third Intermediate Period of Egypt were perhaps the driving force behind the technological diversity in glass-making which emerged at this time. Trading networks, both maritime and land-based, were extensive and the demands for supplies of raw materials and goods by powerful states in the region fuelled the re-establishment of the glass industry in the Near East (sections 9.7.2, 9.7.3).

Glass object types were limited in the Early IA, but glasses were often made in dark colours, such as dark blue or black, or were colourless and weakly coloured, translucent. Traditional colours of turquoise opaque and translucent blues persisted, but were no longer dominant (section 1.2). There is evidence of continuity from the LBA in the soda-lime-silica compositions, the traditional opaque and translucent blue colours, including the use of Co alum, and the use of plant ash as a flux. Innovation is equally evident in the introduction of mineral soda as a flux (section 1.9), the use of different plant ashes, the increased use of sand as a silica source, the strong colours, and the ability to decolour with Sb (sections 1.10.2, 9.6 to 9.8).

Glasses exhibited diverse compositions at this time, and evidence of experimentation is seen in the composition of glasses containing insufficient lime to achieve chemical stability (Chapters 5 to 7, sections 9.2, 9.5). The use of different raw materials required the modification of conventional LBA recipes to be able to produce chemically stable glasses.

Comparisons made in this study between Co blue glasses from IA Nimrud (Group 4) and from contemporary France revealed that they were all coloured using a pigment derived from cobaltiferous alum, probably that used in New Kingdom Egypt, although variations in the sourcing or processing of the alum has resulted in slightly different proportions of the chemical components in the glasses. The Co blue glasses from Egypt, France and Nimrud have elevated levels of magnesia and alumina and share the characteristic that all are low in potash. Coupled with exceptionally low lime content, this characteristic indicates that the alkali of the Nimrud and French glasses was added mainly in the form of natron, with minor or no plant ash, although the relative amounts of natron and plant ash in the New Kingdom glasses remain uncertain (sections 1.10.1, 9.7).

These technological characteristics indicate both continuity, in the form of the use of Co alum, and change, with a shift from plant ash to natron as the prime glass flux. The use of plant ash as a flux continued in the IA, but compositional indicators suggest that the plant ashes were different from those used in the LBA. The Nimrud Group 2 glass compositions are reminiscent of those of the LBA Near Eastern plant ash glasses and, like those from the IA at Pella, might in fact be remnants of LBA production (sections 4.2, 9.2).

Compositional analyses, combined with archaeological evidence, have strongly suggested that the IA glasses from Pella and Nimrud were the products of a Near Eastern region, possibly the Syrian coast (sections 1.2, 6.2.2, 9.6, 9.7). The likely geological origins of the chromite found in the Pella LMLK glasses (sections 5.4.2, 9.6.1), and the link between the 1st millennium BC Nimrud Co blue, low potash glasses and Phoenician ivories (section 6.3) suggests that the IA black and Co blue glasses could have originated in Syria. The possibility of glass-making in Syria, although from an earlier period, has been discussed by several authors based on studies of textual evidence (section 1.3). Taken together, these strands of evidence for a Syrian glass-making centre are compelling and deserving of further research.

As there is no other known source of the glass ingredients, natron and Co alum, it is proposed that during the Third Intermediate Period in Egypt, when the country was under Libyan control, natron and Co alum were sourced from the Western Desert oases and Wadi Natrun by overland routes, either through the oases to the Nile delta and the coast, or to the Nile River and thereby along the river to the coast (section 9.7.2). Phoenician traders are likely to have collected these raw materials as part of their extensive Mediterranean trading networks, and transported these and other goods to the Syrian ports where the glass-makers could acquire them. The association of Phoenician ivories and Co alum glass inlays strengthens the theory for this connection. That Co blue and iron black glasses, closely similar in chemical composition to the Nimrud Co blue glasses and Pella black glasses, were found in France, is likely to be the result of Phoenician trade (section 9.7.3).

The origin of the Nesikhons glasses is uncertain and it would be curious if natron glasses were made in Egypt in the IA, albeit in an experimental fashion (sections 9.5, 9.7), while Co blue glass was not. It is proposed that the specialised Co blue glass-makers had, for some reason perhaps associated with the changed political and religious environment in Egypt at the time, moved north to Syria where their tradition was modified by the use of natron as a flux, instead of plant ash (or a plant ash/ natron mix), but otherwise resembled that of their LBA predecessors.

Natron was sourced from Egypt, but it is not known whether the use of natron as a flux originated in Egypt. LMLK glasses from Pella are contemporary with, and some may even pre-date, those discovered from the 10th C BC in Egypt (section 5.2). On the basis of the unstable composition of the glasses from the tomb of Nesikhons, it is assumed that other, perhaps earlier, LMLK glasses have perished. If the manufacture of non-Co coloured LMLK glasses occurred in Egypt, this would indicate that while the colour-specialist glass-makers had moved on, there were still craftsmen left in Egypt who were perhaps released from the strict state or temple control of the previous periods, giving them the freedom to experiment with new combinations of raw materials. It is not known whether difficulty in the supply of plant ash could have been a factor in this change in practice (sections 1.2, 1.9, 9.5).

The Pella black LMLK glasses are early examples of the production of the new natron flux technology in the Near East. They too were low in lime and were preserved by the high contents of iron oxide, which gave them their colour (section 9.6). They were traded to distant France and eastward to Hasanlu, indicating that the commercial links of the IA Near East spread to far away lands by maritime and overland routes (section 9.7.3).

The glasses of Nimrud Group 3 are the third group of LMLK glasses, probably made with a natron flux, which have been examined in the present study. The low levels of sedimentary contamination in Group 3 glasses indicate that they were made from comparatively pure raw materials (section 6.4). That they include colourless, opaque and translucent blue glasses shows that the colourant and decolourant agents did not contribute significant sedimentary contamination to the glass batch. Either these glasses represent the achievement of a successful mineral soda recipe, which contained a sufficiently high level of lime to ensure the preservation of the glasses, or they were made with a quartz silica and LMLK plant ash combination. It is tempting to conclude that these glasses provide the earliest evidence of the mastery of the natron-fluxed glass technology that later dominated Hellenistic and Roman glass production. That Nimrud Group 1 glasses also included colourless examples, but were of a different plant ash composition, is probably another indication that more than one workshop was operating during this time.

The LBA and IA compositional data from Near Eastern glass, compiled in the present study, have enhanced and facilitated inter-regional and chronological comparisons of glass production in its early stages. The information regarding the important technological developments that has been presented and discussed here, has contributed to our knowledge of the achievements of a sophisticated, but poorly understood, industry. The investigation of the growth, development, organisation and spread of glass-making, including the addition of a substantial chemical data-base, presented in this study, has provided unique insights into ancient glass production and has laid the foundations for future research.

10.3 Future Research

The compositional research undertaken in the present study has raised many new questions in the process of answering those set out in the aims of this study (section 1.16). These complex issues are not easily resolved, for example it is clear that while LBA glass is of comparatively homogeneous composition, there are internal chemical variations which have implications with regard to our understanding of industrial organisation and workshop practices. Although our knowledge of the use and sourcing of raw materials has been deepened and refined by the interpretation of sensitive base glass oxide and trace element analyses, the additional information provided by isotope analyses would significantly enhance our knowledge of LBA and IA glass manufacture.

Isotopes have been used by researchers to identify and distinguish raw materials used in ancient glass-making, thereby clarifying our understanding of the nature of raw material sources employed by early craftsmen. The isotopes of Pb, Sr, oxygen and Nd in particular have been used in glass studies. Examples of isotope analyses in the study of ancient glass include those of Brill (1970b), who used isotopes of oxygen and of Pb as a means of distinguishing between different glasses, and of Brill and Shirahata (1997) who were able to distinguish different origins of glasses and other materials by the use of Pb isotopes. The isotopes of Pb were used by Degryse *et al.* (2005) to distinguish between two silica raw materials in glasses from Sagalassos in Turkey, and to determine that recycled glass was used. The analysis of Sr isotopes confirmed the recycling of glass and was used to suggest the possible origin of the silica raw materials (Degryse *et al.* 2006). Freestone *et al.* (2003) found the ratios of Sr isotopes useful in distinguishing between lime derived from marine and terrestrial sources. Leslie *et al.* (2006) used isotopes of oxygen and Sr to discriminate between sources of lime and silica in ancient glasses. At the time of writing, distinctions between the different sand raw materials used for primary glass production of Roman glasses in the Mediterranean region, using a combination of Nd and Sr isotopes, have been identified by Degryse and Schneider (2008). The isotopes of Nd have been used to characterise Nile River silt as a component which dominates coastal sands of the Egyptian and Levantine region (Freestone *et al.* 2009).

It is suggested that the analyses of a combination of isotopes from the LBA and IA glasses could help to unravel the mystery of the lime concentrations and to clarify the sources of silica. They would also prove useful in discriminating between otherwise homogenous glasses produced with similar raw materials, but which perhaps originated in different workshops and/ or regions.

Chapter 11

Epilogue: Published Research 2009-2019

11.1 Introduction

Since the completion in 2009 of the thesis on which this study is based, new research has continued to build our knowledge of Late Bronze Age to Iron Age glass-making practices, raw materials and provenance. With analysis of more glass, often now including isotopic analyses, glass research continues to answer questions about the technology and development of glass industries within the wider Mediterranean region. By producing new data and refining interpretations of previously collected data, glass research is at once achieving a more intricate understanding of the organisation and processes of the industry, while piecing together the bigger picture of the distribution of its products, and the relationship between the peoples and regions who produced and consumed glass.

The following is a brief review of publications, 2009 to present (mid 2019), and the issues raised in them that are most relevant to the study of LBA and IA glass production as presented in this study. This survey of the last decade of literature, beginning with more general, overarching publications, is arranged in loose chronological order grouped into sections that have particular relevance to the themes and issues tackled here, namely analytical techniques, opacification, cobalt (Co) colourants, provenance, and distribution of raw and finished glass.

11.2 New and Consolidated Studies

New and more extensive chemical analyses continually add to our collected compositional data upon which new interpretive studies are based. These studies find their focus in being site or region specific, comparative within or across datasets, compilations of information from multiple sources, or overviews of research that continue to expand and enrich our knowledge of ancient glass production and distribution. With the publication in 2012 of Brill's third volume of data and interpretations (Brill & Stapleton 2012) a grand collection of analyses has been completed and continues to provide a valuable resource in the ongoing quest to know ancient glass-making.

Isotopic characterisation is being explored and used increasingly in studies of provenance and composition to complement data provided by elemental and oxide analyses. The complex nature of the components, and pyrotechnical journey of vitreous materials makes for particularly challenging study. The potential and limitations of isotopes for distinguishing primary glass sources were explored in a compilation edited by Degryse and co-workers, *Isotopes in Vitreous Materials* (2009). This volume is a collection of studies of 'state-of-the-art' isotopic analyses that investigate the provenance of Syrian plant ash glass and natron glass production, LBA Egyptian pigment sourcing, and compositional characterisation of later Levantine and south-eastern Mediterranean glass. *Modern Methods for Analysing Archaeological Glass, I* (Janssens (Ed.) 2013) presented a wide range of current analytical techniques for determining glass composition and identifying glass components, including studies of glass from specific regions and chronological periods.

Thoughtful overviews bring together what usually amounts to years of research, interpretation and discussion in syntheses such as *Lapis Lazuli from the Kiln,* Shortland's (2012) reflective examination of the impact that scientific analyses have had so far on interpretation of information from all sources on LBA Egyptian and Near Eastern glass, raw materials, and workshops. Henderson, in *Ancient Glass* (2013), presented a broad study spanning the material and chemical properties of glass, and glass provenance from the earliest to Islamic period glass, incorporating data from the Near East, Europe, Africa and Asia, while Rehren and Freestone (2015) took stock of scientific glass research to date as a means to direct research questions into the future. Gliozzo's (2016) useful review of research into the composition of colourless glass from the LBA to the modern period, traced its chemistry and developments in manufacturing technology from naturally colourless to intentionally decoloured with Sb or Mn or a mixture of both. Compilations such as Shortland, Freestone and Rehren's *From Mine to Microscope* (2009) presented a range of papers that place glass technology in its wider pyrotechnical context alongside other vitreous and metallic products of ancient high-temperature technology. Most recently, Schmidt (2019) examined the history of glass in IA Mesopotamia in a multi-disciplinary work that incorporated archaeological and philological evidence with archaeometric analyses. This study of typology, manufacturing techniques, the organisation of production, and the importance of glass to Mesopotamian society, aimed to reconstruct the operational sequence of primary and secondary glass production. Such specialised studies provide the bases for more general overviews with broader popular appeal, such as Wight's *Molten Color* (2011), a catalogue of glass in the J. Paul Getty Museum in Los Angeles.

11.3 Calcium Antimonate Opacification

Advances in the understanding of opacification processes have been made in several studies in the past decade, but scholars have differing interpretations of the evidence, and details of the glass opacification process remain elusive. Studies have focused on whether the opacifier, calcium antimonate, was crystallised before or after addition to the glass melt, and on implications of the levels of lime and antimony in translucent and opacified glasses.

According to Biron and Chopinet (2013: 60-62), there were two main ways to opacify ancient glass: *in situ* crystallisation, or *ex situ* production of opacifying crystals, which were then added to a pre-prepared glass or a glass melt. In a series of studies, Lahlil et al. (2008, 2010a, 2010b, 2011) investigated calcium and lead antimonate opacifiers in ancient glasses. They first concluded that Roman glass was opacified by the process of *in situ* crystallisation for white calcium antimonate, but that, as had been proposed by Shortland (2002), the addition of separately or *ex situ* synthesised lead antimonate opacifier was used to achieve yellow (Lahil et al. 2008), later shown to be the case also in yellow LBA Egyptian glasses (Molina et al. 2014: 172). Further investigation by Lahlil et al. (2010a) suggested that opacity in white Roman glasses was obtained probably by introducing roasted stibnite to the melt, and was influenced by kiln temperature and duration of heat treatment.

Lahlil et al. (2010b) concluded from their experiments on 18th Dynasty Egyptian glass that LBA Egyptian glassmakers, with a sophisticated grasp of high-temperature technology, first synthesised the calcium antimonate opacifier ($CaSb_2O_6$ or $CaSb_2O_7$) *ex situ*, and then added the pre-prepared opacifier to the glass. The authors proposed that this was achieved by mixing Sb_2O_3 or Sb_2O_5 with calcium carbonate and heating between 1000° C and 1100° C forming calcium antimonate with Sb^{V} as the predominant species. The ratio of Sb^{III} to Sb^{V} changes with increase in firing temperature, with Sb^{V} being the major species up to 1250° C, and Sb^{III} thereafter. The predominant presence of Sb^{V} in the glass samples analysed by Lahlil et al. is indicative of *ex situ* production of calcium antimonate. *In situ* production of calcium antimonate exhibits a predominance of Sb^{III} or the presence of a mix of Sb^{III} and Sb^{V} (Lahlil et al. 2010b: 7; 2010a: 686-688). The authors therefore concluded that the oxidation state of Sb in a vitreous matrix appears to be a reliable discriminant between *in situ* production of calcium antimonate and the addition of *ex situ* prepared calcium antimonate crystals in opaque glasses heated below 1200° C.

The analysis by Smirniou and Rehren (2011) of a collection of semi-finished, or incompletely fused, white glass from Amarna identified an intermediate stage of glass-making, previously theorised, but not proven. The presence of a significant quantity of semi-finished glass at this site is important in demonstrating that glass-making, not just glass-working, occurred there (see also Jackson and Nicholson 2013). A comparison with similar material from Qantir found the two groups to be identical. This material is consistent with a two-stage process of glass production employed at both sites.

The procedure for adding antimony for opacification is less clear, and is not investigated by Smirniou and Rehren (2011) who observed that the concentrations of antimony in some semi-finished white glasses are equal to those of opacified finished glass. Lime levels, on the other hand, are lower in the semi-finished phase. They considered this provides evidence for the addition of an Sb-based opacifier during primary glass-making before the addition of colourants at a later stage of production. The authors deduced that second phase remelting at higher temperatures, where semi-finished glass can absorb lime-rich phases such as wollastonite, together with incorporation of lime from the parting layer, applied to the inside of crucibles to facilitate removal of the glass, both contributed to higher lime levels in the finished glass.

Shortland (2012: 110-112) and Shortland et al. (2018: 777-778) observed that lime levels are the same in opacified as in translucent finished LBA Egyptian glasses, and that there is a range of antimony concentrations for all glasses, opaque and translucent, that provides no clear distinction between them. In section 8.4 of this study the lime levels in 283 Egyptian glasses are also shown to form a continuous range that provides no distinction between opacified and translucent glasses. Shortland suggested that antimony alone was added to the glass, effectively producing *in situ* opacification with lime already present in the glass, *contra* Lahlil et al. (2010b) who described *ex situ* opacification of Egyptian glass.

Shortland (2012: 111-112) argued that in the case of LBA Near Eastern glass from Nuzi, antimony and lime contents are clearly different between opaque and translucent glass, in contrast to the Egyptian model. According to Shortland, finished opacified glass at Nuzi contains significantly more Sb and on average nearly twice as much CaO as translucent glass, implying that both lime and antimony were added to glass to achieve opacification, but whether it was added as pre-prepared calcium antimonate, or lime and antimony as separate ingredients is not addressed. It may be of significance that no correlation between the contents of lime and antimony was found in Nuzi glasses in the research undertaken for this study (chapter 8); if such a correlation existed it would be good evidence that lime and antimony were added together.

In their 2018 publication, Shortland and co-workers, also noted the bimodal distribution of antimony in the Nuzi glasses and they suggested that this clearly divides translucent from opaque glass, as opposed to the continuous variation that occurs across both opaque and translucent glass in Egyptian production. Shortland and co-workers therefore concluded that the higher lime content of the Nuzi opaque glass compared with the Nuzi translucent glass suggested that both lime and antimony were added to achieve opacification, presumably *in situ*, but the actual process is not discussed (Shortland et al. 2018; Shortland 2012). According to these studies, antimony was added to glass to achieve opacification (*in situ*) in Egypt, and lime and antimony were both added to opacify Nuzi glass, but whether these components were combined to form calcium antimonate *in situ* or *ex situ* remains unclear (see section 8.4.5 for a discussion of the unlikeliness of the separate addition of lime to Nuzi and Pella glasses). Duckworth and co-workers (2012b) chose the ToF-SIMS (Time of Flight Secondary Mass Spectrometry) analytical technique to identify the chemical composition of opacifiers in LBA glass. They interpreted their results to suggest that a turquoise opaque glass from Tell Brak (14th century BCE) and a white glass from Greece (late 12th century BCE) were made using *in situ* opacification.

The findings in this study (chapter 8) show that while Sb content may be an indicator of opacity, the opaque and translucent Near Eastern glasses do not correspond neatly to the identified high- and low-lime groups, with the conclusion that the two lime groups are not directly related to opacification in the glasses from Nuzi,. The two lime groups also observed in the present study of glass from LBA Pella (section 8.4), are however more closely aligned with opacification, but the picture is less clear when we look at Tell Brak where opacified glasses could have high- or low-lime contents. Anomalous compositions, that place opaque and translucent glasses in both high- and low-lime groups from all three sites, show that these glasses could have high-lime levels and be translucent, or could equally be opacified with a low-lime content. The reasons for this intriguing inconsistency have yet to be determined, but as noted in section 8.4.3, and also observed by Shortland et al. (2018: 778), the differences between the Nuzi opaque and translucent blue glasses could indicate the use of raw materials from two sources, or the use of the same raw materials but processed and heated differently. The significance is that it suggests the translucent and opaque blue colours were made in different workshops (section 8.5; Shortland et al. 2018: 778).

11.4 Cobalt Colourants

The question of Co sources for LBA and IA Near Eastern and Egyptian glasses has been an exciting avenue of research, addressed in this study particularly with the identification of an Egyptian Co colourant in the IA Nimrud glasses made in a period when Egypt was itself apparently not utilising this resource as a glass colourant (Reade et al. 2005; section 9.7).

11.4.1 Late Bronze Age

It has long been known that another, non-Egyptian source of Co was at times used in the Near East (section 9.7). Electron microprobe analyses of samples of blue glass from the LBA, 14th to 13th century BCE, Nippur axes (Clayden 2011: Appendix B) revealed the blue colourant to be a mixture of Cu and Co. While these glasses appear to be of plant ash composition, closer to Near Eastern than Egyptian glass compositions of this period, they were not identical.

This study was followed by the chemical analysis of the Nippur axes by Walton et al. (2012). They identified the glasses as being the first Near Eastern Co glasses known from the LBA, and the Co colourant as being of a previously unidentified Near Eastern origin. It had been thought previously that Co was sourced only from Egypt during the LBA. Further characterisation of Egyptian and Mycenaean Co glass, and more recently identified mixed Cu-Co, or Cu-rich Co blue glass, was undertaken by Smirniou and Rehren (2013). The authors set out to characterise Mycenaean Cu-Co glass, to define its role and significance, and to investigate the geographic and technological relationship between the two glass types. The Cu-Co glasses appear to be Egyptian but are more prevalent in the Mycenaean world. There is no clear answer yet as to why this was so, and many new questions about LBA glass have been posed.

11.4.2 Iron Age

Abe et al. (2012) discussed the decline in glass-making and the absence of Co as a colourant in 19th Dynasty Egypt (early 13th to 12th century BCE), despite its popularity in the preceding 18th Dynasty. They concluded from their preliminary portable XRF study of the cobalt blue colourant of 19th Dynasty Ramesside glasses from the Memphite region that this 19th Dynasty colourant differs from that used in the preceding 18th Dynasty. The authors acknowledged that surface weathering of the samples analysed means that further analysis is needed to confirm the findings. The present study of cobalt colourant in the later, 8th to 7th century BCE, Nimrud glasses shows that cobalt alum colourant, as used in the 18th Dynasty, was being used to colour glass made outside of Egypt at a time when it was apparently no longer used in Egyptian-made glass (sections 9.7, 10.2). Both studies suggest there was a change in the system of Egyptian glass-making at this time (Abe et al. 2012: 1794). The use of Egyptian Co in the IA glasses found at Nimrud, and also identified further afield in Italy and France, for example (Conte et al. 2016a; Arletti et al. 2011; Gratuze and Picon 2006; section 9.7.1) in a period when it was not used in Egypt itself remains puzzling.

While scientific studies of glass composition continue to be published, consideration is also being given to the contemporary value and position of glass and the significance of its colour especially in its ill-defined role vis-à-vis precious stones of the time, principally lapis lazuli and turquoise. Archaeological and textual evidence form the basis of discussions that revolve around how glass was viewed: as a type of stone itself, or imitative of, or competitive with stones (see for example Duckworth 2012a; Nicholson 2012).

11.5 Trade and Provenance

Comparative analyses of chemical compositions can establish connections that provide evidence for the extensive trade of glass from the LBA and IA Near East around the Mediterranean region. In this study, such evidence was provided by comparison of IA Co and green-black iron-rich glasses found at Nimrud and Pella respectively, with closely similar glasses found as far afield as France (section 9.6). Since then, evidence continues to be gathered as more samples of LBA and IA glasses are analysed and added to the comparative pool, not only from around the Mediterranean, but even into northern Europe, greatly increasing the picture of a vast trading network that operated from early times. Much of the ability to make finer differentiations between glass groups, and equally to identify their similarities, has been made possible by analyses of an increasing number of isotopes. The resulting data not only reveal information about glass manufacturing technology and provenance of batch components, but help to trace trade and distribution of raw and finished glass products.

11.5.1 Late Bronze Age

Walton et al. (2009) presented evidence from the chemical analysis of LBA Mycenaean glass beads for the trade in raw glass to the Mycenaean states from the Near East and Egypt (see also Smirniou et al. 2012). This trade had already been strongly suggested by the cargo of the Ulu Burun shipwreck containing glass ingots identified as being made in Egypt with Egyptian Co (Kaczmarczyk 1986; Nicholson et al. 1997; Jackson and Nicholson 2009). For a discussion of the value, meaning and status of glass in the Aegean region, see Jackson and Wager (2011).

The isotopes of Nd are generally considered to have been introduced with the silica component of the glass and are used to provenance sand-based glasses produced around the Mediterranean, while Sr isotopes help to identify lime sources. Both are becoming staples of the chemical analytical regime in glass characterisation and provenancing, along with the increasing investigation into and use of isotopes of Pb, O, Sb, Cu and B (see for example Degryse et al. 2015; Rehren and Freestone 2015; Devulder et al. 2014; Lobo et al. 2014; Brems et al. 2013a; 2013b; Degryse et al. 2010a). Henderson and co-workers (2010) and Degryse et al. (2010) used Nd and Sr isotopic evidence to identify separate Mesopotamian and Egyptian glass manufacturing compositions in the 14th century BCE, following on from the identification by Shortland et al. (2007) of Cr, La, Ti and Zr as elemental discriminants between glasses of these regions. The data presented in this study show that Sr, Cs, Li and Be and their ratios are also useful discriminants between Egyptian and Near Eastern products (section 8.3.3.2).

Degryse and co-workers (2010a) argued that the Nd and Sr isotopic data indicate independent primary production occurred in both Egypt and Mesopotamia (see also Henderson et al. 2010), and that both areas exported glass to Greece (cf Walton et al. 2009; Rehren 2014). Their chemical evidence suggested more than one glass production site can be identified in the LBA Near East, while Egyptian production appears to have been more limited. Analyses and examination of Amarna glass continue to answer questions about production technology and the issue of whether glass was made as well as worked at this site (e.g. Jackson and Nicholson 2013). A study of ceramic vessels used in LBA Egyptian glass-making from Lisht, Amarna and Qantir, combined with replication experiments, provided intriguing new evidence for primary glass-making at all three sites through the identification of a layered structure that developed within the ceramic only as a result of the interaction during firing of the primary glass batch with the ceramic (Smirniou and Rehren 2016), and further, Smirniou et al. (2018) identified primary glass-making at Lisht in a study of production waste.

Research was also undertaken by Degryse and co-workers (2010a) of O, Sr and Nd isotopes of Mesopotamian glass from Nuzi and Tell Brak, alongside a group of Egyptian glasses. Combined Sr and Nd isotope analysis enabled them to distinguish Mesopotamian from Egyptian glass, reinforcing the elemental distinctions already identified. Isotope analysis also allowed these researchers to differentiate at least two Near Eastern primary glass compositions, probably relating to two separate glass-making centres in the region, supporting the view expressed in this study that from available elemental data at the time, Tell Brak and Nuzi glasses appear to exhibit significant differences (sections 8.3.1, 8.3.3). These findings were further refined by a subsequent study of Sr-Nd-Sb isotopic analyses of Nuzi, Nippur, Tell Brak and Egyptian LBA glasses (Degryse et al. 2015), which concluded that while Egyptian glasses were clearly separate from the contemporary Near Eastern glasses, there were two distinct groups of Nuzi-Tell Brak, and Nuzi-Tell Brak-Nippur glasses. The second of these groups had an Nd isotope signature indicative of the use of eastern Mediterranean coastal sands, supporting the theory advanced in this study that Near Eastern glasses could have originated in the Syro-Levantine coastal region (chapters 8, 9). These results prompt further discussion of the likelihood of more than one glass-making site and certainly of multiple glass-making raw material mixtures. Of interest is the finding that glass opacification was achieved using the same Sb source, possibly the Caucasus, in both the Near East and Egypt in the LBA, while a different source was used in Roman times (Degryse et al. 2015). Such indications of provenance from isotopic analysis illustrates how this kind of analysis can provide chemical evidence of LBA exchange within the Near East.

Continuing the investigations into provenance and trade, Jackson and Nicholson revisited the Ulu Burun glass ingots in their 2010 chemical analytical study of the composition of one turquoise and two deep blue ingots. The analyses complemented the work of Walton et al. (2009) by confirming that the ingots were made in Egypt and traded to the Mycenaean world where they were worked into objects. This picture is being extended by the chemical analyses of Mycenaean period glasses such as those found at Pylos in the Peloponnese, identifying their compositions as consistent with both Egyptian and Mesopotamian origins based on the discriminants Ti and Zr (Polikreti et al. 2011). The typological study of glass from the port city of Panaztepe, in western Anatolia, also filled a gap in the map of LBA Mediterranean trade routes (Çinardali-Karaaslan 2012).

Beyond the Mediterranean, the research of Varberg et al. (2015; 2016) has extended our knowledge of LBA trade links to northern Germany, France, Romania and Denmark where glass beads, canes and vessel fragments from 14th to 12th century BCE burials have been analysed and found to have come from both the LBA Near East and Egypt, around the Mediterranean and onwards via European rivers (Varberg 2016: 193). Similarly, Purowski et al. (2016) reported glass beads from both these regions found at 13 mainly burial sites in Poland. It is unsurprising therefore that IA black beads of closely similar composition have been identified in France, Jordan and Iran when wide-ranging trade routes were clearly well established at least as early as the LBA (chapter 8, section 10.1), as were trade routes within the region. The elemental analyses of green glass canes from Amarna (Varberg et al. 2016) revealed two to have been made in the Near East and three in Egypt demonstrating that Egypt imported raw or unformed glass from the Near East at this time.

Recent reinvestigation of LBA glass from Nuzi (Shortland et al. 2018), including Sr and Nd isotope analyses, has extracted more information from these much-studied glasses (chapters 3, 8), and questioned the long-held view that glass object production began in the Near East (Mesopotamia) before it did in Egypt. The authors judged the Egyptian production to be more mature, based on a review of the dating of finds from Nuzi, the processes of glass object manufacture and decoration, and the limited palette of colours compared with Egyptian glass (Shortland et al. 2018: 779; also noted section 8.3.3). However, the mid-16th century BCE glasses from Pella, examined in this study (chapters 4, 8), are of Near Eastern composition closely similar to that of the Nuzi glasses, and predate the Amarna glass production, tipping the scales back in favour of a Near Eastern origin for the earliest glass objects.

11.5.2 Iron Age

Stapleton's (2011) publication of unusual black-coloured iron-rich glasses from Hasanlu, Iran, 11th to 9th century BCE, is of interest because of the discovery at Pella of identical contemporary IA glass (section 9.6). Stapleton concluded from analysis of the Hasanlu glasses that their composition was unlike any other glass in the LBA or IA Near East or Egypt (see also Stapleton 2003). Reade et al. (2009; section 9.6 this volume) have, however, found direct comparison of the Hasanlu glasses with the green-black IA Pella glasses from Pella, Jordan, and with closely similar 'black' glasses from several contemporary sites in France (Gratuze n.d.(a); n.d. (b); 2001a; 2001b), all coloured by iron. The Pella glasses are unusually rich in chromium, thanks to the inclusion of chromite-bearing sand, but a lack of comparative data precludes further comparison between the glasses from these sites.
Comparable dark iron-green natron glass containing chromite, along with Co blue glass made with Egyptian cobaltiferous alum, have however been identified from Campania in southern Italy, dating from the 9th to 8th centuries BCE (Conte et al. 2016a), and iron green-black glass from Bologna, Cumae and Pozzuoli of the 9th to 7th centuries BCE (Arletti et al. 2011; Conte et al. 2016b), once again illustrating the extent of the Mediterranean glass trade. Comparisons with the Pella iron green-black glasses (chapter 5, section 9.6) and the Nimrud Co blue glasses (chapter 7, section 9.7) made by Conte et al. (2016a; 2016b; see also Arletti et al. 2011) are

therefore very interesting, especially considering that this appears to have been a period when the Egyptians themselves were not using Co to colour glass. It shows that the Nimrud blue glasses coloured with Egyptian Co are not an isolated occurrence, and that an IA glass-making industry with access to Egyptian cobaltiferous alum manufactured dark blue glass somewhere most likely on the Levantine coast, or further north during the IA (sections 9.7, 10.2; see also O'Hea 2011), and produced some of the earliest examples of natron glass-making after those of Nesikhons, in the 10th century BCE (Schlick-Nolte and Werthmann 2003).

Conte et al. (2016a: 429-431; 2016b), without comparative trace element data for Nimrud Co glass, or Pella iron green-black glass, suggested that these glasses could both have been imported from Egypt, wanting to demonstrate a continuum of natron glass production first noted in Egypt. The trace element data provided in this volume do not suggest an Egyptian origin for either group, and the puzzling lack of Co or similar iron green-black glass in IA Egypt itself is not encouraging. The green-black glasses from Bologna, Cumae and Pozzuoli exhibited high chemical variability (Conte et al. 2016b) making comparison with the Pella green-black glasses more tenuous. Renewed examination and comparison with recent analyses of other IA glasses is encouraged.

In addition, comparisons were made by Conte et al. (2016a: 423) between the Campanian plant ash glass and Group 1 glass from Nimrud (see chapter 6 this volume; Reade et al. 2005) suggesting that it was not just the Co blue and iron green-black glasses, but others of the IA glass types observed in the Near East (chapter 9), that were also traded widely to the West.

Questions of provenance in discussions of IA trade and exchange were explored, for example, in a detailed stylistic and technical study by O'Hea (2011), who discussed the origins of translucent glass bowls, and dark blue inlays found predominantly at IA Nimrud. The chemical analyses of these glasses, as presented in this study (chapter 9), found that the Co colourant appears to be of Egyptian origin, but the base glass itself does not. O'Hea argued for the disassociation of glass inlay plaques from use within ivory objects that have been attributed to Phoenician production. By their association the same Phoenician origin had previously been suggested for the glass plaques. Instead, O'Hea preferred a Urartian/ north Syrian origin, but could draw no firm conclusion (see also Dardeniz 2015; Ingo et al. 2013). Barag (2011), on the other hand, attributed a single Neo-Assyrian glass bowl fragment, a rare find from Tel 'Aroer, Israel, to Phoenician production of the 8th to 6th centuries BCE.

Conclusion

While scholars continue to examine the evidence for glass technology and raw materials that lies buried in chemical analyses of glass compositions, there are still many questions to answer. Undoubtedly, the advent of isotope analysis will help to elucidate further glass-making processes, and the distribution of raw and finished products around the ancient Mediterranean world. It is only with the contribution of new information from the pursuit of wider and deeper studies that the whole picture will continue to gradually emerge from over two and a half thousand years of silence.

Bibliography 2009 – 2019

Abe, Y., Harimoto, R., Kikugawa, T. Yazawa, K., Nishisaka, A., Kawai, N., Yoshimura, S. & Nakai, I. 2012 Transition in the use of cobalt-blue colorant in the New Kingdom of Egypt. *Journal of Archaeological Science,* 39, 1793-1808.

Arletti, R., Bertoni, E., Vezzalini, G. & Mengoli, D. 2011 Glass beads from Villanovian excavation in Bologna (Italy): an archaeometrical investigation. *European Journal of Mineralogy,* 23, 959–968.

Barag, D. 2011 A Neo-Assyrian glass bowl fragment, in Y. Thareani, I. Taxel, D. Barag, D. Ilan & Applbaum, N. (Eds.), *Tel 'Aroer: The Iron Age II caravan town and the Hellenistic-early Roman settlement.* Annual of the Nelson Glueck School of Biblical Archaeology/ Hebrew Union College – Jewish Institute of Religion, 259-260.

Biron, I. & Chopinet, M-H. 2013 Colouring, decolouring and opacifying of glass, in Janssens, K. H. A. (Ed.). *Modern Methods for Analysing Archaeological and Historical Glass.* John Wiley and Sons Ltd, 49-65.

Brems, D., Ganio, M., Latruwe, K., Balcaen, L., Carremans, M., Gimeno, D., Silvestri, A., Vanhaecke, F., Muchez, P. & Degryse, P. 2013a Isotopes on the beach, part 1: strontium isotope ratios as a provenance indicator for lime raw materials used in Roman glass-making. *Archaeometry,* 55, 2, 214-234.

Brems, D., Ganio, M., Latruwe, K., Balcaen, L., Carremans, M., Gimeno, D., Silvestri, A., Vanhaecke, F., Muchez, P. & Degryse, P. 2013b Isotopes on the beach, part 2: neodymium isotopic analysis for the provenancing of Roman glass-making. Archaeometry, 55, 3, 449-464.

Brill, R. H. & Stapleton, C. P. 2012 *Chemical analyses of early glasses: Volume 3. The Years 2000-2011, Reports, and Essays.* Corning, NY: Corning Museum of Glass.

Çinardali-Karaaslan, N. 2012 The east Mediterranean Late Bronze Age glass trade within the context of the Panaztepe finds. *Oxford Journal of Archaeology,* 31, 2, 121-141.

Clayden, T., with Frame, G. & Kirk, S. 2011 Glass axes of the Kassite Period from Nippur, *Zeitschrift für Orient-Archäologie,* 4, 92-135.

Conte, S., Arletti, R., Mermati, F. & Gratuze, B. 2016a Unravelling the Iron Age glass trade in southern Italy: the first trace-element analyses. *European Journal of Mineralogy,* 28, 409-433.

Conte, S., Arletti, R., Henderson, J., Degryse, P. & Blomme, A. 2016b Different glassmaking technologies in the production of Iron Age black glass from Italy and Slovakia. *Archaeological and Anthropological Science.* DOI 10.1007/s12520-016-0366-4.

Dardeniz, G. 2015 Was ancient Egypt the only supplier of natron? New research. reveals major Anatolian deposits. *Anatolica,* 41, 191-202.

Degryse, P., Henderson, J. & Hodgins, G. (Eds.) 2009 *Isotopes in Vitreous Materials.* Studies in Archaeological Sciences 1. Leuven, Belgium, Leuven University Press.

Degryse, P., Boyce, A., Erb-Satullo, N., Eremin, K., Kirk, S., Scott, R., Shortland, A.J., Schneider, J. & Walton, M. 2010a Isotopic discriminants between late Bronze age glasses from Egypt and the Near East. *Archaeometry,* 52, 380-388.

Degryse, P., Shortland, A., De Muynck, D., Van Heghe, L., Scott, R., Neyt, B. & Vanhaecke, F. 2010b Considerations on the provenance determination of plant ash glasses using strontium isotopes. *Journal of Archaeological Science,* 37, 3129-3135.

Degryse, P., Lobo, L., Shortland, A., Vanhaecke, F., Blomme, A., Painter, J., Gimeno, D., Eremin, K., Greene, J., Kirk, S. & Walton, M. 2015 Isotopic investigation into the raw materials of Late Bronze Age glass making. *Journal of Archaeological Science,* 62, 153-160.

Devulder, V., Vanhaecke, F., Shortland, A., Mattingly, D., Jackson, C. & Degryse, P. 2014 Boron isotopic composition as a provenance indicator for the flux raw material in Roman natron glass. *Journal of Archaeological Science,* 46, 107-113.

Duckworth CN. 2012a Imitation, artificiality and creation: the colour and perception of the earliest glass in New Kingdom Egypt. *Cambridge Archaeological Journal,* 22, 3, 309-327.

Duckworth, C. N., Henderson J., Rutten F. J. M. & Nikita, K. 2012b Opacifiers in Late Bronze Age glasses: the use of ToF-SIMS to identify raw ingredients and production techniques. *Journal of Archaeological Science,* 39, 2143-2152.

Gliozzo, E. 2016 The composition of colourless glass: a review. *Archaeological and Anthropological Science.* DOI 10.1007/s12520-016-0388-y.

Henderson, J., Evans, J.A. & Nikita, K. 2010 Isotopic evidence for the primary production, provenance and trade of Late Bronze Age glass in the Mediterranean. *Mediterranean Archaeology and Archaeometry,* 10, 1, 1-24.

Ingo, G.M., Çilingiroglu, A., Di Carlo, Batmaz, A., De Caro, T., Riccucci, C., Parisi, E.I. & Faraldi, F. 2013 Egyptian Blue cakes from the Ayanis Fortress (Eastern Anatolia, Turkey): micro-chemical and -structural investigations for the identification of manufacturing process and provenance. *Journal of Archaeological Science,* 40, 4283-4290.

Jackson, C.M. & Nicholson, P.T. 2009 The provenance of some glass ingots from the Ulu Burun shipwreck. *Journal of Archaeological Science*, 37, 295-301.

Jackson C.M. & Nicholson P.T. 2010 The provenance of some glass ingots from the Uluburun shipwreck. *Journal of Archaeological Science*, 37, 295–301.

Jackson, C. & Nicholson, P. 2013 Glass at el-Amarna, in Janssens, K. H.vA. (Ed.), *Modern Methods for Analysing Archaeological and Historical Glass.* John Wiley and Sons Ltd, 345-353.

Jackson, C.M., Wager, E.C. 2011 Glass in the Aegean Bronze Age: value, meaning and status. In: Vianello, A. (Ed.), *Exotica in the Prehistoric Mediterranean.* Oxbow Books, Oxford, 115-123.

Janssens, K. (Ed.) 2013 *Modern Methods for Analysing Archaeological and Historical Glass.* Vol. 1. Chichester, West Sussex: John Wiley & Sons.

Lahlil, S., Biron, I., Galoisy, L. & Morin, G. 2008 Rediscovering ancient glass technologies through the examination of opacifier crystals. *Applied Physics A*, 92, 109-116.

Lahlil, S., Biron, I., Cotte, M. & Susini, J. 2010a New insight on the in situ crystallization of calcium antimonate opacified glass during the Roman period. *Applied Physics A*, 100, 683-692.

Lahlil, S., Biron, I., Cotte, M. & Susini, J. 2010b Synthesis of calcium antimonate nano-crystals by the 18th dynasty Egyptian glassmakers. *Applied Physics A*, 98, 1-8.

Lahlil, S., Cotte, M., Biron, I., Szlachetko, J., Menguy, N. & Susini, J. 2011 Synthesizing lead antimonate in ancient and modern opaque glass. *Journal of Analytical Atomic Spectrometry*, 26, 1040-1050.

Lobo, L., Degryse, P., Shortland, A., Eremin, K. & Vanhaecke, F. 2014 Copper and antimony isotopic analysis *via* multi-collector ICP-mass spectrometry for provenancing ancient glass. *Journal of Analytical Atomic Spectrometry*, 29, 58-64.

Molina, G., Odin, G.P., Pradell, T., Shortland, A.J. & Tite, M.S. 2014 Production technology and replication of lead antimonate yellow glass from New Kingdom Egypt and the Roman Empire. *Journal of Archaeological Science*, 41, 171-184.

Nicholson, P. T. 2012 "Stone ... that flows": faience and glass as man-made stones in Egypt. *Journal of Glass Studies*, 54, 11-23.

O'Hea, M. 2011 Another look at the origins of Iron Age II cast glass vessels in the Levant. *Levant*, 43, 153-172.

Polikreti, K., Murphy, J. M. A., Kantarelou, V. & Karydas, A. G. 2011 XRF analysis of glass beads from the Mycenaean palace of Nestor at Pylos, Peloponnesus, Greece: new insight into the LBA glass trade. *Journal of Archaeological Science*, 38, 2889-2896.

Purowski, T., Kępa, L. & Wagner, B. 2016 Glass on the Amber Road: the chemical composition of glass beads from the Bronze Age in Poland. *Archaeological and Anthropological Science.* DOI 10.1007/s12520-016-0443-8.

Rehren, T. 2014 Glass production and consumption between Egypt, Mesopotamia and the Aegean, in Pfälzner, P. Niehr, H., Pernicka, E., Lange, S. & Köster, T, (Eds.), *Contextualising Grave Inventories in the Ancient Near East: Proceedings of a workshop at the London 7th ICAANE in April 2010 and an International Symposium in Tübingen in November 2010, both organised by the Tübingen Post-Graduate School 'Symbols of the Dead'.* Aatna Studien Supplementa, 3. Wiesbaden, Harrassowitz Verlag, 217-223.

Schmidt, K. 2019 *Glass and Glass Production in the Near East During the Iron Age. Evidence from Objects, Texts and Chemical Analysis.* Oxford, Archaeopress.

Shortland, A. J. 2012 *Lapis Lazuli from the Kiln. Glass and Glassmaking in the Late Bronze Age.* Studies in Archaeological Sciences 2, Leuven University Press.

Shortland, A., Freestone, I. & Rehren, T. (Eds.) 2009 *From Mine to Microscope: Advances in the Study of Ancient Technology*, Oxford, Oxbow Books, 115-127.

Shortland, A. J., Kirk, S., Eremin, K., Degryse, P. & Walton, M. 2018 The analysis of Late Bronze Age glass from Nuzi and the question of the origin of glass-making. *Archaeometry*, 60, 764-783.

Smirniou, M. & Rehren, T. 2011 Direct evidence of primary glass production in late Bronze age Amarna, Egypt. *Archaeometry*, 53, 1, 58-80.

Smirniou, M. & Rehren, T. 2013 Shades of blue – cobalt-copper coloured blue glass from New Kingdom Egypt and the Mycenaean world: a matter of production or colourant source? *Journal of Archaeological Science*, 40, 4731-4743.

Smirniou, M. & Rehren, T. 2016 The use of technical ceramics in early Egyptian glass-making. *Journal of Archaeological Science*, 67, 52-63.

Smirniou, M. & Rehren, T. 2018 Lisht as a New Kingdom glass-making site with its own chemical signature. *Archaeometry*, 60, 3, 502-516.

Smirniou, M., Rehren, T., Adrymi-Sismani, V., Asderaki, E. & Gratuze, B. 2012 Mycenaean bead from Kazanaki, Volos: a further node in the LBA glass network, in Ignatiadou, D. & Antonaras, A. (Eds.) *Annales du 18e Congrès de l'Association Internationale pour l'Histoire du Verre. Thessaloniki 2009.* AIHV, Thessaloniki, 11-18.

Stapleton, C. P. 2011 Glass and glaze analysis and technology from Hasanlu, Period IVB, in M. de Schauensee (Ed.), *Peoples and Crafts in Period IVB at Hasanlu, Iran.* Hasanlu Special Studies 4. Philadelphia, PA, University of Pennsylvania, Museum of Archaeology and Anthropology/ University of Pennsylvania Press, 87-102.

Varberg, J., Gratuze, B. & Kaul, F. 2015 Between Egypt, Mesopotamia and Scandinavia: Late Bronze Age glass beads found in Denmark. *Journal of Archaeological Science*, 54, 168-181.

Varberg, J., Gratuze, B., Kaul, F., Haslund Hansen, A., Rotea, M. & Wittenberger, M. 2016 Mesopotamian glass from Late Bronze Age Egypt, Romania,

Germany, and Denmark. *Journal of Archaeological Science,* 74, 184-194.

Walton, M., Eremin, K., Shortland, A., Degryse, P. & Kirk, S. 2012 Analysis of Late Bronze age glass axes from Nippur—a new cobalt colourant. *Archaeometry,* 54, 835–852.

Walton M., Shortland, A., Kirk, S. & Degryse, P. 2009 Evidence for the trade of Mesopotamian and Egyptian glass to Mycenaean Greece. *Journal of Archaeological Science,* 36, 7, 1496-1503.

Wight, K. B. 2011 *Molten Color: Glassmaking in Antiquity.* Los Angeles, CA, J. Paul Getty Museum.

Bibliography

Akkermans, P., M.M.G. and Schwartz, G. M. 2003 *The Archaeology of Syria from Complex Hunter-Gatherers to Early Urban Societies (c. 16,000 - 300 BC)*, Cambridge, Cambridge University Press.

Anderson, W. P. 1990 The Beginnings of Phoenician pottery: vessel shape, style, and ceramic technology in the early phases of the Phoenician Iron Age. *Bulletin of the American Schools of Oriental Research* 279, 35-54.

ACQUIRE 2003 Isotope Research Centre (ACQUIRE) Information Brochure.

Bamford, C. R. 1977 *Colour Generation and Control in Glass*, Amsterdam, Elsevier Scientific Publishing Co.

Bamford, C. R. 1982 Optical properties of flat glass. *Journal of Non-Crystalline Solids* 47, 1-20.

Barag, D. 1962 Mesopotamian glass vessels of the second millennium B.C. *Journal of Glass Studies* 4, 9-27.

Barag, D. 1964 A Note on a Mesopotamian bottle. *Journal of Glass Studies* 6, 9-10.

Barag, D. 1970 Mesopotamian core-formed glass vessels (1500 - 500 B.C.), in Oppenheim, A. L., Brill, R. H., Barag, D. and Von Saldern, A. (eds) *Glass and Glassmaking in Ancient Mesopotamia.* Corning, The Corning Museum of Glass Press, 131-200.

Barag, D. 1985 *Catalogue of Western Asiatic Glass in the British Museum,* London, British Museum Publications Ltd.

Barkoudah, Y. and Henderson, J. 2006 Plant ashes from Syria and the manufacture of ancient glass: ethnographic and scientific aspects. *Journal of Glass Studies* 48, 297-321.

Barnett, R. D. 1963 Hamath and Nimrud. *Iraq* 25, 81-85.

Bartoloni, P. 2001 Commerce and industry, in Moscati, S. (ed.) *The Phoenicians,* London, I.B. Taurus, 92-100.

Bass, G. F. 1986 A Bronze Age shipwreck at Ulu Burun (Kas): 1984 campaign. *American Journal of Archaeology* 90, 269-296.

Bass, G. F. 1987 Oldest known shipwreck reveals splendours of the Bronze Age. *National Geographic* 172, 693-733.

Bass, G. F., Pulak, C., Collon, D. and Weinstein, J. 1989 The Bronze Age shipwreck at Ulu Burun: 1986 campaign. *American Journal of Archaeology* 93, 1-29.

Baxter, M. J., Cool, H. E. M., Heyworth, M. P. and Jackson, C. M. 1995 Compositional variability in colourless Roman vessel glass. *Archaeometry* 37, 129-141.

Baxter, M. J., Cool, H. E. M. and Jackson, C. M. 2006 Comparing glass compositional analyses. *Archaeometry,* 48, 399-414.

Baxter, M. J. and Jackson, C. M. 2001 Variable selection in artefact compositional studies. *Archaeometry* 43, 253-268.

Beck, H. C. 1934 Glass before 1500 B.C. *Ancient Egypt and the East*, 7-21.

Bezborodov, M. A. 1975 *Chemie und Technologie der antiken und mittelatlterlichen Glaser,* Mainz, Verlag Philipp von Zabern.

Bianchi, R. S., Schlick-Nolte, B., Bernheimer, G. M. and Barag, D. (eds) 2002 *Reflections on Ancient Glass from the Borowski Collection,* Mainz, Verlag Philipp von Zabern.

Biek, L. and Bayley, J. 1979 Glass and other vitreous materials. *World Archaeolcgy* 11, 1-25.

Bimson, M. 1987 Opaque red glass: a review, in Bimson, M. and Freestone, I. C. (eds) *Early Vitreous Materials.* London, British Museum, 165-171.

Bimson, M. and Freestone, I. C. 1983 An analytical study of the relationship between the Portland vase and other Roman cameo glasses. *Journal of Glass Studies* 25, 55-64.

Bimson, M. and Freestone, I. C. (eds) 1987 *Early Vitreous Materials,* London, British Museum.

Bimson, M. and Freestone, I. C. 1988 Some Egyptian glasses dated by royal inscriptions. *Journal of Glass Studies* 30, 11-15.

Bjorkman, J. 1999 How to bury a temple: the case of Nuzi's Ishtar Temple A, in Owen, D. I. and Wilhelm, G. (eds) *Nuzi at Seventy-Five.* Bethesda, CDL Press, 103-122.

Boardman, J. 2001 Aspects of "colonization". *Bulletin of the American Schools of Oriental Research* 322, 33-42.

Bourke, S. 1992 First preliminary report on the excavation and study of human remains at Pella, in McNicoll, A. W., Edwards, P. C., Hanbury-Tenison, J., Hennessy, J. B., Potts, T. F., Smith, R. H., Walmsley, A. and Watson, P. 1992 *Pella in Jordan 2. The Second Interim Report of the Joint University of Sydney and College of Wooster Excavations at Pella 1982-1985,* Sydney, Mediterranean Archaeology, 215-226.

Bourke, S. 1997 Pre-classical Pella in Jordan: a conspectus of ten years' work (1985-1995). *Palestine Exploration Quarterly* 129, 94-115.

Bourke, S. 2004 Cult and archaeology at Pella in Jordan: Excavating the Bronze and Iron Age temple precinct (1994-2001). *Journal & Proceedings of the Royal Society of New South Wales* 137, 1-31.

Bourke, S. 2006 Pella and the Jordanian Middle and Late Bronze Ages, in Fischer, P. (ed.) *The Jordan Valley in the Middle and Late Bronze Ages,* Vienna, 243-256.

Bourke, S. forthcoming. Preliminary report on the University of Sydney's twentieth and twenty-first seasons of excavation at Pella (Tabaqat Fahl) in 1999/2001. *Annals of the Department of Antiquities Jordan.*

Bourke, S. and Descoeudres, J.-P. (eds) 1995 *Trade, Contact, and the Movement of Peoples in the Eastern Mediterranean. Studies in Honour of J. Basil Hennessy,* Sydney, Mediterranean Archaeology Supplement 3.

Bourke, S., Sparks, R., McLaren, P., Sowada, K., Mairs, L., Meadows, J., Hikade, T. and Reade, W. J. 2003 Preliminary report on the University of Sydney's eighteenth and nineteenth seasons of excavations at Pella (Tabaqat Fahl) in 1996/97. *Annals of the Department of Antiquities Jordan* 47, 335-388.

Brill, R. H. 1962 A Note on the scientist's definition of glass. *Journal of Glass Studies* 4, 127-138.

Brill, R. H. 1963a Ancient glass. *Scientific American* 209, 120-130.

Brill, R. H. 1963b Glass finds. *Bulletin of the American Schools of Oriental Research* 170, 62-65.

Brill, R. H. 1965 Interlaboratory comparison experiments of the analysis of ancient glass. *Comptes Rendus; VIIe Congrès Internationale du Verre, Bruxelles, 28 juin-3 juillet 1965, Section B* Brussels, 226.1-226.4.

Brill, R. H. 1967 A Great glass slab from ancient Galilee. *Archaeology* 20, 89-95.

Brill, R. H. 1969 The Scientific investigation of ancient glasses. *Eighth International Congress on Glass, London.* Sheffield, The Society of Glass Technology 47-68.

Brill, R. H. 1970a The Chemical interpretation of the texts, in Oppenheim, A. L., Brill, R. H., Barag, D. and Von Saldern, A. (eds) *Glass and Glassmaking in Ancient Mesopotamia.* Corning, The Corning Museum of Glass, 105-128.

Brill, R. H. 1970b Lead and oxygen isotopes in ancient objects. *Philosophical Transactions of the Royal Society London* 269, 143-164.

Brill, R. H. 1971 A Chemical-analytical round-robin on four synthetic ancient glasses. *International Congress on Glass, Versailles, Sept.-Oct. 1971, Artistic and Historical Communications.* Paris, L'Institut du Verre, 93-110.

Brill, R. H. 1978 Some miniature painted glass plaques from Fort Shalmaneser, Nimrud. Part II: Laboratory studies. *Iraq* 40, 23-39.

Brill, R. H. 1979 *The Glassmakers of Herat.* USA, Corning Museum of Glass (video).

Brill, R. H. 1987 Chemical analyses of some early Indian glasses. *Archaeometry of Glass, Proceedings of the Archaeometry Session of the XIVth International Congress on Glass, 1986, New Delhi* New Delhi, Indian Ceramic Society, 1-25.

Brill, R. H. 1988 Scientific investigations of the Jalame glass and related finds, in Weinberg, G. D. (ed.) *Excavations at Jalame, Site of a Glass Factory in Late Roman Palestine.* Columbia, University of Missouri Press, 257-294.

Brill, R. H. 1991 Scientific investigations of some glasses from Sedeinga. *Journal of Glass Studies* 33, 11-28.

Brill, R. H. 1995 Chemical analyses of some glass fragments from Nishapur in The Corning Museum of Glass, in Kroger, J. (ed.) *Nishapur: Glass of the Early Islamic Period.* New York, The Metropolitan Museum of Art, 211-233.

Brill, R. H. 1999a *Chemical Analyses of Early Glasses: Volume I: Catalogue of Samples,* Corning, N.Y., Corning Museum of Glass.

Brill, R. H. 1999b *Chemical Analyses of Early Glasses: Volume II: Tables of Analyses,* Corning, N.Y., Corning Museum of Glass.

Brill, R. H. and Barnes, I. L. 1988 The examination of some Egyptian glass objects, in Rothenberg, B. (ed.) *Researches in the Arabah 1959-1984. The Egyptian Mining Temple at Timna.* London, The Institute for Archaeo-Metallurgical Studies, 217-222.

Brill, R. H. and Moll, S. 1963 The Electron-beam probe microanalysis of ancient glass, in Matson, F. R. and Rindone, G. E. (eds) *Advances in Glass Technology, part 2. History Papers and Discussions of the Technical Papers of the VI International Congress on Glass.* New York, Plenum Press, 293-302.

Brill, R.H. and Shirahata, H. 1997 Laboratory analyses of some glasses and metals from Tell Brak, in Oates, D., Oates, J. and McDonald, H. 1997 *Excavations at Tell Brak. Vol. I: The Mitanni and Old Babylonian periods,* British School of Archaeology in Iraq, 89-94.

Brill, R. H. and Wosinski, J. F. 1965 A Huge slab of glass in the ancient necropolis of Beth She'arim. *Comptes Rendus; VIIe Congrès International du Verre, Bruxelles, 28 juin-3 juillet 1965, Section B* Brussels, 219.1-219.11.

Brothwell, D. R. and Pollard, A. M. (eds) 2001 *Handbook of Archaeological Sciences,* Chichester, John Wiley & Sons Ltd.

Bryan, B. M. 2000 The Eighteenth Dynasty before the Amarna Period, in Shaw, I. (ed.) *The Oxford History of Ancient Egypt.* Oxford, Oxford University Press, 218-271.

Caley, E. R. 1962 *Analysis of Ancient Glasses 1790-1957,* Corning, NY, The Corning Museum of Glass.

Canti, M. G. 2003 Aspects of the chemical and microscopic characteristics of plant ashes found in archaeological soils. *Catena* 54, 339-361.

Caubet, A. and Kaczmarczyk, A. 1987 Bronze Age faience from Ras Shamra, in Bimson, M. and Freestone, I. C. (eds) *Early Vitreous Materials,* London, British Museum, 47-56.

Colinart, S. and Menu, M. (eds) 1998 *La Couleur dans la Peinture et l'Emaillage de l'Egypte Ancienne,* Bari, Edipuglia.

Cosyns, P. and Hurt, V. 2007 Les perles en verre de Neufchateau Le-Sart. *Arduinna* 62, 1-5.

Crowfoot, J. W. and Crowfoot, G. M. 1938 *Early Ivories from Samaria,* London, Palestine Exploration Fund.

Crowfoot, J. W., Crowfoot, G. M. and Kenyon, K. M. 1957 *The Objects from Samaria,* London, Palestine Exploration Fund.

Curtis, J. (ed.) 1982 *Fifty Years of Mesopotamian Discovery. The Work of the British School of Archaeology in Iraq 1932 - 1982,* London, The British School of Archaeology in Iraq (London).

Curtis, J. (ed.) 1993 *Early Mesopotamia and Iran: Contact and Conflict 3500 - 1600 BC*, London, British Museum Press.

Curtis, J. 1999 Glass inlays and Nimrud ivories. *Iraq* 61, 59-69.

Curtis, J., Collon, D. and Green, A. 1993 British Museum excavations at Nimrud and Balawat in 1989. *Iraq* 55, 1-37.

De Schauensee, M. 2001 A note on three glass plaques from Hasanlu. *Iraq* 63, 99-106.

Degryse, P. and Schneider, J. 2008 Pliny the Elder and Sr-Nd isotopes: tracing the provenance of raw materials for Roman glass production. *Journal of Archaeological Science* 35, 1993-2000.

Degryse, P., Schneider, J., Poblome, J., Waelkens, M., Haack, U. and Muchez, P. 2005 A geochemical study of Roman to early Byzantine glass from Sagalassos, south-west Turkey. *Journal of Archaeological Science* 32, 287-299.

Degryse, P., Schneider, J., Haack, U., Lauwers, V., Poblome, J., Waelkens, M. and Muchez, P. 2006 Evidence for glass 'recycling' using Pb and Sr isotopic ratios and Sr-mixing lines: the case of early Byzantine Sagalassos. *Journal of Archaeological Science* 33, 494-501.

Dever, W. G. 1992 The chronology of Syria-Palestine in the second millennium B.C.E.: a review of current issues. *Bulletin of the American Schools of Oriental Research* 288, 1-25.

Donnelly, P. 2001 Chocolate-on-White Ware. Towards a definition 15 years on from Hennessy, in Walmsley, A. (ed.) *Australians Uncovering Ancient Jordan.* Sydney, University of Sydney, 253-258.

Dothan, M. and Porath, Y. 1993 *Ashdod V: Excavation of Area G: the Fourth -Sixth Seasons of Excavations: 1968-1970. Atiqot XXIII.* Jerusalem, Israel Antiquities Authority.

Drews, R. 1993 *The End of the Bronze Age. Changes in Warfare and the Catastrophe ca.1200 B.C.*, Princeton, Princeton University Press.

Drews, R. 2000 Medinet-Habu: oxcarts, ships, and migration theories. *Journal of Near Eastern Studies* 59, 161-190.

Drower, M. S. 1973 Syria c. 1550-1400 B.C., in Edwards, I. E. S. and Al, E. (eds) *The Cambridge Ancient History.* London, Cambridge University Press, 417-420.

Eggins, S. M., Woodhead, J.D., Kinsley, L. P. J., Mortimer, G. E., Sylvester, P., McCulloch, M. T., Hergt, J. M. and Handler, M. R. 1997 A simple method for the precise determination of ≥ 40 trace elements in geological samples by ICPMS using enriched isotope internal standardisation. *Chemical Geology* 134, 311-326.

Eichholz, D. E. (ed.) 1962 *Pliny. Natural History,* Volume X, Cambridge, Massachusetts, Harvard University Press.

Emery, K. G. and Neev, D. 1960 Mediterranean beaches of Israel. *Geological Survey of Israel Bulletin* 26, 1-23.

Eriksson, K. O. 2001 Historical transitions revealed in Tomb 62. Some implications of the Late Cypriot pottery in a Middle/ Late Bronze Age tomb at Pella in Jordan, in Walmsley, A. (ed.) *Australians Uncovering Ancient Jordan.* Sydney, University of Sydney, 239-246.

Farnsworth, M. And Ritchie, P. D. 1938 Spectrographic studies on ancient glass. Egyptian glass, mainly of the eighteenth dynasty, with special reference to its cobalt content. *Technical Studies in the Field of Fine Arts* 6, 155-173.

Fischer, A. 2008 *Hot Pursuit. Integrating Anthropology in Search of Ancient Glass-blowers.* Lanham, Lexington Books.

Fleet, A. J. 1984 Aqueous and sedimentary geochemistry of the rare earth elements, in Henderson, P. (ed.) *Rare Earth Element Geochemistry. Developments in Geochemistry 2.* Amsterdam, Elsevier, 343-373.

Fossing, P. 1940 *Glass Vessels Before Glass-blowing,* Copenhagen, Ejnar Munksgaard.

Foy, D. and Nenna, M.-D. (eds) 2003 *Échanges et Commerce du Verre dans le Monde Antique,* Montagnac, editions monique mergoil.

Freestone, I. C. 1987 Composition and microstructure of early opaque red glass, in Bimson, M. and Freestone, I. C. (eds) *Early Vitreous Materials.* London, British Museum, 173-191.

Freestone, I. C. 1991 Technical examination of Neo-Assyrian glazed wall plaques. *Iraq* 53, 55-58.

Freestone, I. C. 1992 Theophilus and the composition of Medieval glass, in Vandiver, P. B., Druzik, J. R., Wheeler, G. S. and Freestone, I. C. (eds) *Materials Issues in Art and Archaeology III.* San Francisco, Materials Research Society, 739-745.

Freestone, I. C. 2001 Post-depositional changes in archaeological ceramics and glasses, in Brothwell, D. R. and Pollard, A. M. (eds) *Handbook of Archaeological Sciences.* Chichester, John Wiley & Sons, 615-625.

Freestone, I. C. 2002 The relationship between enamelling on ceramics and on glass in the Islamic world. *Archaeometry* 44, 251-255.

Freestone, I. C. 2003 Primary glass sources in the mid-first millennium A.D. *Annales du 15e Congrès de l'Association Internationale pour l'Histoire du Verre.* New York, AIHV, 111-115.

Freestone, I. C. 2005 The provenance of ancient glass through compositional analysis, in Vandiver, P. B., Mass, J. L. and Murray, A. *Materials Issues in Art and Archaeology VII.* The Materials Research Society Symposium Proceedings, 852, Warrendale, PA, 008.1.1-008.1.13.

Freestone, I. C. 2006 Glass production in Late Antiquity and the Early Islamic period: a geochemical perspective, in Maggetti, M. and Messiga, B. (eds) *Geomaterials in Cultural Heritage.* London, The Geological Society of London, Special Publications, 201-216.

Freestone, I. C. and Gorin-Rosen, Y. 1999 The great glass slab at Bet She'arim, Israel: an early Islamic glassmaking experiment? *Journal of Glass Studies* 41, 105-116.

Freestone, I. C., Gorin-Rosen, Y. and Hughes, M. J. 2000 Primary glass from Israel and the production of glass in late antiquity and the early Islamic period, in Nenna, M.-D. (ed.) *La Route du Verre. Ateliers Primaires et Secondaires de Verriers du Second Millenaire av. J.-C. au Moyen-Âge.* Lyon, Travaux de la Maison de l'Orient Méditerranéen no. 33, 65-82.

Freestone, I. C., Greenwood, R. and Gorin-Rosen, Y. 2002 Byzantine and early Islamic glassmaking in the eastern Mediterranean: production and distribution of primary glass, in Kordas, G. (ed.) *Hyalos Vitrum Glass.* Athens, 167-174.

Freestone, I. C., Leslie, K. A., Thirlwall, M. and Gorin-Rosen, Y. 2003 Strontium isotopes in the investigation of early glass production: Byzantine and early Islamic glass from the Near East. *Archaeometry* 45, 19-32.

Freestone, I. C., Ponting, M. and Hughes, M. J. 2002 The origins of Byzantine glass from Maroni Petrera, Cyprus. *Archaeometry* 44, 257-272.

Freestone, I. C., Wolf, S. and Thirwall, M. 2009 Isotopic composition of glass from the Levant and south-eastern Mediterranean region, in Degryse *et al.* (eds) *Isotopes in Vitreous Materials,* Leuven University Press.

Fulford, M. G. 1989 To east and west: the Mediterranean trade of Cyrenaica and Tripolitania in antiquity. *Libyan Studies* 20, 169-191.

Garner, H. 1956a The use of imported and native cobalt in Chinese blue and white. *Oriental Art* 2, 48-50.

Garner, H. 1956b An early piece of glass from Eridu. *Iraq* 18, 147-149.

Gates, M.-H. 1981 Alalakh Levels VI and V: a chronological reassessment. *Syro-Mesopotamian Studies* 4.

Gates, M.-H. 1988 Dialogues between ancient Near Eastern texts and the archaeological record: test cases from Bronze Age Syria. *Bulletin of the American Schools of Oriental Research* 270, 63-91.

Giddy, L. 1987 *Egyptian Oases. Bahariya, Dakhla, Farafra and Kharga During Pharaonic Times,* England, Aris and Phillips Ltd.

Gill, R. (ed.) 1997 *Modern Analytical Geochemistry. An Introduction to Quantitative Chemical Analysis Techniques for Earth, Environmental and Materials Scientists,* England, Longman.

Gitin, S., Mazar, A. and Stern, E. (eds) 1998 *Mediterranean Peoples in Transition. Thirteenth to Early Tenth Centuries BCE,* Jerusalem, Israel Exploration Society.

Goldstein, S. M. 1979 *Pre-Roman and Early Roman Glass in The Corning Museum of Glass,* Corning, The Corning Museum of Glass.

Goldstein, J .I., Joy, D. C., Lefshin, E., Lyman, C. E., Michael, J. R., Newbury, D. E. and Sawyer, L. C. 2003. *Scanning Electron Microscopy and X-Ray Microanalysis.* 3rd Edition. New York: Kluwer Academic.

Gorin-Rosen, Y. 2000 The ancient glass industry in Israel. Summary of the finds and new discoveries, in Nenna, M.-D. (ed.) *La Route du Verre. Ateliers Primaires et Secondaires du Second Millénaire av. J.-C. au Moyen Âge* Lyon, Maison de l'Orient Méditerranéen-Jean Pouilloux no. 33, 49-63.

Gratuze, B. 1999 Étude des perles protohistoriques en verre de l'inhumation de Marmilhat (Lempdes 63). Orléans, CNRS.

Gratuze, B. 2000a Étude des perles protohistoriques en verre des dolmens de la Planaise de Sampzon (07) conservées au Museum d'Histoire Naturelle de Lyon. Orléans, Centre National de la Recherche Scientifique.

Gratuze, B. 2000b Étude de perles en verre provenant des niveaux de l'Âge du Bronze et de l'Âge du Fer du site de Roche Noire (Montagnieu, 01). Orléans, IRAMAT, Centre Ernest Babelon, C.N.R.S.

Gratuze, B. 2001a Étude des perles protohistoriques en verre des domens d'Eyne - les Pascarets et la Borda - et d'Enveitg - Bragnoli (Pyrenées Orientales). IRAMAT, Centre E. Babelon, Internal Analytical Report.

Gratuze, B. 2001b Étude des perles protohistoriques en verre de tumulus de Mons (Saint-Georges, 15). IRAMAT, Centre E. Babelon, Internal Analytical Report.

Gratuze, B. 2001c Étude de la composition d'un écarteur de collier en verre de l'Aveyron (Collection B. Tournier, Musée de Gap). Orleans, IRAMAT, Centre Ernest Babelon, C.N.R.S.

Gratuze, B. 2002 Étude d'éléments de parure provenant du site protohistorique de Mez-Notariou (Ouessant, 29). IRAMAT, Centre E. Babelon, Internal Analytical Report.

Gratuze, B. 2004 Étude des perles protohistoriques en verre du tumulus de Courtesoult (Haute-Saone). Orléans, IRAMAT, Centre Ernest Babelon, C.N.R.S.

Gratuze, B. 2005 Les perles protohistoriques en verre du Puech de Mus (Sainte-Eulalie-de-Cernon, Aveyron). Orleans, IRAMAT, Centre Ernest Babelon , C.N.R.S.

Gratuze, B. n.d. (a) Étude des perles protohistoriques en verre conservées au Musée de Chambéry. IRAMAT, Centre E. Babelon, C.N.R.S., Internal Analytical Report.

Gratuze, B. n.d. (b) Étude des perles protohistoriques en verre conservées au Musée d'archéologie de Lons-le-Saunier: - Grotte des Planches (Arbois) - Champ-de-Mont (Quitigny). IRAMAT, Centre E. Babelon, C.N.R.S., Internal Analytical Report.

Gratuze, B. n.d. (c) Étude des perles protohistoriques en verre de la nécropole de Ventavon (tumulus 7 des Mollards). Orléans, IRAMAT, C.N.R.S.

Gratuze, B. n.d. (d) Étude des perles protohistoriques de l'Aveyron. Orléans, IRAMAT, Centre Ernest Babelon, C.N.R.S.

Gratuze, B. and Barrandon, J.-N. 1990 Islamic glass weights and stamps: analysis using nuclear techniques. *Archaeometry* 32, 155-162.

Gratuze, B. and Billaud, Y. 2003 La circulation des perles en verre dans le Bassin Méditerranéen, de l'Âge du Bronze moyen jusqu'au Hallstatt, in Foy, D. and Nenna, M.-D. (eds) *Échanges et Commerce du verre dans le Monde Antique.* Montagnac, éditions monique mergoil, 11-15.

Gratuze, B. and Cosyns, P. 2007 La composition chimique des perles en verre de la tombe à char de la nécropole latenienne de Neufchâteau-Le Sart par LA-ICP-MS. *Arduinna* 63, 1-7.

Gratuze, B. and Picon, M. 2005 Utilisation par l'industrie verrière des sels d'aluns des oasis égyptiennnes au début du premier millénaire avant notre ère. in Borgard, P., Brun, J.-P. and Picon, M. (eds) *L'Alun de Méditerranée.* Centres Jean Berard et Camille Jullian, Collection du Centre Jean Berard, 23, Naples/ Aix-en-Provence, 269-276.

Gratuze, B., Soulier, I., Barrandon, J.-N. and Foy, D. 1992 De l'origine du cobalt dans les verres. *Révue d'Archéometrie* 16, 97-108.

Groot, N. C. E., Dik, J., Van Der Kooli, G., Alkemade, P. F. A., Sivel, V. G. M. and Tichelaar, F. D. 2006 Dark and shiny: the discovery of chromite in Bronze Age faience. *Archaeometry* 48, 229-236.

Grose, D. F. 1989 *Early Ancient Glass,* New York, Hudson Hills Press and The Toledo Museum of Art.

Gunter, A. C. 1990 Artists and ancient Near Eastern art, in Gunter, A. C. (ed.) *Investigating Artistic Environments in the Ancient Near East.* Washington, D.C., Smithsonian Institution, 9-17.

Hadidi, A. (ed.) 1987 *Studies in the History and Archaeology of Jordan III,* London, Department of Antiquities, Amman, Hashemite Kingdom of Jordan and Routledge & Kegan Paul.

Harden, D. B. 1956 Glass and glazes, in Singer, C. (ed.) *A History of Technology, II.* Oxford, 319.

Harden, D. B. 1968 Ancient glass, I: Pre-Roman. *Archaeological Journal* 125, 46-72.

Harden, D. B. 1984 Study and research on ancient glass: past and future. *Journal of Glass Studies* 26, 9-24.

Hartmann, G., Kappel, I., Grote, K. and Arndt, B. 1997 Chemistry and technology of prehistoric glass from Lower Saxony and Hess. *Journal of Archaeological Science* 24, 547-559.

Haskin, L., A., Haskin, M. A., Frey, F., A. and Wildeman, T., R. 1968 Relative and absolute terrestrial abundances of the rare earths, in Ahrens, L. H. (ed.) *Origin and Distribution of the Elements.* Oxford, Pergamon Press, 889-911.

Healy, J. F. 1999 *Pliny the Elder on Science and Technology,* Oxford, Oxford University Press.

Henderson, J. 1985a The raw materials of early glass production. *Oxford Journal of Archaeology* 4, 267-291.

Henderson, J. 1985b Interaction of changes in technology and style - progress or stagnation?, in Kingery, W. D. (ed.) *Technology and Style. Ceramics and Civilization* Columbus, Ohio, The American Ceramic Society, Inc., 21-32.

Henderson, J. 1988a Electron probe microanalysis of mixed-alkali glasses. *Archaeometry* 30, 77-91.

Henderson, J. 1988b Opacifiers in ancient glass: some analytical considerations, in Slater, E. A. and Tate, J., O. (eds) *Science and Archaeology Glasgow 1987. Part i. Proceedings of a conference on the application of scientific techniques to archaeology.* Glasgow, BAR British Series 196(i), 153-154.

Henderson, J. 1989a The use of colorants in mixed-alkali and soda-lime-silica prehistoric glass, in Maniatis, Y. (ed.) *Archaeometry. Proceedings of the 25th International Symposium.* Amsterdam, Elsevier, 217-229.

Henderson, J. (ed.) 1989b *Scientific Analysis in Archaeology and its Interpretation,* Oxford, Oxford University Committee for Archaeology, Monograph No. 19. UCLA Institute of Archaeology, Archaeological Research Tools 5.

Henderson, J. 1989c The scientific analysis of ancient glass and its archaeological interpretation, in Henderson, J. (ed.) *Scientific Analysis in Archaeology and its Interpretation.* Oxford, Oxford University Committee for Archaeology, Monograph No. 19. UCLA Institute of Archaeology, Archaeological Research Tools 5, 30-62.

Henderson, J. 1995 Colour - controlling composition to obtain the desired shade and hue. *Glass Technology* 36, 68-69.

Henderson, J. 1997 Scientific analysis of glass and glaze from Tell Brak and its archaeological implications, in Oates, D., Oates, J. and McDonald, H. (eds) *Excavations at Tell Brak. Vol. I: The Mitanni and Old Babylonian periods* British School of Archaeology in Iraq, 94-100.

Henderson, J. 1999 Archaeological and scientific evidence for the production of early Islamic glass in al-Raqqa, Syria. *Levant* 31, 225-240.

Henderson, J. 2000 *The Science and Archaeology of Materials. An Investigation of Inorganic Materials,* London, Routledge.

Henderson, J. 2001 Glass and glazes, in Brothwell, D. R. and Pollard, A. M. (eds) *Handbook of Archaeological Sciences.* Chichester, John Wiley & Sons Ltd, 471-481.

Henderson, J. 2002 Tradition and experiment in first millennium A.D. glass production - the emergence of early Islamic glass technology in Late Antiquity. *Accounts of Chemical Research* 35, 594-602.

Henderson, J. 2003 Glass trade and chemical analysis: a possible model for Islamic glass production, in Foy, D. and Nenna, M.-D. (eds) *Échanges et Commerce du Verre dans le Monde Antique.* Montagnac, éditions monique mergoil, 109-123.

Henderson, J., McLoughlin, S. D. and McPhail, D. S. 2004 Radical changes in Islamic glass technology: evidence for conservatism and experimentation with new glass recipes from Early and Middle Islamic Raqqa, Syria. *Archaeometry* 46, 439-468.

Henderson, J., Wood, N. and Tregear, M. 1989 The relationship between glass, enamel and glaze technologies: two case studies, in McGovern, P. E., Notis, M. D. and Kingery, W. D. (eds) *Cross-craft and Cross-cultural Interactions in Ceramics. Ceramics and Civilizations.* Westerville, Ohio, The American Ceramic Society, Inc., 315-346.

Henderson, P. 1982 *Inorganic Geochemistry,* Oxford, Pergamon Press.

Henderson, P. (ed.) 1984a *Rare Earth Element Geochemistry. Developments in Geochemistry 2.,* Amsterdam, Elsevier.

Henderson, P. 1984b General geochemical properties and abundances of the rare earth elements, in Henderson, P. (ed.) *Rare Earth Element Geochemistry. Developments in Geochemistry 2.* Amsterdam, Elsevier, 1-32.

Hendrickx, S. and Vermeersch, P. 2000 Prehistory from the Palaeolithic to the Badarian Culture, in Shaw, I. (ed.) *The Oxford History of Ancient Egypt.* Oxford, Oxford University Press, 17-43.

Hennessy, J. B. 2001 Introduction. The Pella Excavations. *Pella in Jordan 1979-1990.* Sydney, Adapa, 1-4.

Herr, L. G. and Najjar, M. 2001 The Iron Age, in MacDonald, B., Adams, R. and Bienkowski, P. (eds) *The Archaeology of Jordan.* Sheffield, Sheffield Academic Press, 323-345.

Herrmann, G. 2002 The Nimrud Ivories 5: the Ornate Group, in Al-Gailani Werr, L., Curtis, J., Martin, H., McMahon, A., Oates, J. and Reade, J. (eds) *Of Pots and Plans. Papers on the Archaeology and History of Mesopotamia and Syria presented to David Oates in Honour of his 75th Birthday.* London, Nabu Publications, 128-142.

Heyworth, M. P., Hunter, J. R. and Warren, S. E. 1989 The role of inductively coupled plasma spectrometry in glass provenance studies, in Maniatis, Y. (ed.) *Archaeometry. Proceedings of the 25th International Symposium.* Elsevier, 661-669.

Horden, P. and Purcell, N. 2000 *The Corrupting Sea. A Study of Mediterranean History,* Oxford, Blackwell Publishers Ltd.

Hussein, M. M. 2002 Excavations of the Department of Antiquities and Heritage at Nimrud, 1988-1993, in Al-Gailani Werr, L., Curtis, J., Martin, H., McMahon, A., Oates, J. and Reade, J. (eds) *Of Pots and Plans. Papers on the Archaeology and History of Mesopotamia and Syria presented to David Oates in Honour of his 75th Birthday.* London, Nabu Publications, 143-157.

Hussein, M. M. and Suleiman, A. 2000 *Nimrud. A City of Golden Treasures,* Baghdad, Ministry of Culture and Information, Directorate of Antiquities and Heritage.

Jackson, C. M. 1992 A Compositional Analysis of Roman and Early Post-Roman Glass and Glassworking Waste from Selected British Sites. *Department of Archaeological Sciences.* Bradford, University of Bradford.

Jackson, C. M. 2005a Glassmaking in Bronze-Age Egypt. *Science* 308, 1750-1752.

Jackson, C. M. 2005b Making colourless glass in the Roman period. *Archaeometry* 47, 763-780.

Jackson, C. M., Baxter, M. J. and Cool, H. E. M. 2003 Identifying group and meaning: an investigation of Roman colourless glass, in Foy, D. and Nenna, M.-D. (eds) *Échanges et Commerce du Verre dans le Monde Antique.* Montagnac, éditions monique mergoil, 33-39.

Jackson, C. M., Booth, C. A. and Smedley, J. W. 2005 Glass by design? Raw materials, recipes and compositional data. *Archaeometry* 47, 781-795.

Jackson, C. M., Joyner, L., Booth, C. A., Day, P. M. and Wager, E. C. W. 2003 Roman glass-making at Coppergate, York? Analytical evidence for the nature of production. *Archaeometry* 45, 435-456.

Jackson, C. M. and Nicholson, P. T. 2007 Compositional analysis of the vitreous materials found at Amarna, in Nicholson, P. T. (ed.) *Brilliant Things for Akhenaten. The Production of Glass, Vitreous Materials and Pottery at Amarna Site 045.1.* London, Egypt Exploration Society, 101-116.

Jackson, C. M., Nicholson, P. T. and Gneisinger, W. 1998 Glassmaking at Tell el-Amarna: an integrated approach. *Journal of Glass Studies* 40, 11-23.

Jackson, C. M. and Smedley, J. W. 2004 Medieval and post-medieval glass technology: melting characteristics of some glasses melted from vegetable ash and sand mixtures. *Glass Technology* 45, 36-42.

Jackson, C. M. and Smedley, J. W. 2008 Medieval and post-medieval glass technology: seasonal changes in the composition of bracken ashes from different habitats through a growing season. *Glass Technology* 49, 240-245.

James, F. and McGovern, P. E. 1993 *The Late Bronze Egyptian Garrison at Beth Shan: A Study of Levels VII and VIII, Volumes I & II,* The University Museum, University of Pennsylvania with The University of Mississippi.

Jarvis, K. E. 1989 Determination of the rare earth elements in geological samples by inductively coupled plasma mass spectrometry. *Journal of Analytical Atomic Spectrometry* 4, 563-570.

Jarvis, K. E., Gray, A. L. and Houk, R. S. 1992 *Handbook of Inductively Coupled Plasma Mass Spectrometry,* Glasgow, Blackie.

Kaczmarczyk, A. 1986 The source of cobalt in ancient Egyptian pigments, in Olin, J. S. and Blackman, M. J. (eds) *Proceedings of the 24th International Archaeometry Symposium.* Washington, DC, Smithsonian Institution Press, 369-376.

Kaczmarczyk, A. and Hedges, R. E. M. 1983 *Ancient Egyptian Faience,* Warminster, Aris and Phillips Ltd.

Kamber, B. S., Greig, A., Schoenberg, R. and Collerson, K. D. 2003 A refined solution to Earth's hidden niobium:

implications for evolution of continental crust and mode of core formation. *Precambrian Research*, 126, 289-308.

Kamber, B. S., Greig, A. and Collerson, K. D. 2005 A new estimate for the composition of weathered young upper continental crust from alluvial sediments, Queensland, Australia. *Geochimica et Cosmochimica Acta*, 69, 1041-1058.

Kitchen, K. A. 1986 *The Third Intermediate Period in Egypt (1100 - 650 B.C.)*, Warminster, Aris & Phillips Ltd.

Klengel, H. 1992 *Syria 3000 to 300 B.C. A Handbook of Political History*, Berlin, Akademie Verlag.

Knapp, A. B. 1993 *Society and Polity at Bronze Age Pella: an Annales Perspective*, JSOT/ASOR Monograph Series No. 6. Sheffield, Sheffield Academic Press.

Kühne, C. 1999 Imperial Mittani: an attempt at historical reconstruction, in Owen, D. I. and Wilhelm, G. (eds) *Nuzi at Seventy-Five*. Bethesda, CDL Press, 203-221.

Kuhrt, A. 1995 *The Ancient Near East c.3000-330 BC, Volume I*, London, Routledge.

Lambert, J. B., Johnson, S. C., Parkhurst, R. T. and Bronson, B. 1996 Analysis of 9th century Thai glass. *American Chemical Society Symposium Series* 625. Westerville, Ohio: American Chemical Society, 10-22.

Larsen, M. 1987 Commercial networks in the Ancient Near East, in Rowlands, M., Larsen, M. and Kristiansen, K. (eds) *Centre and Periphery in the Ancient World*. Cambridge, Cambridge University Press, 47-56.

Lawler, A. 2008 Who were the Hurrians? *Archaeology*, July/August, 46-52.

Leahy, A. 1985 The Libyan period in Egypt: an essay in interpretation. *Libyan Studies* 16, 51-65.

Leslie, K. A., Freestone, I. C., Lowry, D. and Thirlwall, M. 2006 The Provenance and technology of Near Eastern glass: oxygen isotopes by laser fluorination as a complement to strontium. *Archaeometry* 48, 253-270.

Liebowitz, H. 1987 Late Bronze II ivory work in Palestine: evidence of a cultural highpoint. *Bulletin of the American Schools of Oriental Research* 265, 3-24.

Lilyquist, C. 1998 The use of ivories as interpreters of political history. *Bulletin of the American Schools of Oriental Research* 310, 25-33.

Lilyquist, C. and Brill, R. H. 1996 A collaborative study of early glassmaking in Egypt c. 1500 BC. *Annales du 13e Congrès de l'Association Internationale pour l'Histoire du Verre*. Lochem, AIHV, 1-9.

Lilyquist, C., Brill, R. H. with Wypyski, M. T. 1993 *Studies in Early Egyptian Glass*, New York, Metropolitan Museum of Art.

Liverani, M. 1987 The collapse of the Near Eastern regional system at the end of the Bronze Age: the case of Syria, in ROWLANDS, M., LARSEN, M. and KRISTIANSEN, K. (eds) *Centre and Periphery in the Ancient World*. Cambridge, Cambridge University Press, 66-73.

Lucas, A. and Harris, J. R. 1989 *Ancient Egyptian Materials and Industries*, London, Histories & Mysteries of Man Ltd.

Macdonald, B., Adams, R. and Bienkowski, P. (eds) 2001 *The Archaeology of Jordan*, Sheffield, Sheffield Academic Press.

Mallowan, M. E. L. 1954 The excavations at Nimrud (Kalhu) 1953. *Iraq* 16, 59-163.

Mallowan, M. E. L. 1966 *Nimrud and Its Remains. Volumes I-II*, London, Collins.

Mallowan, M. E. L. 1978 *The Nimrud Ivories*, London, British Museum Publications, Ltd.

Mallowan, M. E. L. and Hermann, G. 1974 *Ivories from Nimrud III: furniture from SW7 Fort Shalmaneser*. London, British School of Archaeology in Iraq.

Marfoe, L. 1987 Cedar forest to silver mountain: social change and the development of long-distance trade in early Near Eastern societies, in Rowlands, M., Larsen, M. and Kristiansen, K. (eds) *Centre and Periphery in the Ancient World*. Cambridge, Cambridge University Press, 25-35.

Markoe, G. E. 1990 The emergence of Phoenician art. *Bulletin of the American Schools of Oriental Research* 279, 13-26.

Markoe, G. E. 2000 *Phoenicians*. London, British Museum Press.

Mass, J. L., Stone, R. E. and Wypyski, M. T. 1998 The mineralogical and metallurgical origins of Roman opaque colored glasses, in Kingery, W. D. and McCray, P. (eds) *The Prehistory and History of Glassmaking Technology. Ceramics and Civilization*. Westerville, Ohio, The American Ceramics Society, 121-144.

Mass, J. L., Wypyski, M. T. and Stone, R. E. 2001 Evidence for the metallurgical origins of glass at two ancient Egyptian glass factories. *MRS Bulletin* January, 38-43.

Mass, J. L., Wypyski, M. T. and Stone, R. E. 2002 Malkata and Lisht glassmaking technologies: towards a specific link between second millennium BC metallurgists and glassmakers. *Archaeometry* 44, 67-82.

Master, D. M. 2003 Trade and politics: Ashkelon's balancing act in the seventh century B.C.E. *Bulletin of the American Schools of Oriental Research* 330, 47-64.

Matoïan, V. 1999 L'art des objets en matériaux vitreux. *Le Monde de la Bible* 120, 56-57.

Matoïan, V. 2000a Données nouvelles sur le verre en Syrie au IIe millénaire av. J.-C. Le cas de Ras Shamra-Ougarit, in Nenna, M.-D. (ed.) *La Route du Verre. Ateliers primaires et secondaires du second milllénaire av. J.-C. au Moyen Âge* Lyon, Maison de l'Orient Méditerranéen-Jean Pouilloux 23-47.

Matoïan, V. 2000b Matières premieres - matériaux vitreux: données récentes. *Orient-Express*, 2, 41-42.

Matson, F. R. 1951 The composition and working properties of ancient glass. *Journal of Chemical Education*, 28, 82-87.

Mccray, P. and Kingery, W. D. (eds) 1998 *The Prehistory and History of Glassmaking Technology. Ceramics and*

Civilization, Westerville, Ohio, The American Ceramic Society.

McClellan, T. L. 1989 The chronology and ceramic assemblages of Alalakh, in Leonard, A. and Williams, B.B. (eds) *Essays in Ancient Civilization Presented to Helen J. Kantor.* The Oriental Institute, University of Chicago, 181-212.

McDonough, W. F. and Frey, F. A. 1989 Rare earth elements in upper mantle rocks, in Lipin, B. R. and McKay, G. A. (eds) *Reviews in Mineralogy. Geochemistry and Mineralogy of Rare Earth Elements.* Washington D.C., Mineralogical Society of America, 99-145.

McGinnis, C. E., Jain, J. C. and Neal, C. R. 1997 Characterisation of memory effects and development of an effective wash protocol for the measurement of petrogenetically critical trace elements in geological samples by ICP-MS. *Geostandard Newsletter* 21, 289-305.

McGovern, P. E. 1980 *Ornamental and Amuletic Jewelry Pendants of Late Bronze Age Palestine: An Archaeological Study,* Ann Arbor, University Microfilms International.

McGovern, P. E. 1985 *Late Bronze Palestinian Pendants: Innovation in a Cosmopolitan Age,* Sheffield, JSOT Press.

McGovern, P. E. 1986 *The Late Bronze and Early Iron Ages of Central Transjordan: the Baq'ah Valley Project, 1977-1981,* Philadelphia, University Museum.

McGovern, P. E. 1987a Central Transjordan in the Late Bronze and Early Iron Ages: an alternative hypothesis of socio-economic transformation and collapse, in Hadidi, A. (ed.) *Studies in the History and Archaeology of Jordan III.* London, Department of Antiquities, Amman, Hashemite Kingdom of Jordan and Routledge & Kegan Paul.

McGovern, P. E. 1987b Silicate industries of Late Bronze-Early Iron Palestine: technological interaction between Egypt and the Levant, in Bimson, M. and Freestone, I. C. (eds) *Early Vitreous Materials.* London, British Museum, 91-114.

McGovern, P. E. 1995 Technological innovation and artistic achievement in the Late Bronze and Iron Ages of Central Transjordan, in 'Amr, K., Zayadine, F. and Zaghloul, M. (eds) *Studies in the History and Archaeology of Jordan.* Amman, Department of Antiquities Jordan, 29-37.

McGovern, P. E., Fleming, S. J. And Swann, C. 1991 The beads from Tomb B10a B27 at Dinkha Tepe and the beginnings of glassmaking in the ancient Near East. *American Journal of Archaeology* 95, 395-402.

McGovern, P. E., Fleming, S. J. And Swann, C. P. 1993 The Late Bronze Egyptian garrison at Beth Shan: glass and faience production and importation in the Late New Kingdom. *Bulletin of the American Schools of Oriental Research* 290/291, 1-27.

McLennan, S. M. 1989 Rare earth elements in sedimentary rocks: influence of provenance and sedimentary processes, in Lipin, B. R. And McKay, G. A. (eds) *Reviews in Mineralogy. Geochemistry and Mineralogy of Rare Earth Elements.* Washington D.C., Mineralogical Society of America, 169-200.

McNicoll, A. W., Edwards, P. C., Hanbury-Tenison, J., Hennessy, J. B., Potts, T. F., Smith, R. H., Walmsley, A. And Watson, P. 1992 *Pella in Jordan 2. The Second Interim Report of the Joint University of Sydney and College of Wooster Excavations at Pella 1982-1985,* Sydney, Mediterranean Archaeology.

McNicoll, A. W., Hennessy, J. B. And Smith, R. H. 1980 The 1979 season at Pella of the Decapolis. *Bulletin of the American Schools of Oriental Research* 240, 63-84.

McNicoll, A. W., Smith, R. H. And Hennessy, J. B. 1982 *Pella in Jordan I.* Canberra, Australian National Gallery.

Mirti, P., Pace, M., Negro Ponzi, M. M. and Aceto, M. 2008 ICP-MS analysis of glass fragments of Parthian and Sasanian epoch from Seleucia and Veh Ardasir (central Iraq). *Archaeometry* 50, 429-450.

Misra, M. K., Ragland, K. W. and Baker, A. J. 1993 Wood ash composition as a function of furnace temperature. *Biomass and Bioenergy* 4, 103-116.

Moorey, P. R. S. 1985 *Materials and Manufacture in Ancient Mesopotamia: The Evidence of Archaeology and Art. Metals and metalwork, glazed materials and glass.* Oxford, BAR International Series 237.

Moorey, P. R. S. 1989 The Hurrians, the Mittani and technological innovation, in De Meyer, L. and Haerinck, E. (eds) *Archaeologia Iranica et Orientalis. Miscellanea in Honorem Louis Vanden Berghe.* Gent, Peeters Press, 273-286.

Moorey, P. R. S. 1999 *Ancient Mesopotamian Materials and Industries. The Archaeological Evidence,* Oxford, Oxford University Press.

Moorey, P. R. S. 2001 The mobility of artisans and opportunities for technology transfer between Western Asia and Egypt in the Late Bronze Age, in Shortland, A. (ed.) *The Social Context of Technological Change. Egypt and the Near East, 1650-1550 BC.* Oxford, Oxbow Books, 1-14.

Moran, W. L. (ed.) 1992 *The Amarna Letters,* Baltimore, The Johns Hopkins University Press.

Moscati, S. (ed.) 2001 *The Phoenicians,* London, I.B. Taurus.

Nenna, M.-D., Picon, M. And Vichy, M. 2000 Ateliers primaires et secondaires en Égypte a l'époque greco-romaine, in Nenna, M.-D. (ed.) *La Route du Verre: Ateliers Primaires et Secondaires de Verriers du Second Millénaire av. J.-C. au Moyen-Âge.* Lyon, Travaux de la Maison de l'Orient Méditerrnaéen - Jean Pouilloux no. 33, 97-112.

Neumann, B. 1927 Der babylonisch-assyrische kunstliche Lasurstein. *Chemische Zeitung* 51, 1013-1015.

Newton, R. G. 1980 Recent views on ancient glasses (review paper). *Glass Technology* 21, 173-183.

Newton, R. G. and Davison, S. 1989 *Conservation of Glass,* London, Butterworths.

Nicholson, P. T. 1993 *Egyptian Faience and Glass,* Buckinghamshire, Shire Publications Ltd.
Nicholson, P. T. 1995a Glass making/ working at Amarna: some new work. *Journal of Glass Studies* 37, 11-19.
Nicholson, P. T. 1995b Recent excavations at an ancient Egyptian glassworks: Tell el-Amarna 1993. *Glass Technology* 36, 125-128.
Nicholson, P. T. 1996 New evidence for glass and glazing at Tell el-Amarna (Egypt). *Annales du 13e Congrès de l'Association Internationale pour l'Histoire du Verre.* Netherlands, AIHV, 11-19.
Nicholson, P. T. 2007 *Brilliant Things for Akhenaten. The Production of Glass, Vitreous Materials and Pottery at Amarna Site 045.1,* London, Egypt Exploration Society.
Nicholson, P. T. and Henderson, J. 2000 Glass, in Nicholson, P. T. and Shaw, I. (eds) *Ancient Egyptian Materials and Technology.* Cambridge, Cambridge University Press, 195-224.
Nicholson, P. T. and Jackson, C. M. 1998 Kind of blue: glass of the Amarna period replicated, in McCray, P. and Kingery, W. D. (eds) *The Prehistory and History of Glass and Glassmaking Technology.* Ceramics and Civilisation, 105-120.
Nicholson, P. T. and Jackson, C. M. 2000 Tell el-Amarna and the glassmakers' workshop of the second millennium BC, in Nenna, M.-D. (ed.) *La Route du Verre. Ateliers primaires et secondaires du second millénaire av. J.-C. au Moyen Âge* Lyon, Maison de l'Orient Méditerranéen-Jean Pouilloux 12-21.
Nicholson, P. T. and Jackson, C. M. 2007 The Furnace Experiment, in Nicholson, P. T. (ed.) *Brilliant Things for Akhenaten. The Production of Glass, Vitreous Materials and Pottery at Amarna Site 045.1.* London, Egypt Exploration Society, 83-99.
Nicholson, P. T., Jackson, C. M. and Trott, K. M. 1997 The Ulu Burun glass ingots, cylindrical vessels and Egyptian glass. *Journal of Egyptian Archaeology* 83, 143-153.
Nicholson, P. T. and Shaw, I. (eds) 2000 *Ancient Egyptian Materials and Technology,* Cambridge, Cambridge University Press.
Niu, Y. and Batiza, R. 1997 Trace element evidence from seamounts for recycled oceanic crust in the Eastern Pacific mantle. *Earth Planetary Science Letters* 148, 471-483.
Novak, M. 1999 The Architecture of Nuzi and its significance in the architectural history of Mesopotamia, in Owen, D. I. and Wilhelm, G. (eds) *Nuzi at Seventy-Five.* Bethesda, CDL Press, 123-140.
Oates, D., Oates, J. and McDonald, H. 1997 *Excavations at Tell Brak. Vol. I: The Mitanni and Old Babylonian periods,* British School of Archaeology in Iraq.
Oates, J. and Oates, D. 2001 *Nimrud. An Assyrian Imperial City Revealed,* London, The British School of Archaeology in Iraq.
Oppenheim, A. L. 1973a Towards a history of glass in the ancient Near East. *Journal of the American Oriental Society* 93, 259-266.
Oppenheim, A. L. 1973b A note on research in Mesopotamian glass. *Journal of Glass Studies* 15, 9-11.
Oppenheim, A. L., Brill, R. H., Barag, D. and Von Saldern, A. 1970 *Glass and Glassmaking in Ancient Mesopotamia,* Corning, NY, The Corning Museum of Glass.
Orchard, J. J. and Brill, R.H. 1978 Some miniature painted glass plaques from Fort Shalmaneser, Nimrud. Part I: Description and a restoration. *Iraq* 40, 1-22.
Owen, D. I. and Wilhelm, G. (eds) 1999 *Studies on the Civilization and Culture of Nuzi and the Hurrians. Nuzi at Seventy-Five.,* Bethesa, CDL Press.
Peltenburg, E. J. 1987 Early faience: recent studies, origins and relations with glass, in Bimson, M. and Freestone, I. C. (eds) *Early Vitreous Materials.* London, British Museum, 5-29.
Petrie, W. M. F. 1926 Glass in the early ages. *Journal of the Society of Glass Technology* 10, 229-234.
Pollard, A. M. and Heron, C. 1996 *Archaeological Chemistry,* Cambridge, The Royal Society of Chemistry.
Pollard, A. M. and Moorey, P. R. S. 1982 Some analyses of Middle Assyrian faience and related materials from Tell al-Riman in Iraq, *Archaeometry* 14, 45-50.
Pollard, A. M., Batt, C., Stern, B. and Young, S. M. M. 2007 *Analytical Chemistry in Archaeology.* Cambridge, Cambridge University Press.
Potts, T. F., Colledge, S. M. and Edwards, P. C. 1985 Preliminary report on a sixth season of excavation by the University of Sydney at Pella in Jordan (1983/84). *Annual of the Department of Antiquities of Jordan* 29, 181-210.
Pulak, C. 1988 The Bronze Age shipwreck at Ulu Burun, Turkey: 1985 campaign. *American Journal of Archaeology* 92, 1-37.
Pusch, E. and Rehren, T. (eds) 2007 *Hochtemperatur-Technologie in der Ramses-Stadt - Rubinglas für den Pharao,* Hildesheim, Gerstenberg-Verlag.
Reade, J. 1982 Nimrud, in Curtis, J. (ed.) *Fifty Years of Mesopotamian Discovery.* London, British School of Archaeology in Iraq, 99-112.
Reade, J. 1987 Field observations of glass and glazed materials, in Bimson, M. and Freestone, I. C. (eds) *Early Vitreous Materials.* London, British Museum, 31-38.
Reade, W. J., Freestone, I. C. and Bourke, S. 2009 Innovation and continuity in Bronze and Iron Age glass from Pella in Jordan. *Annales du 17e Congrès de l'Association Internationale pour l'Histoire du Verre.* Antwerp, AIHV.
Reade, W. J., Freestone, I. C. and Simpson, S. J. 2005 Innovation or continuity? Early first millennium BCE glass in the Near East: the cobalt blue glasses from Assyrian Nimrud. *Annales du 16e Congrès de l'Association Internationale pour l'Histoire du Verre.* London, AIHV, 23-27.
Rehren, T. 2000a Rationales in Old World base glass compositions. *Journal of Archaeological Science* 27, 1225-1234.

Rehren, T. 2000b New aspects of ancient Egyptian glassmaking. *Journal of Glass Studies* 42, 13-24.

Rehren, T. 2001 Aspects of the production of cobalt-blue glass in Egypt. *Archaeometry* 43, 483-489.

Rehren, T. 2003 Comments on J.L. Mass, M.T. Wypyski and R.E. Stone, 'Malkata and Lisht glassmaking technologies: towards a specific link between second millennium BC metallurgists and glassmakers', *Archaeometry* 44(1) (2002), 67-82, and reply. *Archaeometry* 45, 185-198.

Rehren, T. 2008 A Review of factors affecting the composition of early Egyptian glasses and faience: alkali and alkali earth oxides. *Journal of Archaeological Science* 35, 1345-1354.

Rehren, T. and Pusch, E. 1997 New Kingdom glass-melting crucibles from Qantir-Piramesses. *Journal of Egyptian Archaeology* 83, 127-141.

Rehren, T. and Pusch, E. 1999 Glass and glass making at Qantir-Piramesses and beyond. *Agypten und Levante* IX, 171-179.

Rehren, T. and Pusch, E. 2005 Late Bronze Age glass production at Qantir-Piramesses, Egypt. *Science* 308, 1756-1758.

Rehren, T. and Pusch, E. 2007 Glas für den Pharao - Glasherstellung in der Spätbronzezeit des Nahen Ostens, in Wagner, G. A. (ed.) *Eihführung in die Archäometrie.* Berlin, Springer-Verlag, 215-235.

Rehren, T. and Pusch, E. 2008 Crushed rock and molten salt? Some aspects of the primary glass production at Qantir/Pi-Ramesse, in Jackson, C. M. and Wager, E. C. W. (eds) *Vitreous Materials in the Late Bronze Age Aegean.* Sheffield Studies in Aegean Archaeology. Oxford, Oxbow Books, 14-33.

Rehren, T., Pusch, E. and Herold, A. 1998 Glass coloring works within a copper-centered industrial complex in Late Bronze Age Egypt, in McCray, P. and Kingery, W. D. (eds) *The Prehistory and History of Glassmaking Technology.* The American Ceramic Society, 227-250.

Rehren, T., Pusch, E. and Herold, A. 2001 Qantir-Piramesses and the organisation of the Egyptian glass industry, in Shortland, A. J. (ed.) *The Social Context of Technological Change. Egypt and the Near East, 1650-1550 BC.* Oxford, Oxbow, 223-238.

Richards, F. V. 1992 *Scarab Seals from a Middle to Late Bronze Age Tomb at Pella in Jordan*, Biblical Institute of the University of Fribourg Switzerland Vandenhoeck & Ruprecht Gottingen.

Rising, B. 1999 Analytical procedures, in Brill, R. H. *Chemical Analyses of Early Glasses.* Volume 2, Corning, N.Y., Corning Museum of Glass, 529-544.

Robertson, A. H. F. 2002 Overview of the genesis and emplacement of Mesozoic ophiolites in the Eastern Mediterranean Tethyan region. *Lithos* 65, 1-67.

Robson, E. 2001 Society and technology in the Late Bronze Age: a guided tour of the cuneiform sources, in Shortland, A. J. (ed.) *The Social Context of Technological Change: Egypt and the Near East 1650-1550 BC.* Oxford, Oxbow Books.

Roehrig, C. H. 2005 Glass, in Roehrig, C. H., Dreyfuss, R. and Keller, C. A. (eds) *Hatshepsut from Queen to Pharaoh.* New York, Metropolitan Museum of Art, 67-69.

Rollinson, H. 1993 *Using Geochemical Data: Evaluation, Presentation, Interpretation,* Harlow, Longman Group Ltd.

Rooksby, H. P. 1959 An investigation of ancient opal glasses with special reference to the Portland vase. *Journal of the Society of Glass Technology* 43, 285T-288T.

Rooksby, H. P. 1962 Opacifiers in opal glasses. *Journal of Science and Technology* 29, 20-26.

Saleh, S. A., George, A. W. and Helm, T. M. 1972 Study of glass and glass-making at Wadi el-Natrun. *Studies in Conservation* 17, 143-172.

Sanderson, D. C. W. and Hunter, J. R. 1981 Composition variability in vegetable ash. *Science and Archaeology* 23, 27-30.

Santagostino Barbone, A., Gliozzo, E., D'Acapito, F., Memmi Turbanti, I., Turchiano, M. and Volpe, G. 2008 The *Sectilia* panels of Faragola (Ascoli Satriano, Southern Italy): a multi-analytical study of the red, orange and yellow glass slabs. *Archaeometry* 50, 451-473.

Santopadre, P. and Verità, M. 2000 Analyses of the production technologies of Italian vitreous materials of the Bronze Age. *Journal of Glass Studies* 42, 25-40.

Sauer, J. A. 1986 Transjordan in the Bronze and Iron Ages: a critique of Glueck's synthesis. *Bulletin of the American Schools of Oriental Research* 263, 1-26.

Sayre, E. V. 1963 The intentional use of antimony and manganese in ancient glasses, in Matson, F. R. and Rindone, G. E. (eds) *VI International Congress on Glass: Advances in Glass Technology, Part 2.* New York, Plenum Press, 263-282.

Sayre, E. V. 1964 Some ancient glass specimens with compositions of particular archaeological significance. New York, Brookhaven National Laboratory.

Sayre, E. V. 1965 Summary of the Brookhaven program of analysis of ancient glass. *The Application of Science in the Examination of Works of Art.* Boston, MA., Museum of Fine Arts, 145-154.

Sayre, E. V. and Smith, R. H. 1961 Compositional categories of ancient glass. *Science* 133, 1824-1826.

Sayre, E. V. and Smith, R. H. 1967 Some materials of glass manufacturing in antiquity, in Levey, M. (ed.) *Archaeological Chemistry. A Symposium.* Philadelphia, University of Pennsylvania Press, 279-311.

Sayre, E. V. and Smith, R. W. 1974 Analytical studies of ancient Egyptian glass, in Bishay, A. (ed.) *Recent Advances in Science and Technology of Materials.* New York, Plenum Press, 47-70.

Schiffer, M. B. 2005 The Devil is in the details: the cascade model of invention processes. *American Antiquity* 70.3, 485.

Schlick-Nolte, B. and Lierke, R. 2002 From silica to glass. On the track of the ancient glass artisans, in Bianchi, R. S., Schlick-Nolte, B., Bernheimer, G. M. and Barag, D. (eds) *Reflections on Ancient Glass in the Borowski Collection.* Mainz, Verlag Philipp von Zabern, 11-40.

Schlick-Nolte, B. and Werthmann, R. 2003 Glass vessels from the burial of Nesikhons. *Journal of Glass Studies* 45, 11-34.

Schoer, B. and Rehren, T. 2007 The composition of glass and associated ceramics from Qantir, in Pusch, E. and Rehren, T. (eds) *Hochtemperatur-Technologie in der Ramses-Stadt - Rubinglas für den Pharao.* Hildesheim, Gerstenberg-Verlag, 171-199.

Schreurs, J. W. H. and Brill, R. H. 1984 Iron and sulfur related colours in ancient glasses. *Archaeometry* 26, 199-209.

Shaw, I. (ed.) 2000 *The Oxford History of Ancient Egypt,* Oxford, Oxford University Press.

Shaystehfar, M. R., Tourchi, M. and Ghorbani, M. 2001 Investigation of texture and structure of Soughan Chromite Mine, Esphandagheh (genetic model of texture and structure). *20th Symposium of Geosciences,* Geological Survey of Iran, Tehran, Iran, 69-70.

Shortland, A. J. 2000a The number, extent and distribution of the vitreous materials workshops at Amarna. *Oxford Journal of Archaeology* 19, 115-134.

Shortland, A. J. 2000 *Vitreous Materials at Amarna: the Production of Glass and Faience in 18th Dynasty Egypt,* Oxford, British Archaeological Reports International Series, Archaeopress.

Shortland, A. J. 2000c Depictions of glass vessels in two Theban tombs and their role in the dating of early glass. *Journal of Egyptian Archaeology* 86, 159-161.

Shortland, A. J. 2001 Social influences on the development and spread of glass, in Shortland, A. J. (ed.) *The Social Context of Technological Change. Egypt and the Near East, 1650-1550 BC.* Oxford, Oxbow Books, 211-222.

Shortland, A. J. 2002 The use and origin of antimonate colorants in early Egyptian glass. *Archaeometry* 44, 517-530.

Shortland, A. J. 2004 Evaporites of the Wadi Natrun: seasonal and annual variation and its implication for ancient exploitation. *Archaeometry* 46, 497-516.

Shortland, A. J. 2005 The Raw materials of early glasses: the implications of new LA-ICPMS analyses. *Annales du 16e Congrès de l'Association Internationale pour l'Histoire du Verre, London 2003.* London, AIHV, 1-5.

Shortland, A. J. 2006 Application of lead isotope analysis to a wide range of Late Bronze Age Egyptian materials. *Archaeometry* 48, 657-669.

Shortland, A. J. 2007 Who were the glassmakers? Status, theory and method in mid-second millennium glass production. *Oxford Journal of Archaeology* 26, 261-274.

Shortland, A. J. 2008 Cuneiform glass texts: a question of meaning, in Martinon-Torres, M. and Rehren, T., *Archaeology, History and Science: Integrating Approaches to Ancient Materials,* London, UCL Institute of Archaeology, 61-75.

Shortland, A. J. and Eremin, K. 2006 The analysis of second millennium glass from Egypt and Mesopotamia, part 1: new WDS analyses. *Archaeometry* 48, 581-603.

Shortland, A. J., Nicholson, P. T. and Jackson, C. M. 2001 Glass and faience at Amarna: different methods of both supply for production, and subsequent distribution, in SHORTLAND, A. (ed.) *The Social Context of Technological Change. Egypt and the Near East, 1650-1550 BC.* Oxford, Oxbow Books, 147-160.

Shortland, A. J., Rogers, N. and Eremin, K. 2007 Trace element discriminants between Egyptian and Mesopotamian Late Bronze Age glasses. *Journal of Archaeological Science* 34, 781-789.

Shortland, A. J., Schachner, L., Freestone, I. C. and Tite, M. S. 2006 Natron as a flux in the early vitreous materials industry: sources, beginnings and reasons for decline. *Journal of Archaeological Science* 33, 521-530.

Shortland, A. J. and Tite, M. S. 1998 The interdependence of glass and vitreous faience production at Amarna, in McCray, P. and Kingery, W. D. (eds) *The Prehistory and History of Glassmaking Technology.* The American Ceramic Society, 251-268.

Shortland, A. J. and Tite, M. S. 2000 Raw materials of glass from Amarna and implications for the origins of Egyptian glass. *Archaeometry* 42, 141-151.

Shortland, A. J., Tite, M. S. and Ewart, I. 2006 Ancient exploitation and use of cobalt alums from the western oases of Egypt. *Archaeometry* 48, 153-168.

Shugar, A. and Rehren, T. 2002 Formation and composition of glass as a function of firing temperature. *Glass Technology 43C, Proceedings XIX International Congress on Glass, Edinburgh, 1-6 July 2001.* Sheffield, Society of Glass Technology, 145-150.

Sillar, B. and Tite, M. S. 2000 The challenge of 'technological choices' for materials science approaches to archaeology. *Archaeometry* 42, 2-20.

Smedley, J. W. and Jackson, C. M. 2002 Medieval and post-medieval glass technology: batch measuring practices. *Glass Technology* 43, 22-27.

Smedley, J. W. and Jackson, C. M. 2006 Medieval and post-medieval glass technology: bracken as a sustainable resource for glassmaking. *Glass Technology* 47, 39-47.

Smedley, J. W., Jackson, C. M. and Booth, C. A. 1998 Back to the roots: the raw materials, glass recipes and glassmaking practices of Theophilus, in McCray, P. and Kingery, W. D. (eds) *The Prehistory and History of Glass and Glassmaking Technology. Ceramics and Civilisation Volume VIII.* American Ceramic Society, 145-165.

Smith, R. H. 1982 Preliminary report on the 1981 season of the Sydney/Wooster Joint Expedition to Pella (Spring session). *Annual of the Department of Antiquities of Jordan* 26, 323-334.

Smith, R. H. 1987 Trade in the life of Pella of the Decapolis, in HADIDI, A. (ed.) *Studies in the History and Archaeology of Jordan III.* London, Department of Antiquities, Amman, Hashemite Kingdom of Jordan and Routledge & Kegan Paul, 53-55.

Smith, R. H., McNicoll, A. W. and Hennessy, J. B. 1981 The 1980 season at Pella of the Decapolis. *Bulletin of American Schools of Oriental Research* 243, 1-30.

Smith, R., H. and Potts, T. 1992 The Middle and Late Bronze Ages, in McNicoll, A. W., Edwards, P. C., Hanbury-Tenison, J., Hennessy, J. B., Potts, T. F., Smith, R. H., Walmsley, A. And Watson, P. 1992 *Pella in Jordan 2. The Second Interim Report of the Joint University of Sydney and College of Wooster Excavations at Pella 1982-1985,* Sydney, Mediterranean Archaeology, 35-82.

Smith, R., H. and Potts, T. 1992 The Iron Age, in McNicoll, A. W., Edwards, P. C., Hanbury-Tenison, J., Hennessy, J. B., Potts, T. F., Smith, R. H., Walmsley, A. and Watson, P. 1992 *Pella in Jordan 2. The Second Interim Report of the Joint University of Sydney and College of Wooster Excavations at Pella 1982-1985,* Sydney, Mediterranean Archaeology, 83-102.

Smith, R. W. 1963 Archaeological evaluation of analyses of ancient glass, in Matson, F. R. and Rindone, G. E. (eds) *Advances in Glass Technology Part 2.* New York, Plenum Press, 283-290.

Spaer, M. 1984 On some of our earliest glass finds. *The Israel Museum Journal* 3, 14-16.

Spaer, M. 2005 Reconsidering the Iron Age glass inlays found in association with carved ivories. *Annales du 16e Congrès de l'Association Internationale pour l'Histoire du Verre, London 2003.* London, AIHV, 28-31.

Sparks, R. 2001 Stone vessel workshops in the Levant: luxury products of a cosmopolitan age, in Shortland, A. (ed.) *The Social Context of Technological Change. Egypt and the Near East, 1650-1550 BC.* Oxford, Oxbow Books, 93-112.

Speiser, E. A. 1929 A Letter of Saushshatar and the date of the Kirkuk Tablets. *Journal of the American Oriental Society* 49, 269-275.

Stapleton, C., P. 2003 *Geochemical Analysis of Glass and Glaze from Hasanlu, Northwestern Iran: Constraints on Manufacturing Technology,* The University of Georgia, Ph.D. Dissertation.

Stapleton, C., P. and Swanson, S. E. 2002a Chemical analysis of glass artifacts from Iron Age levels at Hasanlu, Northwestern Iran. *Journal of Glass Technology* 43C, 151-157.

Stapleton, C., P. and Swanson, S. E. 2002b Batch material processing and glassmaking technology of 9th century B.C. artifacts excavated from the site of Hasanlu, Northwest Iran, in Vandiver, P., Goodway, M. and Mass, J. L. (eds) *Materials Issues in Art and Archaeology VI.* Materials Research Society Symposium Proceedings, 315-321.

Starr, R. F. S. 1939 *Nuzi: Report on the Excavations at Yorgan Tepa Near Kirkuk, Iraq,* Cambridge Massachusetts, Harvard University Press.

Stein, D. L. 1989 A Reappraisal of the "Saustatar Letter" from Nuzi. *Zeitschrift für Assyriologie und Vorderasiatische Archaologie*, 36-60.

Stein, D. L. 1997 Nuzi, in Meyers, E. M. (ed.) *The Oxford Encyclopedia of Archaeology in the Near East.* Oxford, Oxford University Press, 171-175.

Stern, M. E. 1993 Glass working before glass blowing. *Annales du 12e Congrès de l'Association Internationale pour l'Histoire du Verre.* Amsterdam, AIHV, 21-29.

Stern, M. E. and Schlick-Nolte, B. 1994 *Early Glass of the Ancient World 1600 B.C. - A.D. 50,* Ostfildern, Verlag Gerd Hatje.

Sun, S. And McDonough, W. F. 1989 Chemical and isotopic systematics of oceanic basalts: implications for mantle composition and processes, in Saunders, A. D. and Norry, M. J. (eds) *Magmatism in the Ocean Basins.* 42 ed. Oxford, Blackwell Scientific Publications,

Swann, C. P., McGovern, P. E. and Fleming, S. J. 1989 Colorants in glasses from ancient Syro-Palestine: specialized studies using PIXE spectrometry. *Nuclear Instruments and Methods in Physics Research* B40/41, 615-619.

Swann, C. P., McGovern, P. E. and Fleming, S. J. 1990 Colorants used in ancient Egyptian glassmaking: specialized studies using PIXE spectrometry. *Nuclear Instruments and Methods in Physics Research* B45, 311-314.

Tanimoto, S. and Rehren, T. 2008 Interactions between silicate and salt melts in LBA glassmaking. *Journal of Archaeological Science* 35, 2566-2573.

Tatton-Brown, V. and Andrews, C. 1991 Before the Invention of Glassblowing. *Five Thousand Years of Glass,* London, The British Museum Press, 26-61.

Taylor, J. 2000 The Third Intermediate Period, in Shaw, I. (ed.) *The Oxford History of Ancient Egypt.* Oxford, Oxford University Press, 330-368.

Taylor, S. R. and McLennan, S., M. 1985 *The Continental Crust: its Composition and Evolution. An Examination of the Geochemical Record Preserved in Sedimentary Rocks,* Oxford, Blackwell Scientific Publications.

Thureau-Dangin, F., Barrois, A., Dossin, G. and Dunand, M. 1931 *Arslan-Tash,* Paris, Librairie Orientaliste Paul Geuthner.

Tite, M. S. 1987 Characterisation of early vitreous materials. *Archaeometry* 29, 21-34.

Tite, M. S. and Bimson, M. 1987 Identification of early vitreous materials, in Black, J. (ed.) *Recent Advances in the Conservation and Analysis of Artifacts.* London, Summer Schools Press, University of London, 81-85.

Tite, M. S. and Shortland, A. J. 2003 Production technology for copper- and cobalt-blue vitreous materials from the New Kingdom site of Amarna - a reappraisal. *Archaeometry,* 45, 285-312.

Tite, M. S., Shortland, A. J., Maniatis, Y., Kavoussanaki, D. and Harris, S. A. 2006 The composition of the soda-rich and mixed alkali plant ashes used in the production of glass. *Journal of Archaeological Science* 33, 1284-1292.

Tite, M. S., Shortland, A. J., Nicholson, P. T. and Jackson, C. M. 1998 The use of copper and cobalt colorants in vitreous materials in ancient Egypt, in Colinart, S. and Menu, M. (eds) *La Couleur dans la Peinture at l'Emaillage de l'Egypte Ancienne.* Bari, Edipuglia, 111-120.

Tite, M. S., Shortland, A. J. and Paynter, S. 2002 The beginnings of vitreous materials in the Near East and Egypt. *Accounts of Chemical Research* 35, 585-593.

Towle, A., Henderson, J., Bellintani, P. and Gambacurta, G. 2001 Frattesina and Adria: report of scientific analyses of early glass from the Veneto. *Padusa* XXXVII, 7-68.

Turner, W. E. S. 1930 The scientific basis of glass melting. *Journal of the Society of Glass Technology* 30, 369-393.

Turner, W. E. S. 1954 Studies of ancient glasses and glass-making processes. Part II. The composition, weathering characteristics and historical significance of some Assyrian glasses of the eighth to sixth centuries B.C. from Nimrud. *Journal of the Society of Glass Technology* 38, 445T-456T.

Turner, W. E. S. 1955 Glass fragments from Nimrud of the eighth to the sixth century B.C. *Iraq* 17, 57-68.

Turner, W. E. S. 1956a Studies in ancient glasses and glassmaking processes. Part III. The chronology of the glassmaking constituents. *Journal of the Society of Glass Technology* 40, 39T-52T.

Turner, W. E. S. 1956b Studies in ancient glasses and glassmaking processes. Part IV. The chemical composition of ancient glasses. *Journal of the Society of Glass Technology* 40, 162T-186T.

Turner, W. E. S. 1956c Studies in ancient glasses and glassmaking processes. Part V. Raw materials and melting processes. *Journal of the Society of Glass Technology* 40, 277T-300T.

Turner, W. E. S. 1956d Studies of ancient glasses and glass-making processes. Part II. The composition, weathering characteristics and historical significance of some Assyrian glasses of the eighth to sixth centuries B.C. from Nimrud. *Journal of the Society of Glass Technology* 38, 445T-456T.

Turner, W. E. S. 1963 The precision attainable in the chemical analyses of ancient glasses, in Matson, F. R. and Rindone, G. E. (eds) *Advances in Glass Technology Part 2. VI International Congress on Glass,* Washington, D.C., Plenum Press, 384-387.

Turner, W. E. S. and Rooksby, H. P. 1959 A study of opalizing agents in ancient glasses throughout three thousand four hundred years, part 1. *Glastechnische Berichte* 32 K 17, 17-28.

Turner, W. E. S. and Rooksby, H. P. 1961 Further historical studies based on x-ray diffraction methods of the reagents employed in making opal and opaque glasses. *Jahrbuch des Romisch-Germanischen Zentralmuseums* 8, 1-16.

Turner, W. E. S. and Rooksby, H. P. 1963 A Study of the opalizing agents in ancient glasses throughout 3400 years, Part II, in Matson, F. R. and Rindone, G. E. (eds) *Advances in Glass Technology Part 2. VI International Congress on Glass.* New York, Plenum Press, 306-307.

Van De Mieroop, M. 2004 *A History of the Ancient Near East ca. 3000 - 323 BC,* Oxford, Blackwell Publishing.

Vandiver, P. 1982 Second millennium soda-lime-silicate technology at Nuzi (Iraq), in Wertime, T. A. and Wertime, S. F. (eds) *Early Pyrotechnology.* Washington, DC, Smithsonian Institution Press, 73-92.

Vandiver, P. 1983 Glass technology at the mid-second-millennium B.C. Hurrian site of Nuzi. *Journal of Glass Studies* 25, 239-247.

Vandiver, P., Swann, C. and Cranmer, D. 1991 A Review of mid-second millennium B.C. Egyptian glass technology at Tell el-Amarna. *Materials. Research. Society. Symposium. Proceedings,* 185, 609-616.

Velde, B. 1990 Alumina and calcium oxide content of glass found in western and northern Europe, First to Ninth centuries. *Oxford Journal of Archaeology* 9, 105-117.

Velde, B. and Gendron, C. 1980 Chemical composition of some Gallo-Roman glass fragments from central western France. *Archaeometry* 22, 183-187.

Verità, M., Basso, R., Wypyski, M. T. and Koestler, R. J. 1994 X-ray microanalysis of ancient glass materials: a comparative study of wavelength dispersive and energy dispersive techniques. *Archaeometry* 36, 241-251.

Vicenzi, E. P., Eggins, S., Logan, A. and Wyzoczanski, R. 2002 Microbeam characterization of Corning archaeological reference glasses: new additions to the Smithsonian Microbeam standard collection. *Journal of Research of the National Institute Standards and Technology* 107, 719-727.

Von Saldern, A. 1959 Glass finds at Gordion. *Journal of Glass Studies* 1, 23-25.

Von Saldern, A. 1965 Recent excavations at Nimrud, Iraq, and their importance for the history of glass of the first millennium B.C. *Proceedings of the VIIth International Congress on Glass, Brussels 1965.* Brussels, AIHV, paper 241.

Von Saldern, A. 1966a Glass, in Mallowan, M. E. L. (ed.) *Nimrud and Its Remains. Volume II.* London, Collins, 623-634.

Von Saldern, A. 1966b Mosaic glass from Hasanlu, Marlik, and Tell Al-Rimah. *Journal of Glass Studies* 8, 9-25.

Von Saldern, A. 1970 Other Mesopotamian glass vessels (1500-600 B.C.), in Oppenheim, A. L., Brill, R. H., Barag, D. and Von Saldern, A. (eds) *Glass and Glassmaking in Ancient Mesopotamia.* Corning, NY, The Corning Museum of Glass, 203-228.

Von Saldern, A. 1980 *Ancient and Byzantine Glass from Sardis,* Cambridge, Massachusetts, Harvard University Press.

Von Saldern, A. 2004 *Antikes Glas.* Verlag C.H. Beck, München.

Walmsley, A. (ed.) 2001 *Australians Uncovering Ancient Jordan: Fifty Years of Middle Eastern Archaeology,* Sydney, The University of Sydney.

Webb, V. E. S. 1987 Aegean glass: continuity or discontinuity?, in Bimson, M. and Freestone, I. C. (eds) *Early Vitreous Materials.* London, British Museum, 145-150.

Wedepohl, K. H. 1997 Chemical composition of medieval glass from excavations in West Germany. *Glastechnische Berichte Glass Science and Technology* 70, 246-255.

Weyl, W. A. 1953 *Coloured Glasses,* New York, Corning Museum of Glass.

White, D. 1994 Before the Greeks came: a survey of the current archaeological evidence for the pre-Greek Libyans. *Libyan Studies* 25, 31-44.

Wieppert, M. 1987 The relations of the states east of the Jordan with the Mesopotamian powers during the first millennium BC, in Hadidi, A. (ed.) *Studies in the History and Archaeology of Jordan III.* London, Department of Antiquities, Amman, Hashemite Kingdom of Jordan and Routledge & Kegan Paul., 97-105.

Wilhelm, G. 1989 *The Hurrians*, Warminster.

Wilhelm, G. 1999 Introduction: recent trends in Nuzi and Hurrian studies, in Owen, D. I. and Wilhelm, G. (eds) *Nuzi at Seventy-Five.* Bethesda, CDL Press, 3-11.

Winter, I. J. 1981 Is there a south Syrian style of ivory carving in the early First Millennium B.C.? *Iraq* 43, 101-130.

Zaccagnini, C. 1983 Patterns of mobility among ancient Near Eastern craftsmen. *Journal of Near Eastern Studies,* 42, 245-264.

Zaccagnini, C. 1987 Aspects of ceremonial exchange in the Near East during the late second millennium BC, in Rowlands, M., Larsen, M. and Kristiansen, K. (eds) *Centre and Periphery in the Ancient World.* Cambridge, Cambridge University Press, 57-65.

Appendix I

Model for Compositional Investigation of Ancient Glasses

INSPIRATION THINKING RESEARCH	1. Decide on archaeological objectives by experience, observation, enquiry, discussion and conducting literature survey 2. Design investigation to achieve objectives, including survey of methods of chemical analysis and interpretation, including statistical approaches
ACTION PREPARATION	3. Find academic supervisors and advisors 4. Determine availability of samples from a range of sites in a defined geographic region, and from a defined chronological context by contacting individuals and collecting institutions 5. Obtain quotations for travel and analysis and raise funds
SAMPLING RECORDING	6. Study collections and seek permission to sample 7. Collect minimal samples with least damage to objects 8. Create catalogue of samples with full detail of objects from which they were taken, including site, date, colour, type, registration numbers, excavation information, photographs and/or illustrations where possible
CHEMICAL ANALYSIS	9. Identify appropriate analytical methods (SEM/ EPMA and ICP-MS) to obtain largest possible suite of elements with highest quality analyses 10. Prepare samples for both analytical methods 11. Create concordance of sample analysis numbers with registration numbers 12. Analyse standard reference glasses and glass samples
DATA ANALYSIS INTERPRETATION	13. Assess quality of sampling and analysis 14. Tabulate and analyse data, normalising & graphing to aid interpretation 15. Consider results in relation to archaeological information from excavations, ancient texts and literature, and in the light of current knowledge of ancient glass technology, raw materials & geochemistry
COMPARISONS	16. Collect available data from other researchers 17. Compare all compositional data to gain a better understanding of glass production as a whole in regional and chronological contexts, revealing patterns, similarities and differences

CONCLUSIONS	18. Re-examine literature and data with better defined focus.
	19. Draw conclusions about development of glass technology with regard to recipes, raw material sources, production centres and their organisation, trade and distribution of glass and raw materials, continuity and innovation over time and across geographical space.
WRITING	20. Final writing of dissertation including aims, literature review, methodology, results, discussion, selecting appropriate graphs and illustrations, followed by conclusions, recommendations for further research and bibliography.
PUBLICATION	21. Process of refinement and reassessment, further research.

The idea fcr this model for the compositional investigation of ancient glasses was adapted from Gill (1997: Fig. 2.1), and was developed to summarise the stages of this research. Note that the steps from Action to Analysis cannot usually be repeated due to impracticality relating to costs and limited amount of sample available. These steps must therefore be very carefully planned and executed. During Data Analysis, Comparisons, Interpretation and Conclusions, the initial aims and objectives of the project may be reviewed, reassessed and modified as data are analysed and new information comes to light. The process of writing inevitably necessitates the review of data and research, and the incorporation of new insights gained from the intense focus required in the ultimate stage of writing a doctoral dissertation. Discussions with supervisors, colleagues and specialists from various fields was a constant thread throughout this enterprise

.

Appendix II

NUZI:
Results of Compositional Analyses (Chapter 3)

Table 1 Raw values NUZI base glass oxides (wt%)

NUZI Base Glass Oxides (wt%)															
Sample	**Colour**	**Na2O**	**MgO**	**Al2O3**	**SiO2**	**P2O5**	**K2O**	**CaO**	**TiO2**	**MnO**	**FeO**	**CoO**	**CuO**	**Sb2O3**	**Sum**
16	turq.O	15.97	4.4	0.5	62.05	0.11	2.48	11.14	0.03	0.04	0.25	0.02	1	2.07	100.06
17	blue T	15.69	5.04	0.32	66.7	0.15	2.88	7.08	0.04	0	0.12	0	1.7	0.16	99.88
18	green T	17.87	3.5	0.95	66.35	0.16	2.25	9.09	0.07	0.05	0.4	0	0.06	0.14	100.89
19	turq.O	14.06	3.08	0.47	67.07	0.23	2.74	8.48	0.03	0.03	0.26	0.01	1.21	1.82	99.49
20	blue T	15.55	5.92	0.59	64.71	0.15	4.08	6.79	0.06	0.03	0.4	0	0.93	0.15	99.36
21	blue T	16.51	3.48	0.29	66.05	0.14	1.87	9.73	0.02	0.03	0.14	0.01	0.95	0.12	99.34
22	blue T	14.29	3.77	0.49	67.96	0.16	1.97	8.7	0.01	0.02	0.24	0.02	1.48	0.1	99.21
23	green T	17.18	3.48	0.93	65.94	0.18	2.29	9.08	0.07	0.04	0.42	0.01	0.09	0.12	99.83
26	blue T	17.25	4.94	0.35	66.97	0.21	2.82	5.18	0.03	0.03	0.2	0.01	1.43	0.16	99.58
28	blue T	17.05	4.83	0.83	67.13	0.15	3.09	5	0.04	0.03	0.38	0	1.18	0.21	99.92
30	blue T	16.97	4.7	0.83	66.97	0.15	3.05	4.97	0.04	0.06	0.38	0.01	1.17	0.22	99.52
31	blue T	16.93	4.7	0.81	67.08	0.15	3.15	4.98	0.04	0.05	0.35	0.01	1.17	0.17	99.59
32	blue T	13.64	3.42	0.49	72.32	0.23	2.7	4.46	0.01	0.03	0.23	0	1.22	0.16	98.91
33	blue T	13.98	3.65	0.36	67.01	0.15	2.96	8.24	0.02	0.03	0.17	0.01	2.09	0.16	98.83
34	blue T	16.75	6.21	0.66	63.18	0.21	4.77	6.04	0.03	0.02	0.22	0	1.09	0.23	99.41
27	green T	15.15	3.41	0.78	64.99	0.17	2.38	5.45	0.01	0.11	4.24	0.01	2.4	0.2	99.3

Table 2 Raw values NUZI rare earth elements (ppb)

NUZI Rare Earth Elements (ppb)														
Sample	La	Ce	Pr	Nd	Sm	Eu	Gd	Tb	Dy	Ho	Er	Tm	Yb	Lu
16	3414	2783	340	1283	282	65	272	44	258	53	146	22	136	20
17	1262	2836	303	1181	262	67	267	42	233	46	120	17	106	15
18	2613	5207	624	2433	550	140	565	88	504	101	276	41	252	37
19	2544	2577	299	1140	240	58	245	40	230	49	134	21	132	18
20	1703	3517	413	1605	360	82	353	57	323	67	180	27	165	25
21	1200	2418	286	1095	259	66	266	43	217	43	112	16	95	14
22	1459	2989	352	1389	323	84	335	49	274	56	146	21	129	18
23	2540	5099	617	2348	550	136	556	87	495	101	275	41	250	37
26	1043	2057	239	926	200	49	209	35	192	40	112	17	105	15
28	2044	4179	504	1966	456	130	503	80	452	91	246	37	220	33
30	2033	4182	495	1966	476	128	501	80	448	89	243	36	226	32
31	2090	4238	505	2016	491	132	511	82	451	92	241	37	223	31
32	1325	2670	322	1254	334	102	477	74	389	79	206	30	178	26
33	1135	2352	271	1024	226	56	230	35	192	39	110	15	95	14
34	2093	4188	488	1882	405	96	427	70	391	79	220	32	199	29
27	1850	3486	561	2432	562	130	600	90	469	99	263	41	275	43

Table 3 Raw values for NUZI sediment-related elements (ppb)

NUZI Sediment-Related Elements (ppb)														
Sample	Sc	Ti	V	Cr	Ga	Y	Zr	Nb	Mo	Hf	Ta	W	Tl	Th	U
16	582	204,015	5,796	16,762	697	1,419	8,880	496	820	230	25	30	20	337	166
17	-102	129,026	3,116	9,966	354	1,279	6,179	389	311	156	24	17	4	221	76
18	905	374,543	10,381	34,251	1,059	2,815	15,216	1,093	806	386	72	95	31	571	234
19	315	196,396	5,150	19,024	525	1,363	10,149	345	339	250	22	15	19	323	140
20	597	245,491	6,609	12,654	736	1,858	9,501	686	755	237	44	23	10	386	185
21	656	94,026	3,038	8,156	344	1,203	4,354	296	228	112	21	19	6	186	71
22	266	160,521	5,513	6,814	542	1,582	6,384	482	428	165	41	18	6	276	107
23	1,378	364,952	10,369	35,259	1,066	2,807	16,197	1,090	847	409	56	99	27	561	217
26	127	160,219	4,121	10,777	460	1,118	6,077	413	187	152	26	15	128	241	130
28	761	241,683	7,852	20,623	957	2,580	10,971	852	645	293	39	68	17	524	190
30	791	239,639	7,806	21,394	966	2,565	12,456	859	877	342	71	68	15	528	189
31	586	245,215	7,970	21,570	946	2,583	11,319	853	642	305	44	67	24	530	185
32	2,011	151,845	5,625	11,352	547	2,113	6,837	473	144	174	32	18	11	298	152
33	201	135,765	3,734	13,893	411	1,125	5,851	418	431	147	20	20	22	230	89
34	431	188,110	4,883	16,203	784	2,312	10,688	762	504	285	43	33	40	509	175
27	2,055	164,524	24,973	760,304	622	2,869	8,476	580	201	216	39	295	7	296	268

Table 4 Raw values for NUZI alkali and alkaline earth elements (ppb)

NUZI Alkali and Alkaline Earth Elements (ppb)						
Sample	Li	Be	Rb	Sr	Cs	Ba
16	17,219	105	21,110	384,443	1,042	52,640
17	36,095	74	8,801	393,142	193	22,893
18	13,253	166	13,170	403,941	562	58,317
19	18,919	72	22,525	446,444	760	35,931
20	20,375	98	10,537	447,350	269	31,568
21	16,314	73	10,503	543,693	382	26,414
22	19,408	87	9,532	447,529	231	35,618
23	13,421	166	13,444	404,047	544	58,767
26	9,973	55	12,586	324,809	266	27,238
28	21,079	148	9,286	337,444	285	38,652
30	21,426	152	9,098	343,388	286	40,209
31	20,303	151	8,565	341,975	275	38,887
32	17,860	80	16,511	237,625	346	35,503
33	27,878	63	13,700	453,546	454	22,140
34	39,787	117	13,021	411,468	377	57,516
27	25094	186	16707	351903	386	24943

Table 5 Raw values for NUZI colourant and related elements (ppb)

NUZI Colourant and Related Elements (ppb)									
Sample	Colour	Co	Ni	Cu	Zn	Cd	Sn	Sb	Pb
16	turq. O	3,537	28,145	8,384,503	19,250	67	87,517	12,797,342	69,710
19	turq. O	4,588	43,349	8,778,512	15,917	44	8,787	9,160,017	98,965
17	blue T	1,600	11,664	12,267,899	14,160	37	2,603	16,445	5,339
20	blue T	2,023	12,423	7,177,741	15,255	77	215	7,224	3,075
21	blue T	3,104	7,327	7,350,490	13,974	12	685	2,169	2,715
22	blue T	2,415	15,384	10,120,834	17,039	49	595	2,069	3,245
26	blue T	1,436	15,103	9,472,465	13,549	52	10,592	37,855	10,507
28	blue T	3,947	15,740	8,109,356	18,010	66	10,784	4,299	8,559
30	blue T	4,057	16,028	8,195,514	17,793	64	10,874	4,308	8,243
31	blue T	4,024	16,245	8,060,622	18,084	70	10,981	4,299	8,381
32	blue T	5,723	12,782	9,334,967	18,447	38	12,131	30,947	7,371
33	blue T	4,395	11,338	14,417,283	13,393	89	4,104	12,892	6,307
34	blue T	2,593	14,042	7,404,483	15,579	62	10,246	2,241	6,749
18	green T	2,121	12,802	1,984,839	18,562	20	9,098	1,939	2,790
23	green T	2,389	11,761	1,756,525	17,385	20	8,760	1,271	2,583
27	green T	56,452	494,320	15,262,020	47,917	73	8,742	357,071	7,935

Appendix III

PELLA:
Results of Compositional Analyses (Chapter 4)

Table 6 Raw values for PELLA HMHK base glass oxides (wt %)

PELLA HMHK Base Glass Oxides (wt%)																				
Sample	Colour	Na2O	MgO	Al2O3	SiO2	P2O5	SO3	Cl	K2O	CaO	TiO2	MnO	FeO	CoO	CuO	ZnO	SnO2	Sb2O3	BaO	PbO	Sum
74	black	19.28	5.84	1.04	61.49	0.08	0.41	0.61	2.13	8.05	0.01	0.01	0.42	0.03	0.09	-0.04	-0.19	-0.04	0.05	-0.01	99.27
77	blue T	18.63	4.33	2.24	61.91	0.09	0.37	0.83	3.88	6.24	0.09	-0.03	0.47	0.04	1.09	0.02	0	-0.10	-0.04	-0.07	100.00
79	turq. O	13.31	2.79	0.37	68.26	0.05	0.29	0.92	2.39	8.06	0.02	-0.02	0.16	0.04	1.33	-0.01	-0.03	1.90	-0.04	-0.09	99.71
86	blue T	15.4	2.9	0.38	72.08	0.18	0.17	0.83	2.82	3.15	0.02	-0.01	0.2	0.03	1.69	-0.04	0.03	-0.16	-0.01	-0.02	99.65
93	turq. O	14.37	3.78	0.35	62.14	0.13	0.33	0.73	2.47	9.4	0.06	0.05	0.17	-0.05	2.08	0.03	-0.04	2.55	-0.08	-0.05	98.42
95	blue T	15.49	2.88	0.41	71.96	0.25	0.20	0.83	2.88	2.78	0	-0.04	0.21	0	1.72	-0.01	-0.03	-0.01	0.01	-0.11	99.42
98	turq. O	14.29	2.82	0.53	63.66	0.16	0.50	0.88	2.69	9.27	-0.05	0.05	0.23	-0.04	0.9	0	-0.03	1.83	0.05	-0.1	97.63
119	colourless	18.29	5.25	1.60	62.64	0.1	0.41	0.69	3.41	5.99	0.1	0.1	0.51	0	-0.11	0.02	-0.06	-0.11	-0.08		98.74
127	blue T	16.16	4.56	0.83	66.44	0.16	0.51	0.48	2.77	4.72	0	0.07	0.44	-0.01	1.15	-0.05	-0.1	-0.11	-0.05	0.02	97.99
136	blue T	18.43	4.94	1.93	62.70	0.15	0.33	0.8	3.19	4.79	0.04	0.04	0.31	0.01	1.12	0.05	0.04	-0.02	0.04	-0.08	98.81
144	blue T	16.77	4.59	0.95	66.86	0.09	0.38	0.69	2.96	4.82	0.07	0	0.41	-0.03	0.82	0.04	-0.24	-0.15	-0.01	0.03	99.04
187	blue T	14.39	3.52	0.73	67.26	0.18	0.24	0.85	3.54	4.92	0.02	0.06	0.35	-0.04	1.08	0.07	-0.28	-0.04	0.03	-0.07	96.81
188	turq. O	17.08	3.38	0.54	60.97	0.18	0.38	0.9	3.18	8.54	0.11	0.09	0.27	-0.02	0.74	-0.05	-0.33	1.79	0.01	-0.15	97.60
189	blue T	17.93	4.57	0.61	63.56	0.03	0.34	0.8	2.93	6.26	0.04	0.07	0.53	-0.01	0.77	-0.03	-0.1	0.08	0.03	0.1	98.50
190	blue T	16.64	3.73	0.60	67.10	0.18	0.18	0.94	3.74	5.05	-0.07	0.02	0.4	-0.02	1.4	-0.04	-0.06	-0.07	0.08	-0.08	99.73
192	turq. O	12.23	2.14	0.29	69.00	0.14	0.23	0.88	2.83	8.78	0.02	0	0.16	0.01	1.36	0	-0.12	1.84	-0.07	0.04	99.77
193	turq. O	16.48	3.96	0.44	63.45	0.11	0.29	0.97	2.94	8.54	0	0.07	0.28	0.01	0.6	0	-0.15	1.26	-0.03	0.05	99.27
194	blue T	13.45	4.02	0.19	68.03	0.23	0.21	0.94	2.68	8.79	0.02	0.01	0.11	-0.02	1.65	0.07	-0.05	-0.11	0.05	-0.09	100.19
195	turq. O	17.67	3.66	0.88	60.84	0	0.44	0.84	2.63	9.74	0.09	0.06	0.45	-0.02	0.89	-0.01	-0.12	1.29	0.04	-0.07	99.28
196	turq. O	14.84	2.67	0.66	65.15	0.13	0.39	0.74	2.8	9.02	0.03	0.01	0.37	0	0.81	0.03	-0.04	1.26	-0.03	-0.13	98.71
197	blue T	16.45	2.95	0.52	68.96	0.12	0.28	0.84	2.29	4.8	0.01	0.02	0.26	0.01	1.58	-0.12	-0.11	-0.06	0.11	-0.03	98.88
198	blue T	15.34	3.91	1.32	66.32	0.2	0.36	0.66	4.26	6.67	0.05	0.04	0.76	0.04	1	-0.02	0.12	0.12	-0.01	-0.03	101.11
199	turq. O	15.86	3.35	0.59	65.44	0.27	0.33	0.89	2.51	8.82	0.02	-0.04	0.33	-0.03	0.68	-0.04	0.01	1.51	-0.07	-0.06	100.36
200	turq. O	17.41	4.01	1.01	62.87	0.12	0.31	0.91	2.95	9.32	0.07	-0.01	0.62	-0.01	0.66	-0.03	-0.21	0.99	-0.04	0.04	100.99
201	blue T	15.53	3.76	0.73	69.67	0.18	0.28	0.84	2.67	4.49	0.07	0.08	0.35	0	1.34	0.02	-0.01	-0.02	-0.04	0.01	99.95
202	blue T	19.65	5.73	0.73	60.83	0.17	0.31	0.93	3.37	6.42	0.08	0.08	0.53	-0.04	0.93	0.01	0.14	0.09	-0.02	0.08	100.01
203	blue T	17.87	4.22	0.46	67.29	0.12	0.23	0.98	2.76	5.87	0.04	-0.05	0.43	-0.03	1.16	0.05	0.05	0.08	-0.02	0.04	101.54

PELLA HMHK Base Glass Oxides (wt%)																				
Sample	Colour	Na2O	MgO	Al2O3	SiO2	P2O5	SO3	Cl	K2O	CaO	TiO2	MnO	FeO	CoO	CuO	ZnO	SnO2	Sb2O3	BaO	PbO	Sum
204	blue T	19.5	5.3	0.56	61.64	0.09	0.29	0.87	4.58	6.04	0.06	0.09	0.47	-0.01	1.03	0.03	0.02	0.15	-0.03	-0.11	100.58
205	blue T	17.99	4.75	0.50	62.32	0.08	0.26	0.92	2.27	7.38	0.02	0.02	0.54	0.03	1.12	-0.05	-0.05	-0.15	-0.02	0	97.93
206	blue T	18.49	4.79	0.58	63.27	0.07	0.28	0.91	2.35	7.58	0.04	0.07	0.55	0.04	1.1	0.03	0.09	0.11	-0.12	0.06	100.28
207	blue T	15.54	6.27	0.95	64.25	0.1	0.26	0.72	3.88	6.97	0.1	0.02	0.5	-0.01	1.12	-0.02	0.19	-0.02	-0.06	-0.08	100.67
208	blue T	20.18	5.23	0.56	61.37	0.06	0.26	0.84	4.57	6.11	-0.05	0.04	0.48	-0.01	0.95	0.06	-0.09	0.08	0.12	0	100.76
209	blue T	14.37	2.95	0.84	63.62	0.1	0.35	0.68	2.47	9.88	0.06	0.05	0.56	0.05	1.16	0.01	-0.05	2.05	0.04	-0.03	99.16
210	blue T	14.99	3.45	0.40	71.15	0.11	0.56	0.74	2.6	4.05	0.05	0.02	0.18	0	2.1	-0.03	-0.06	0.15	0.03	0.06	100.55
211	turq. O	14.57	3	0.67	64.88	0.13	0.34	0.76	2.77	8.7	0.09	0.06	0.28	0.03	1.12	0.03	-0.16	1.63	-0.08	-0.01	98.82
212	blue T	18.33	3.71	0.90	65.75	0.07	0.31	0.9	2.52	5.69	0.05	0.09	0.55	-0.01	1.08	0.03	-0.13	-0.13	-0.03	0.02	99.69
213	turq. O	14.46	2.98	0.82	64.03	0.15	0.41	0.71	2.5	9.72	0.11	0.05	0.56	-0.06	1.09	-0.01	0.05	1.61	-0.05	-0.07	99.06
214	blue T	18.52	3.69	0.92	65.89	0.09	0.29	0.91	2.55	5.8	0.1	0.06	0.49	0.01	1.11	0.1	0.05	0.17	-0.11	0.01	100.66
215	turq. O	14.56	3.07	0.78	65.01	0.1	0.41	0.72	2.44	9.99	0.01	0.03	0.56	0.04	1.04	0.07	-0.07	1.29	0.07	-0.1	100.01
216	blue T	14.57	3.42	0.44	70.58	0.31	0.28	0.77	2.66	4.09	0.05	0.03	0.14	-0.05	2.15	0.09	-0.08	0.23	-0.02	0.04	99.70
218	turq. O	12.94	3.02	0.31	67.36	0.28	0.30	0.79	2.42	8.9	0.01	0.01	0.18	-0.02	1.47	0.03	0.08	2.21	-0.05	0	100.24
219	blue T	14.61	2.72	0.64	70.33	0.07	0.27	0.77	2.68	5.19	0.03	0.04	0.21	-0.02	1.96	0.11	0	-0.04	0.02	0.05	99.64
222	turq. O	17.17	4.04	0.60	63.91	0.1	0.32	0.92	2.73	6.94	0.09	0	0.32	0.05	0.46	-0.07	-0.14	1.42	-0.11	0.08	98.83

Table 7 Raw values for PELLA HMHK rare earth elements (ppb)

PELLA HMHK Rare Earth Elements (ppb)														
Sample	La	Ce	Pr	Nd	Sm	Eu	Gd	Tb	Dy	Ho	Er	Tm	Yb	Lu
74	2,173	4,205	501	1,925	408	102	415	64	375	79	221	33	209	31
77	4,029	8,418	977	3,726	844	177	813	129	757	151	421	63	397	57
79	840	1,762	207	803	174	41	163	26	156	31	90	13	83	12
86	981	1,783	237	960	217	60	225	31	174	32	84	11	69	10
93	1,070	1,957	223	859	184	43	179	28	168	35	96	14	88	13
95	930	1,738	222	907	210	59	211	31	169	32	78	11	66	9
98	1,122	2,275	262	1,008	238	55	227	35	219	46	125	19	122	18
119	3,189	6,614	769	3,007	712	172	710	107	615	123	337	49	309	45
127	1,984	3,992	479	1,864	436	121	448	70	396	79	215	31	195	28
136	3,773	7,676	891	3,411	734	150	707	108	638	126	346	51	321	45
144	2,003	4,123	484	1,866	401	101	408	62	381	79	217	32	202	30
187	1,877	3,918	445	1,742	364	84	351	55	337	70	201	28	182	27
188	1,267	2,677	311	1,235	263	67	268	42	254	55	146	22	137	20
189	1,805	3,714	428	1,693	352	91	374	57	339	69	193	28	176	26
190	1,578	3,211	380	1,448	309	79	311	49	286	59	166	25	155	23
192	872	1,751	199	784	159	37	152	24	146	29	80	12	71	11
193	1,369	2,754	318	1,259	274	65	273	43	248	52	144	21	135	20
194	1,194	2,920	283	1,118	236	57	240	35	203	39	109	15	94	14
195	2,025	4,087	467	1,789	376	93	392	60	365	75	210	30	194	29
196	1,401	2,772	324	1,251	255	67	264	42	250	53	149	22	135	20
197	1,279	2,441	301	1,178	249	60	246	38	225	48	133	19	121	18
198	2,333	4,790	558	2,170	452	118	466	74	459	95	268	38	253	37
199	1,355	2,756	314	1,221	260	66	273	43	260	53	148	22	139	20
200	1,699	3,467	392	1,511	311	78	330	52	312	65	183	27	171	24
201	1,544	3,167	369	1,426	302	73	290	44	284	58	160	24	151	22
202	1,972	3,950	462	1,789	379	92	381	59	357	73	203	30	183	27
203	2,018	4,209	470	1,800	394	93	385	59	337	68	181	26	160	23
204	1,699	3,544	408	1,562	330	81	331	51	304	62	167	24	151	22
205	1,649	3,469	393	1,510	323	76	318	50	302	62	170	24	152	23
206	1,661	3,514	395	1,542	321	79	329	51	301	62	172	25	157	23
207	2,257	4,436	528	2,049	417	101	412	65	382	80	223	33	206	30
208	1,702	3,411	402	1,602	340	83	337	52	310	61	165	24	152	22
209	1,820	3,512	412	1,586	364	86	345	51	327	62	167	26	148	22
210	1,181	2,574	285	1,106	245	57	231	37	215	45	122	18	114	17
211	1,410	2,865	332	1,281	282	71	282	44	270	56	156	23	147	22
212	1,902	3,857	443	1,716	367	93	372	59	358	75	215	31	200	29
213	2,127	4,174	493	1,905	387	101	396	65	386	81	225	33	213	32
214	1,975	4,087	462	1,783	364	96	386	62	364	77	213	32	199	30
215	1,651	3,329	382	1,472	290	81	325	52	303	65	178	27	169	25
216	1,160	2,536	286	1,129	223	54	240	36	211	44	121	18	113	16
218	966	2,063	234	907	196	44	199	33	190	39	111	16	96	15
219	1,403	2,773	333	1,292	269	69	260	42	238	51	141	22	127	19
222	1,257	2,466	293	1,137	240	62	249	39	232	49	132	20	128	18

Table 8 Raw values for PELLA HMHK sediment-related elements (ppb)

PELLA HMHK Sediment-Related Elements (ppb)

Sample	Sc	Ti	V	Cr	Ga	Y	Zr	Nb	Mo	Hf	Ta	W	Tl	Th	U
74	1,361	366,220	10,624	33,896	1,060	2,186	13,247	990	814	321	64	79	72	493	385
77	1,034	408,821	9,596	26,000	2,962	4,268	35,149	2,834	638	981	343	123	96	1,735	1,109
79	482	132,376	3,640	14,770	380	887	5,246	374	525	128	24	21	27	185	135
86	425	197,653	4,199	6,375	480	870	4,761	635	403	114	29	23	56	172	1,333
93	467	125,519	3,515	11,594	392	954	6,422	244	557	157	6	32	44	208	115
95	413	195,471	3,751	6,010	461	835	4,816	617	433	114	40	19	35	170	674
98	750	182,915	5,364	13,855	640	1,222	6,659	517	538	164	27	80	8	270	154
119	1,243	376,029	9,830	26,300	1,913	3,450	21,635	1,732	795	570	164	85	418	1,008	1,859
127	802	224,166	6,952	39,264	862	2,257	10,897	774	824	277	58	55	23	464	209
136	741	340,788	6,646	11,386	2,526	3,485	25,408	2,407	557	740	306	112	349	1,520	809
144	1,183	337,674	8,744	35,377	1,026	2,136	11,749	925	529	288	62	73	18	436	296
187	1,094	291,039	8,404	18,502	809	1,869	14,714	803	455	356	47	33	278	519	197
188	868	211,477	6,867	15,119	631	1,423	7,525	568	480	185	39	33	1,427	308	418
189	916	255,705	7,613	21,500	887	1,923	10,719	720	1,022	257	46	40	179	358	248
190	860	246,609	6,928	15,973	769	1,625	10,901	691	403	260	47	479	11	354	147
192	393	98,308	2,549	10,566	320	800	5,223	253	602	125	9	53	19	179	102
193	767	192,138	6,034	14,162	656	1,470	8,938	544	472	210	36	38	88	296	153
194	330	85,818	3,208	6,445	249	1,112	6,456	253	504	123	18	708	12	161	303
195	1,139	298,440	8,891	22,618	879	2,072	12,375	816	837	289	45	1,389	38	436	185
196	909	221,896	7,510	20,998	671	1,452	8,037	605	1,055	189	32	864	6	316	172
197	783	187,488	6,178	15,950	541	1,318	7,229	510	483	176	35	28	65	291	128
198	1,885	488,599	13,701	40,761	1,457	2,561	14,600	1,206	1,613	356	77	4,349	24	553	224
199	804	217,475	6,409	16,409	674	1,503	7,960	575	742	189	37	325	21	306	165
200	1,111	282,240	8,614	24,385	910	1,756	10,012	718	2,127	242	35	204	13	379	189
201	918	241,779	7,109	79,454	692	1,562	8,810	682	396	216	42	53	13	353	219
202	1,040	280,414	8,047	14,099	964	2,034	10,540	799	1,210	258	53	1,372	66	395	188
203	648	183,620	5,420	12,753	645	1,903	11,215	545	2,440	272	36	802	40	279	154
204	834	229,771	6,255	13,538	761	1,700	9,938	653	864	237	42	353	235	375	200
205	661	195,815	6,777	18,795	716	1,709	9,850	551	1,649	236	22	647	45	304	229
206	747	199,689	7,010	18,817	704	1,722	10,596	557	1,584	253	38	125	21	307	196
207	1,103	333,168	8,915	23,473	1,011	2,230	14,432	986	540	352	61	199	21	503	277
208	804	229,882	6,372	15,225	784	1,725	9,997	647	844	234	41	24	344	345	199
209	962	229,952	6,411	16,198	835	1,700	9,701	637	863	236	42	59	370	353	199
210	559	162,658	4,181	14,550	486	1,231	7,976	470	399	191	30	27	20	295	100
211	899	270,828	7,558	22,204	693	1,512	9,813	657	531	235	35	694	89	349	198
212	1,272	323,585	9,743	28,211	1,045	2,102	10,547	855	1,465	254	55	42	58	416	260
213	1,260	346,377	10,385	22,952	1,084	2,189	13,904	898	1,290	331	30	747	16	493	205
214	1,281	336,871	10,108	25,410	1,064	2,082	10,848	873	1,426	260	52	42	57	436	190
215	1,087	283,713	8,670	24,468	948	1,772	10,306	609	2,175	247	13	266	11	379	179
216	534	159,795	4,104	13,162	474	1,199	7,875	465	382	193	30	34	26	293	120
218	933	124,926	3,251	11,610	384	1,088	7,339	263	545	182	6	25	15	253	98
219	817	209,282	5,773	29,946	605	1,443	8,377	607	497	202	30	86	12	322	121
222	699	190,915	5,704	14,789	561	1,352	7,446	334	548	178	6	29	205	278	113

Table 9 Raw values for PELLA HMHK alkali and alkaline earth elements (ppb)

PELLA HMHK Alkali and Alkaline Earth Elements (ppb)						
Sample	**Li**	**Be**	**Rb**	**Sr**	**Cs**	**Ba**
74	6,672	138	13,047	357,394	234	54,654
77	18,386	502	22,090	521,328	478	67,805
79	11,950	55	12,524	434,189	348	39,432
86	9,545	75	11,928	168,138	371	17,109
93	20,563	65	16,988	368,303	674	36,278
95	13,819	81	19,400	246,516	695	19,464
98	11,605	71	16,537	348,393	503	37,609
119	22,157	390	15,515	474,127	426	56,410
127	25,718	144	8,949	384,562	277	38,650
136	22,421	502	22,198	448,008	446	58,481
144	15,480	125	18,896	358,353	383	42,595
187	10,672	130	16,935	284,399	337	52,046
188	10,507	76	16,299	306,434	395	33,805
189	17,562	86	23,090	464,574	741	58,129
190	12,942	101	16,188	339,874	279	37,995
192	19,282	44	15,997	465,745	605	46,063
193	13,878	73	12,852	412,865	325	42,589
194	10,781	86	13,471	616,751	553	32,699
195	11,448	153	18,142	434,000	519	54,643
196	12,308	104	12,699	276,323	323	73,772
197	14,013	89	8,693	298,718	277	52,880
198	8,378	172	26,628	400,222	464	51,589
199	13,878	100	15,511	309,832	506	38,968
200	9,676	117	15,476	385,675	630	54,170
201	11,474	91	11,853	267,981	315	45,159
202	21,470	117	14,806	496,343	475	47,408
203	21,924	104	15,010	682,726	356	45,992
204	28,741	91	23,477	472,038	323	37,649
205	25,712	103	13,112	484,381	378	55,581
206	25,335	85	11,775	494,434	323	56,379
207	36,547	133	12,105	452,293	346	50,477
208	20,934	94	23,307	432,487	349	38,485
209	20,684	110	23,473	429,400	352	39,545
210	17,907	83	17,624	237,464	525	31,694
211	11,164	93	13,663	265,711	418	33,770
212	14,414	135	13,551	301,392	373	40,641
213	15,275	131	21,139	378,672	679	63,966
214	12,896	126	15,678	310,093	449	43,236
215	9,366	100	15,233	401,765	539	56,592
216	22,222	79	22,801	232,369	644	31,385
218	14,235	57	18,374	344,784	686	38,760
219	15,180	69	16,222	305,800	440	97,228
222	13,979	69	14,517	309,692	277	39,899

Table 10 Raw values for PELLA HMHK colourant and related elements (ppb)

PELLA HMHK Colourant and Related Elements (ppb)										
Sample	Colour	Co	Ni	Cu	Zn	As	Cd	Sn	Sb	Pb
77	blue T	6,490	13,446	7,332,865	27,136	17,589	172	3,803	2,900	9,633
86	blue T	7,756	10,372	12,253,423	20,014	40,144	112	3,019	13,328	4,428
95	blue T	7,956	10,414	12,114,583	22,151	39,661	137	2,921	14,806	6,483
127	blue T	5,162	13,423	7,825,720	19,273	12,265	200	6,111	2,929	6,127
136	blue T	8,107	11,750	9,127,957	19,490	17,805	139	3,204	3,761	9,182
144	blue T	3,090	16,887	5,224,972	22,179	4,630	143	686	25,851	3,857
187	blue T	4,353	29,702	9,286,915	34,326	81,106	143	19,737	18,280	42,735
189	blue T	19,976	17,685	6,262,899	22,396	12,348	119	1,587	6,011	6,819
190	blue T	8,877	22,426	10,305,502	21,250	19,141	135	24,703	46,001	11,312
194	blue T	1,745	12,029	13,264,232	17,935	39,933	93	8,140	50,865	8,048
197	blue T	6,954	34,586	12,388,041	14,035	71,097	152	6,349	6,388	9,688
198	blue T	10,152	26,551	7,958,364	29,716	14,384	180	4,984	25,237	10,772
201	blue T	2,565	45,500	10,254,593	16,648	92,088	93	216,082	28,046	27,443
202	blue T	11,888	16,165	7,668,293	39,718	18,232	162	1,187	6,089	4,987
203	blue T	12,758	13,183	6,956,245	23,942	23,953	159	779	30,743	4,306
204	blue T	10,073	13,951	7,873,415	33,755	15,196	188	2,468	3,713	4,431
205	blue T	14,297	11,356	7,750,470	20,460	25,112	169	3,138	1,167	5,756
206	blue T	14,558	10,700	9,299,735	19,731	24,885	124	2,867	1,187	4,742
207	blue T	30,851	15,835	7,156,203	17,224	3,814	204	701	12,618	3,147
208	blue T	10,030	14,454	8,030,307	33,723	15,577	196	2,406	3,464	5,515
210	blue T	1,422	21,774	15,164,703	18,210	86,755	121	893	1,830,281	1,550,714
212	blue T	5,798	18,790	7,427,910	20,924	4,098	149	629	29,617	3,432
214	blue T	5,977	19,272	8,270,625	19,354	3,966	140	532	24,379	5,195
216	blue T	1,398	20,406	15,627,245	17,957	77,943	132	699	1,824,080	1,580,367
219	blue T	6,186	67,080	13,301,629	12,761	64,011	42	23,355	219,137	44,834
79	turq. O	6,079	25,486	8,813,346	14,801	40,768	48	6,192	15,390,049	21,233
93	turq. O	17,124	114,383	14,296,653	16,908	61,589	75	5,088	23,001,878	20,013
98	turq. O	2,324	13,481	6,285,856	18,145	6,368	136	686	14,620,193	7,233
188	turq. O	3,129	33,036	6,833,584	30,092	53,069	617	167,151	16,240,862	17,998
192	turq. O	4,506	25,889	11,372,808	18,684	33,561	126	20,853	11,135,912	22,331
193	turq. O	3,859	12,086	4,313,178	30,378	82,244	287	3,432	11,971,957	6,867
195	turq. O	9,042	19,227	6,842,318	17,429	8,752	123	11,568	11,559,227	7,750
196	turq. O	3,013	33,582	7,366,517	16,903	74,555	50	320,461	12,233,737	142,829
199	turq. O	2,482	14,634	4,098,562	17,156	12,937	119	4,907	12,706,721	8,162
200	turq. O	5,667	18,044	8,831,055	19,070	4,772	85	557	12,697,945	4,453
209	turq. O	10,428	17,472	8,624,407	49,709	16,164	218	3,176	4,237	6,100
211	turq. O	5,621	15,479	7,536,896	18,837	48,256	184	5,864	14,321,406	11,814
213	turq. O	5,359	25,028	4,672,050	19,917	10,887	111	11,335	5,575,014	89,895
215	turq. O	5,647	17,749	7,569,598	14,709	4,567	87	486	8,783,490	3,134
218	turq. O	1,235	19,277	10,767,085	18,614	80,648	94	1,700	9,833,311	27,740
222	turq. O	2,250	17,050	3,587,724	13,514	35,084	77	7,155	8,517,666	21,077
74	black	3,171	23,637	244,951	26,888	2,035	41	5,089	1,456,141	3,159
119	colourless	1,762	13,390	12,661	26,438	1,451	134	381	640	6,630

Appendix IV

PELLA:
Results of Compositional Analyses (Chapter 5)

Table 11 Raw values for PELLA LMLK base glass oxides (wt%)

PELLA LMLK Base Glass Oxides (wt%)																				
Sample	**Colour**	**Na2O**	**MgO**	**Al2O3**	**SiO2**	**P2O5**	**SO3**	**Cl**	**K2O**	**CaO**	**TiO2**	**MnO**	**FeO**	**CoO**	**CuO**	**ZnO**	**SnO2**	**Sb2O3**	**BaO**	**PbO**	**Sum**
46	black	17.68	2.65	2.48	62.10	0.15	0.32	0.46	1.37	3.77	0.23	-0.02	6.04	0.01	0.05	-0.01	0.02	-0.01	0.08	-0.09	97.28
52	black	16.39	2.05	1.61	64.08	0.16	0.26	0.51	1.24	3.39	0.19	0	7.81	0.05	0.11	0.04	0.05	0.01	-0.09	-0.05	97.81
53	black	15.64	1.27	1.46	70.28	0.12	0.26	0.52	1.28	2.04	0.2	0.02	5.46	0.02	0.01	0.04	0.14	0.09	0	-0.01	98.84
54	black	16.8	1.5	1.69	67.02	0.13	0.31	0.57	1.46	2.25	0.23	-0.06	6.13	0.08	0.02	0.03	0.01	0.03	-0.08	-0.06	98.06
55	black	15.95	1.62	1.44	63.99	0.22	0.29	0.49	1.2	2.56	0.18	0.02	10.71	0.02	-0.08	0.02	0.1	0.12	-0.06	-0.01	98.79
56	black	18.25	1.4	1.27	64.87	0.26	0.30	0.62	1.16	2.43	0.07	-0.01	7.95	0.01	-0.01	0.02	0	0.08	-0.08		98.59
57	black	15.85	1.34	1.60	63.99	0.22	0.17	0.44	1.39	2.35	0.19	0.01	11.27	0.02	0.07	-0.08	-0.09	-0.04	0.09	-0.02	98.77
58	black	15.7	1.31	0.96	60.19	0.18	0.27	0.42	1.18	2.21	0.13	0.03	15.91	0.1	-0.04	-0.04	-0.01	0.04	-0.08	-0.01	98.45
59	black	17.05	1.92	2.45	63.48	0.15	0.56	0.47	1.41	2.38	0.18	-0.02	9.21	-0.05	0.01	0.12	-0.2	-0.11	0.06	-0.1	98.97
60	black	17.37	1.49	1.69	67.11	0.19	0.28	0.61	1.38	2.29	0.19	-0.01	6.25	0.02	-0.01	0.01	0.02	0.15	-0.03	-0.08	98.93
61	black	17.07	2	2.26	65.15	0.19	0.65	0.53	1.41	2.46	0.13	0	7.14	-0.01	0.04	-0.07	-0.19	-0.09	0.03	-0.25	98.45
62	black	15.73	1.44	1.62	65.93	0.19	0.23	0.51	1.37	2.51	0.1	0.01	8.98	0.05	-0.02	-0.03	-0.08	-0.04	-0.02	-0.04	98.45
63	black	16.37	1.21	1.58	65.23	0.14	0.20	0.62	1.22	1.96	0.15	0.07	10.17	0.04	0.03	0.03	-0.06	-0.03	0.03	-0.08	98.88
64	black	17.07	1.25	1.20	65.77	0.26	0.23	0.53	1.09	2.35	0.19	0.04	9.59	-0.07	0.08	0.09	-0.09	0.04	-0.07	0.01	99.55
65	black	16.4	1.82	1.03	66.72	0.27	0.26	0.44	1.17	3.02	0.32	0.01	8.48	0.11	0.01	0	0.08	0.04	0.02	-0.07	100.12
66	black	17.27	1.4	1.73	63.44	0.17	0.41	0.73	1.82	2.54	0.1	-0.03	8.81	0.06	-0.03	-0.01	0.04	-0.12	0.03	0.13	98.49
67	black	17.39	1.3	2.00	67.22	0.16	0.26	0.56	1.38	2.24	0.25	-0.02	6.81	0	0.06	0.01	-0.08	-0.19	-0.01	-0.07	99.28
132B	black	13.27	1.9	2.12	62.31	0.09	0.22	0.35	1.04	2.13	0.2	-0.03	15.19	0.06	-0.05	0.02	-0.02	-0.06	-0.07	-0.07	98.60
159	black	16.51	1.43	1.52	64.85	0.15	0.29	0.47	1.14	2.19	0.18	0.01	11.11	-0.01	0.07	0.07	0.04	0.07	-0.04	-0.09	99.96
223	black	14.25	0.79	1.54	64.73	0.05	0.13	0.8	1.94	0.92	0.07	0.05	13.87	0.13	0.08	0.13	0	0.05	0.03	-0.01	99.54
224	black	18.66	0.72	1.09	64.09	0.09	0.16	0.48	1.24	0.75	0.25	-0.04	10.66	-0.04	0.04	0	0.09	-0.01	-0.03	0	98.20
225	black	19.15	0.73	1.10	65.69	0.13	0.16	0.5	1.21	0.68	0.22	-0.01	11.03	0.05	0.05	-0.11	0.03	-0.05	0.03	-0.11	100.48
226	black	18.77	0.87	1.03	65.76	0.08	0.23	0.37	1.04	1.48	0.21	-0.01	10.84	0.03	0	0.05	0	0.03	0.07	0.01	100.85

Table 12 Raw values for PELLA LMLK rare earth elements (ppb)

PELLA LMLK Rare Earth Elements (ppb)														
Sample	La	Ce	Pr	Nd	Sm	Eu	Gd	Tb	Dy	Ho	Er	Tm	Yb	Lu
46	9,855	19,757	2,247	7,977	1,440	182	1,110	170	1,007	215	618	98	631	98
52	8,675	17,705	2,114	7,703	1,437	186	1,296	206	1,259	269	776	118	740	111
53	9,538	19,685	2,317	8,550	1,603	226	1,382	215	1,300	272	783	119	735	111
54	9,438	19,624	2,317	8,600	1,653	225	1,481	232	1,371	284	809	120	738	112
55	8,445	17,003	2,059	7,739	1,516	249	1,376	216	1,301	271	779	117	716	108
56	10,224	24,696	2,966	11,570	2,213	343	1,859	284	1,572	324	903	139	844	127
57	10,382	22,784	2,687	10,280	1,923	287	1,723	268	1,530	319	894	133	838	121
58	7,681	15,369	1,839	6,867	1,368	207	1,191	180	1,041	217	608	92	574	89
59	8,074	16,209	1,928	7,180	1,416	222	1,449	234	1,411	304	878	134	843	126
60	9,240	19,002	2,234	8,170	1,576	216	1,367	214	1,243	265	756	114	708	107
61	6,249	12,714	1,476	5,525	1,081	174	1,079	181	1,088	236	674	103	649	99
62	9,575	19,323	2,367	9,134	2,005	321	2,102	323	1,896	387	1,069	158	952	140
63	8,286	17,121	2,123	8,017	1,611	278	1,564	238	1,334	265	716	105	643	97
64	7,190	14,608	1,867	7,427	1,706	305	1,899	284	1,595	324	859	123	757	110
65	10,250	21,107	2,601	10,064	2,260	398	2,241	341	1,957	404	1,105	162	1,000	149
66	5,914	11,747	1,436	5,276	1,014	191	941	152	951	207	604	95	616	93
67	13,580	27,365	3,259	11,715	2,113	304	1,946	314	1,856	396	1,112	169	1,057	158
132B	5,831	12,321	1,530	6,207	1,482	304	1,609	246	1,360	273	705	99	578	84
159	12,052	26,421	3,147	11,951	2,306	364	2,125	332	1,942	408	1,125	168	1,033	149
223	8,122	16,456	1,953	7,305	1,439	235	1,362	224	1,341	280	798	117	730	107
224	8,002	16,381	1,955	7,430	1,423	272	1,317	194	1,088	220	585	85	515	78
225	8,049	16,411	1,979	7,501	1,443	281	1,345	197	1,104	222	597	86	522	79
226	8,318	16,528	1,826	6,081	962	137	783	139	875	196	591	94	612	96

Table 13 Raw values PELLA LMLK sediment-related elements (ppb)

PELLA LMLK Sediment-Related Elements (ppb)														
Sample	Sc	Ti	V	Cr	Ga	Y	Zr	Nb	Mo	Hf	Ta	W	Tl	Th	U
46	2,704	1,068,205	29,679	108,551	2,394	5,792	65,693	4,441	5,507	1,530	294	648	11	3,241	1,401
52	2,675	760,922	86,934	59,718	2,124	7,535	60,774	3,823	4,494	1,495	251	579	16	3,286	5,259
53	1,844	1,162,135	51,210	59,932	2,268	7,434	66,398	4,973	10,410	1,624	346	627	25	3,202	1,332
54	1,823	1,152,733	50,083	64,311	2,292	7,645	75,039	4,780	9,743	1,859	333	604	25	3,307	1,288
55	1,352	719,894	64,927	33,725	1,978	7,664	55,578	3,598	2,484	1,326	244	555	11	2,614	2,590
56	1,301	792,875	75,839	42,501	1,860	8,692	75,072	4,038	20,838	1,824	270	624	36	4,671	2,640
57	1,532	1,128,937	59,215	51,366	2,143	8,831	71,810	4,662	20,466	1,764	314	596	6	3,747	1,983
58	1,162	653,411	83,673	61,865	1,481	6,061	68,031	2,666	20,412	1,636	171	473	14	2,978	2,100
59	4,061	1,251,785	32,211	199,049	3,145	8,358	80,270	5,220	1,089	1,955	351	609	21	2,962	1,132
60	1,743	1,125,040	49,715	57,790	2,279	7,188	71,752	4,754	10,088	1,736	313	619	25	3,068	1,306
61	3,572	1,079,265	30,470	167,948	2,730	6,323	60,117	4,480	1,040	1,459	315	563	19	2,416	1,495
62	1,084	612,477	49,039	25,109	2,563	10,788	68,677	4,742	4,704	1,656	321	690	13	3,227	2,101
63	1,041	1,195,415	48,850	17,868	2,185	7,277	53,406	3,459	4,876	1,290	242	663	23	2,695	1,365
64	1,043	721,169	44,886	22,976	2,059	8,757	58,794	3,609	6,740	1,398	238	536	14	2,040	1,832
65	1,278	1,994,379	36,589	47,466	1,479	11,124	133,232	5,771	14,880	3,368	394	611	7	5,124	2,250
66	1,271	660,227	76,849	69,486	2,085	5,726	43,809	2,873	4,117	1,075	221	548	38	2,180	3,407
67	1,687	1,397,002	47,367	84,139	2,730	10,634	92,796	5,914	19,273	2,282	409	685	20	4,868	2,280

PELLA LMLK Sediment-Related Elements (ppb)															
Sample	Sc	Ti	V	Cr	Ga	Y	Zr	Nb	Mo	Hf	Ta	W	Tl	Th	U
132B	4,035	849,690	135,861	432,285	2,297	7,494	43,478	3,002	2,939	1,052	196	416	14	1,583	953
159	1,345	1,093,445	91,710	30,720	2,200	11,116	81,019	4,807	5,917	2,004	336	681	13	4,167	2,059
223	1,334	729,704	56,179	57,759	2,408	8,061	50,725	3,543	5,417	1,250	247	3,838	41	2,408	1,809
224	2,830	1,623,864	38,249	323,933	1,523	6,098	74,025	4,258	1,762	1,708	276	524	74	3,282	1,316
225	2,824	1,637,135	35,965	221,797	1,476	6,171	74,029	4,305	1,957	1,724	279	2,980	161	3,312	1,362
226	1,318	1,191,443	200,801	30,378	1,705	5,545	98,337	4,114	2,954	2,446	277	1,880	22	1,763	1,415

Table 14 Raw values for PELLA LMLK alkali and alkaline earth elements (ppb)

PELLA LMLK Alkali and Alkaline Earth Elements (ppb)						
Sample	Li	Be	Rb	Sr	Cs	Ba
46	69,681	331	19,692	202,862	556	123,469
52	55,732	358	18,255	182,390	421	167,018
53	38,749	323	18,708	113,978	444	37,431
54	32,013	306	17,691	110,448	438	36,004
55	49,214	378	16,532	124,284	435	135,257
56	44,050	381	17,976	121,619	415	66,069
57	42,384	427	17,507	102,265	408	131,176
58	39,177	248	12,501	108,534	324	151,979
59	51,965	423	21,935	115,207	589	43,309
60	37,789	351	18,142	112,700	433	35,558
61	52,009	317	20,499	124,196	484	44,296
62	51,019	530	22,757	123,901	669	72,798
63	51,602	357	20,178	99,796	479	85,043
64	46,488	450	16,273	117,368	541	249,062
65	52,501	329	15,111	145,907	373	104,181
66	34,485	338	20,516	119,219	379	198,705
67	52,076	493	22,387	109,217	552	67,255
132B	33,516	219	12,983	60,327	384	90,015
159	45,310	492	19,072	116,074	551	180,274
223	45,684	431	19,726	50,779	469	75,128
224	43,396	89	12,392	41,437	225	27,784
225	44,070	96	12,638	42,283	230	27,382
226	39,077	405	11,858	54,990	264	28,156

Table 15 Raw values for PELLA LMLK colourant and related elements (ppb)

PELLA LMLK Colourant and Related Elements (ppb)

Sample	Colour	Co	Ni	Cu	Zn	As	Cd	Sn	Sb	Pb
46	black	4,662	60,852	49,603	12,233	32,579	45	794	155,225	4,710
52	black	5,849	130,184	79,636	147,136	227,295	55	972	221,780	13,460
53	black	5,754	58,330	64,875	24,972	209,449	76	757	797,173	11,498
54	black	5,789	59,021	63,953	26,114	212,589	105	733	1,026,115	11,992
55	black	4,223	122,698	109,107	112,632	479,120	225	1,057	710,247	16,866
56	black	2,716	98,627	29,412	28,317	42,967	173	605	26,145	3,061
57	black	4,091	107,923	236,391	53,487	23,337	143	704	46,052	4,927
58	black	4,012	149,987	161,910	124,460	31,649	219	376	13,854	14,099
59	black	5,630	74,259	64,332	14,605	162,235	57	1,011	186,400	8,580
60	black	5,602	57,247	70,963	28,223	191,224	181	842	650,500	10,945
61	black	4,781	64,929	52,277	15,126	14,608	185	615	5,947	3,461
62	black	3,442	96,554	152,280	105,789	152,670	143	1,046	693,795	16,581
63	black	2,965	87,651	355,916	104,374	38,770	79	1,342	140,699	18,001
64	black	3,627	108,036	190,748	162,566	26,106	51	784	41,793	18,643
65	black	4,409	101,441	55,849	15,915	160,361	84	856	470,375	5,240
66	black	3,308	53,278	41,060	59,950	31,548	40	724	4,674	6,958
67	black	3,194	79,311	89,954	23,969	24,940	116	1,065	24,913	3,747
132B	black	10,446	232,880	96,932	18,137	40,286	43	2,499	581,327	48,967
159	black	4,414	58,069	59,133	50,196	205,241	100	885	1,169,105	9,264
223	black	6,153	178,163	373,617	240,234	14,321	143	2,131	305,539	356,355
224	black	6,102	80,126	75,487	20,509	13,493	65	522	22,974	6,924
225	black	6,013	78,992	103,905	22,731	12,299	88	552	19,975	6,700
226	black	3,034	69,177	111,786	109,064	174,997	74	779	286,437	52,267

Appendix V

NIMRUD: Results of Compositional Analyses (Chapters 6 and 7)

Table 16 Raw values for NIMRUD base glass oxides (wt %)

NIMRUD Base Glass Oxides (wt%)																	
Sample	Colour	Na2O	MgO	Al2O3	SiO2	P2O5	SO2	K2O	CaO	TiO2	MnO	FeO	CoO	CuO	ZnO	SnO2	Sb2O3	Sum
GROUP 1																		
5Y	colourless	13.58	3.02	0.90	71.13	0.19	0.46	1.46	8.20	0.13	0.06	0.43	0.02	0.07	0.06	0.07	0.12	99.90
7N	colourless	13.67	3.08	0.94	69.23	0.20	0.50	1.55	9.01	0.09	0.04	0.41	0.02	0.06	0.03	0.16	0.49	99.48
10P	colourless	14.37	2.96	0.95	69.61	0.15	0.59	1.47	8.11	0.17	0.01	0.43	0.01	0.05	0.06	0.07	0.73	99.74
12W	colourless	13.76	3.11	0.97	69.58	0.13	0.51	1.52	9.06	0.16	0.03	0.46	0.00	0.01	0.03	0.01	0.30	99.64
22L	purple T	12.99	3.36	1.22	69.02	0.09	0.56	1.31	8.84	0.17	1.53	0.59	0.02	0.02	0.03	0.01	0.06	99.82
23N	colourless	14.11	3.15	1.06	69.53	0.14	0.46	1.44	8.62	0.12	0.06	0.43	0.01	0.04	0.07	0.07	0.20	99.51
36M	colourless	14.56	3.13	1.00	69.21	0.15	0.66	1.37	8.61	0.17	0.19	0.54	0.01	0.01	0.03	0.06	0.19	99.89
38R	pink T	13.81	2.69	0.94	68.39	0.17	0.48	1.56	8.61	0.11	1.53	0.54	0.01	0.00	0.09	0.02	0.41	99.36
GROUP 2																		
34X	turq. O	16.54	5.03	0.32	62.76	0.08	0.56	2.86	4.85	0.00	0.01	0.21	0.05	2.99	0.00	0.09	3.14	99.49
40Q	turq. O	14.67	3.96	1.51	58.19	0.34	0.44	3.61	6.56	0.06	0.03	0.91	0.05	3.87	0.07	0.20	4.79	99.26
48Y	blue T	16.49	6.46	1.18	60.59	0.13	0.54	3.67	7.46	0.11	0.09	0.59	0.04	1.99	0.17	0.03	0.10	99.64
UNGROUPED																		
2P	blue T	17.22	2.96	0.93	65.66	0.15	0.45	1.45	8.59	0.08	0.28	0.44	0.02	1.08	0.02	0.14	0.17	99.64
8Q	blue T	17.27	2.89	0.63	66.08	0.26	0.57	1.69	7.54	0.04	0.02	0.40	0.04	1.94	0.01	0.03	0.08	99.49
43K	turq. O	13.02	1.56	1.77	68.63	0.25	0.40	2.03	4.04	0.06	0.16	0.68	0.02	2.14	0.00	0.31	5.07	100.14
45P	colourless	15.43	4.15	0.80	67.97	0.18	0.42	1.54	7.69	0.06	0.01	0.50	0.07	0.02	0.02	0.06	0.78	99.70
GROUP 3																		
1M	blue T	17.06	1.66	0.68	70.30	0.10	0.46	0.73	5.29	0.06	0.25	0.33	0.03	0.94	0.04	0.14	0.13	98.20
3R	blue T	15.88	1.90	0.68	72.85	0.03	0.40	0.54	4.49	0.07	0.00	0.33	0.03	1.83	0.01	0.16	0.09	99.29
4W	turq. O	16.76	2.01	0.57	67.55	0.10	0.73	0.96	5.59	0.02	0.07	0.41	0.04	2.10	0.01	0.00	2.55	99.47
6L	colourless	19.07	1.40	0.53	70.31	0.03	0.54	0.30	6.01	0.06	0.00	0.28	0.00	0.04	0.05	0.02	0.72	99.36
9T	blue T	17.57	1.91	0.47	69.94	0.20	0.56	1.07	5.70	0.05	0.03	0.22	0.01	1.58	0.03	0.08	0.09	99.51
11R	blue T	18.59	1.11	0.51	69.66	0.03	0.69	0.69	5.99	0.03	0.03	0.25	0.02	1.80	0.00	0.09	0.04	99.53
19K	blue T	15.91	2.17	0.60	70.49	0.09	0.50	0.82	6.74		0.02	0.29	0.01	1.62	0.10	0.23	0.05	99.69
20W	colourless	16.96	1.92	0.37	72.66	0.02	0.39	1.02	5.43	0.02	0.29	0.15	0.01	0.03	0.01	0.11	0.06	99.45
26X	blue T	17.73	2.04	0.45	69.70	0.17	0.54	1.10	5.69	0.05	0.00	0.24	0.04	1.60	0.01	0.01	0.03	99.40
27K	blue T	17.81	1.49	0.47	70.72	0.22	0.34	0.79	5.65	0.02	0.03	0.22	0.06	1.75	0.02	0.08	0.06	99.73
35K	colourless	15.73	2.16	0.76	72.62	0.02	0.45	0.77	6.06	0.12	0.11	0.39	0.03	0.04	0.01	0.01	0.25	99.53
37P	blue T	16.61	1.62	0.42	68.20	0.18	0.76	1.83	7.18	0.06	0.04	0.19	0.04	2.09	0.01	0.05	0.22	99.50
39W	colourless	17.72	1.73	0.55	72.52	0.06	0.29	1.27	4.21	0.02	0.01	0.30	0.01	0.01	0.03	0.06	0.83	99.62
44M	blue T	15.92	1.48	0.44	69.90	0.14	0.57	1.40	7.15	0.01	0.40	0.13	0.03	1.62	0.09	0.04	0.04	99.36
46R	blue T	16.12	1.50	0.43	70.05	0.11	0.59	1.39	7.14	0.04	0.42	0.19	0.02	1.53	0.04	0.03	0.07	99.67
49L	turq. O	15.60	1.29	0.67	61.71	0.04	0.59	0.63	7.67	0.00	0.01	0.44	0.01	5.55	0.03	0.17	5.21	99.62

NIMRUD Base Glass Oxides (wt%)																		
Sample	**Colour**	**Na2O**	**MgO**	**Al2O3**	**SiO2**	**P2O5**	**SO2**	**K2O**	**CaO**	**TiO2**	**MnO**	**FeO**	**CoO**	**CuO**	**ZnO**	**SnO2**	**Sb2O3**	**Sum**
50X	turq. O	15.39	1.34	0.66	62.04	0.09	0.69	0.68	8.02	0.05	0.02	0.42	0.01	5.41	0.05	0.28	5.25	100.40
GROUP 4																		
13Y	dk blue	16.54	2.73	5.81	62.50	0.06	0.51	0.64	5.31	0.04	0.46	1.52	0.05	0.02	0.08	0.15	3.17	99.59
14L	dk blue	20.30	3.62	6.05	63.68	0.02	0.48	0.38	2.08	0.11	0.31	0.71	0.06	0.01	0.01	0.08	1.65	99.55
15N	dk blue	16.68	3.28	6.01	65.40	0.10	0.59	0.85	2.98	0.06	0.34	1.04	0.15	0.09	0.08	0.05	2.35	100.05
16Q	dk blue	17.00	1.88	3.94	68.81	0.00	0.48	0.18	4.01	0.09	0.28	0.73	0.14	0.03	0.05	0.03	2.10	99.75
17T	dk blue	18.63	4.17	6.44	63.67	0.06	0.48	0.12	2.59	0.06	0.73	0.70	0.19	0.05	0.18	0.16	2.22	100.45
18X	dk blue	19.36	1.84	4.25	64.34	0.03	0.58	0.21	3.63	0.02	0.61	0.97	0.12	0.06	0.17	0.04	3.37	99.60
21Y	dk blue	17.23	2.33	4.89	66.26	0.06	0.53	0.20	5.39	0.07	0.35	0.65	0.09	0.00	0.05	0.02	1.69	99.81
24Q	dk blue	17.65	2.38	5.07	70.64	0.08	0.54	0.74	1.14	0.11	0.30	0.62	0.28	0.16	0.19	0.03	0.06	99.99
25T	dk blue	16.96	2.72	5.60	70.41	0.01	0.58	0.69	1.09	0.10	0.36	0.64	0.30	0.07	0.26	0.05	0.04	99.88
29P	dk blue	17.59	3.03	7.17	59.16	0.05	0.46	0.49	3.82	0.04	0.40	1.40	0.16	0.05	0.23	0.07	1.80	95.92
30L	dk blue	20.07	2.64	6.10	62.08	0.07	1.08	0.32	4.23	0.07	0.45	0.77	0.11	0.06	0.07	0.07	1.95	100.14
31N	dk blue	18.22	2.75	5.31	69.38	0.08	0.47	0.49	1.18	0.07	0.35	0.78	0.22	0.14	0.21	0.01	0.05	99.71
32Q	dk blue	18.96	2.22	4.88	67.19	0.01	0.49	0.18	1.51	0.02	0.10	0.74	0.10	0.01	0.03	0.04	3.21	99.69

Table 17 Raw values for NIMRUD base glass oxides (wt %)

NIMRUD Base Glass Oxides (wt%)																				
(Additional glasses from a separate analysis)																				
sample	**Colour**	**Na2O**	**MgO**	**Al2O3**	**SiO2**	**P2O5**	**SO3**	**Cl**	**K2O**	**CaO**	**TiO2**	**MnO**	**FeO**	**CoO**	**CuO**	**ZnO**	**SnO2**	**Sb2O3**	**BaO**	**PbO**	**Sum**
Group 4																					
3A BM	dk blue	16.55	1.73	4.25	68.40	0.00	0.32	0.52	0.15	3.94	0.08	0.23	0.69	0.11	0.01	0.17	-0.04	2.22	0.04	-0.01	98.31
3B BM	dk blue	17.53	2.04	3.58	67.30	-0.07	0.44	0.49	0.24	2.97	0.01	0.42	0.36	0.17	-0.09	0.33	0.02	3.20	-0.02	-0.06	98.04
4A BM	dk blue	18.34	3.8	6.65	61.39	0.06	0.35	0.54	0.47	2.34	-0.04	0.38	0.94	0.14	-0.08	0.3	0.08	2.10	0.15	1.36	97.97
4B BM	dk blue	19.1	4.1	8.32	59.81	0.01	0.55	0.37	0.5	2.9	0	0.54	0.43	0.13	0.01	-0.04	0.01	1.24	-0.03	0	96.47
Group 2																					
5BM	dk blue	12.53	4.88	1.22	69.43	0.08	0.26	0.64	2.37	6.58	0.04	0.76	1.19	0.03	0.32	0.03	-0.22	-0.19	0	0.01	99.18

Table 18 Raw values for NIMRUD rare earth elements (ppb)

NIMRUD	Rare Earth Elements (ppb)													
Sample	La	Ce	Pr	Nd	Sm	Eu	Gd	Tb	Dy	Ho	Er	Tm	Yb	Lu
GROUP 1														
5Y	3,090	6,128	754	2,910	574	126	578	89	544	115	331	51	335	54
7N	3,229	6,367	790	3,034	610	129	591	91	564	119	343	54	351	55
10P	3,494	6,669	837	3,288	651	143	636	97	608	126	369	58	372	58
12W	3,220	6,307	785	3,023	612	129	559	90	571	119	340	53	339	54
22L	4,602	8,111	1,105	4,259	846	179	820	129	811	175	515	83	528	87
23N	3,663	7,350	898	3,463	689	146	651	101	631	135	391	62	398	61
36M	3,755	7,404	919	3,544	719	154	701	106	674	142	425	65	420	68
38R	3,853	6,789	906	3,437	693	151	672	105	665	143	410	65	431	70
GROUP 2														
34X	1,471	3,573	391	1,464	300	74	315	50	280	55	151	22	129	19
40Q	2,652	5,022	754	2,845	618	156	622	102	621	129	359	54	330	52
48Y	3,165	6,235	751	2,908	609	141	585	92	537	111	313	46	282	43
5BM	3,363	6,873	842	3,245	664	155	647	101	606	124	355	51	327	50
UNGROUPED														
8Q	2,541	4,188	603	2,373	489	110	459	66	371	76	212	31	187	29
43K	3,450	7,665	885	3,324	698	165	669	106	639	131	372	59	361	55
45P	2,767	7,749	673	2,608	511	120	484	76	447	89	245	37	222	34
GROUP 3														
3R	1,735	3,284	418	1,607	320	73	300	48	279	57	166	26	155	23
4W	1,629	3,825	442	1,762	373	79	321	56	323	67	189	31	192	28
6L	1,275	2,672	316	1,245	262	61	251	39	224	46	126	20	117	18
9T	1,311	2,607	319	1,238	278	58	235	37	225	45	126	19	126	18
11R	1,222	2,321	300	1,197	248	57	246	37	207	44	116	19	111	17
19K	1,732	3,245	412	1,572	336	71	296	45	281	56	155	23	153	22

Table 18 Raw values for NIMRUD rare earth elements (ppb) (continued)

NIMRUD Rare Earth Elements (ppb)														
Sample	La	Ce	Pr	Nd	Sm	Eu	Gd	Tb	Dy	Ho	Er	Tm	Yb	Lu
26X	1,425	2,847	342	1,311	280	62	248	38	223	45	126	18	115	19
27K	1,202	2,423	292	1,102	242	53	216	35	198	41	121	17	106	16
35K	2,117	3,989	509	1,926	381	88	365	59	328	68	192	27	172	26
37P	1,338	2,564	311	1,225	246	56	251	38	215	45	124	18	114	18
39W	1,687	3,227	385	1,495	321	72	306	48	272	57	164	24	153	22
46R	1,192	2,336	274	1,059	218	55	228	36	222	45	126	18	116	19
49L	1,823	2,558	330	1,243	262	59	258	42	252	51	145	20	133	20
GROUP 4														
3A BM	3,571	8,054	1,320	6,029	1,492	421	1,879	291	1,710	357	996	144	858	135
3B BM	1,785	5,101	778	3,902	1,355	403	1,704	243	1,263	224	543	71	384	57
4A BM	1,782	5,681	1,094	6,214	1,930	526	2,068	293	1,581	297	747	100	562	81
13Y	3,846	9,876	1,729	8,256	2,120	585	2,498	394	2,373	492	1,303	199	1,170	175
15N	3,564	10,087	1,579	7,103	1,743	456	1,853	271	1,515	303	775	114	698	105
16Q	4,050	7,977	1,307	5,974	1,527	409	1,868	289	1,709	361	979	150	854	132
17T	1,548	3,015	447	2,146	516	143	600	87	474	91	235	35	204	31
18X	6,939	17,829	2,677	12,946	4,049	1,423	8,634	1,530	9,183	1,833	4,738	661	3,784	576
21Y	3,314	8,412	1,340	6,105	1,565	430	1,779	243	1,326	254	642	97	548	82
24Q	4,084	12,439	2,082	9,747	2,402	591	2,177	280	1,411	256	637	88	519	74
25T	3,694	11,867	2,037	9,577	2,423	594	2,191	277	1,395	255	632	89	503	76
29P	2,636	6,322	982	4,486	1,127	296	1,278	184	1,083	221	613	92	557	82
30L	2,608	7,095	1,247	6,320	1,825	498	2,117	302	1,733	347	908	133	774	117
31N	3,234	11,337	2,105	10,414	2,658	653	2,580	330	1,755	319	781	110	636	91
32Q	1,622	3,465	530	2,426	594	148	587	78	443	85	235	36	220	32

Table 19 Raw values for NIMRUD sediment-related elements (ppb)

NIMRUD Sediment-Related Elements (ppb)															
Sample	Sc	Ti	V	Cr	Ga	Y	Zr	Nb	Mo	Hf	Ta	W	Tl	Th	U
GROUP 1															
5Y	2,442	664,869	9,627	13,008	1,204	3,234	71,986	2,019	359	1,706	131	31	4	860	1,513
7N	1,970	701,084	10,078	16,408	1,263	3,318	74,256	2,081	443	1,747	137	36	3	894	1,591
10P	5,463	737,931	9,631	12,000	1,195	3,557	87,020	2,135	281	2,019	134	35	2	930	838
12W	12,217	677,381	9,542	12,202	1,299	3,285	76,645	2,090	418	1,812	138	33	3	883	1,581
22L	20,893	908,501	19,847	15,593	2,636	4,885	118,834	2,257	56	2,799	96	1	1	1,249	2,150
23N	1,197	788,715	11,973	13,502	1,306	3,797	91,662	2,355	417	2,140	136	55	4	1,039	1,483
36M	36,152	795,247	12,342	13,767	1,531	4,003	95,634	2,376	438	2,286	156	61	3	1,072	2,859
38R	1,068	691,844	19,898	12,508	2,424	3,961	76,792	178	109	1,828	3	2	3	788	1,638
GROUP 2															
34X	23,679	145,068	3,104	6,917	478	1,521	7,957	425	277	209	28	33	37	276	376
40Q	3,039	638,902	15,530	48,588	1,958	3,428	18,965	1,684	472	474	81	52	66	860	992
48Y	21,641	418,946	10,754	51,907	1,296	3,088	21,551	1,272	834	538	86	125	3	697	276
5BM	1,791	553,149	10,772	39,286	2,934	3,379	28,638	1,344	460	708	14	67	6	965	325

NIMRUD Sediment-Related Elements (ppb)															
Sample	Sc	Ti	V	Cr	Ga	Y	Zr	Nb	Mo	Hf	Ta	W	Tl	Th	U
UNGROUPED															
8Q	3,790	318,478	7,683	39,664	936	2,376	16,904	1,139	1,006	386	72	55	3	413	2,275
43K	5,016	618,401	12,762	46,526	2,006	3,528	31,341	1,934	292	786	135	65	12	1,208	404
45P	45,683	360,860	9,123	12,337	1,154	2,646	16,989	1,466	393	412	92	17	2	536	1,244
	844	259,151	6,061	19,229	1,550	2,776	33,149	1,104	1,091	883	83	286	20	830	433
GROUP 3															
3R	1,339	280,405	6,320	6,907	921	1,564	17,403	1,168	482	411	80	22	4	382	3,043
4W	476	332,917	6,510	7,580	818	1,909	31,770	1,071	309	739	66	19	4	428	1,812
6L	703	203,516	5,444	6,561	652	1,302	11,502	790	847	256	53	2,438	4	305	3,897
9T	509	255,930	4,652	6,781	907	1,304	22,340	1,055	250	461	68	9	5	371	792
11R	26,580	193,390	6,279	5,553	538	1,240	12,698	705	270	282	47	29	6	288	2,463
19K	12,959	262,454	9,776	7,188	805	1,597	18,380	1,131	643	444	71	351	1	361	2,658
26X	19,086	218,832	4,998	6,625	576	1,356	15,174	770	267	341	47	11	4	305	746
27K	4,771	198,242	4,310	5,786	568	1,170	13,990	744	253	312	42	29	5	288	484
35K	1,023	324,458	10,473	10,184	1,087	1,861	22,136	1,391	986	519	87	32	2	430	2,765
37P	521	194,058	4,570	6,562	613	1,361	10,800	737	413	261	56	19	4	272	758
39W	652	233,376	6,060	40,028	677	1,553	16,617	744	789	394	49	49	2	476	1,302
46R	550	162,912	7,503	4,881	994	1,300	17,115	571	260	392	26	115	8	233	644
49L	27,796	253,131	6,130	7,390	783	1,449	14,451	836	455	340	42	24	8	362	1,944
GROUP 4															
3A BM	2,054	611,630	8,052	12,426	998	9,702	90,976	1,685	2,056	50	39	4	1,256	920	105
3B BM	1,219	220,857	4,619	5,231	832	5,578	13,568	114	309	3	45	4	710	714	114
4A BM	1,449	244,216	8,175	7,663	943	8,498	19,183	247	442	3	7	48	665	3,658	143
13Y	1,814	516,563	10,646	13,539	1,747	14,407	61,133	1,567	187	1,346	72	21	1	936	1,372
15N	2,305	400,914	9,590	18,994	1,857	9,005	31,987	1,427	214	597	89	76	12	1,144	5,665
16Q	11,821	605,592	8,021	12,896	1,307	9,896	97,580	2,024	95	2,139	130	36	3	1,253	935
17T	31,455	175,493	5,188	8,842	1,528	2,602	11,797	539	24	266	28	1	0	466	3,962
18X	5,728	144,027	6,535	22,500	1,519	54,272	13,481	506	52	298	33	4	0	815	4,185
21Y	9,779	325,553	5,896	21,378	1,282	6,898	47,762	1,096	158	1,058	69	281	4	1,336	1,015
24Q	1,966	440,676	6,756	19,928	1,225	6,569	41,309	1,517	333	920	99	224	3	1,607	9,808
25T	1,702	358,396	6,154	15,874	1,157	6,501	34,380	1,263	369	764	84	192	4	1,387	9,907
29P	3,585	313,977	8,602	9,769	1,536	6,798	27,462	995	230	635	62	24	61	748	5,733
30L	12,947	267,246	6,790	8,222	1,455	10,022	21,376	945	121	471	59	5	1	795	5,689
31N	1,739	405,827	7,306	12,922	1,499	8,511	31,817	1,266	332	727	85	153	4	1,323	9,081
32Q	3,212	273,527	5,161	6,924	934	2,307	27,466	860	205	580	53	14	5	965	4,476

Table 20 Raw values for NIMRUD alkali and alkaline earth elements (ppb)

NIMRUD Alkali and Alkaline Earth Elements (ppb)						
Sample	Li	Be	Rb	Sr	Cs	Ba
GROUP 1						
5Y	4,296	145	7,118	459,080	123	81,694
7N	4,449	162	7,386	476,456	127	88,120
10P	4,493	149	7,146	388,653	130	92,154
12W	4,387	163	7,289	468,930	125	86,615
22L	5,913	214	7,409	439,301	163	139,338
23N	4,609	168	7,835	417,998	156	102,146
36M	8,793	173	7,463	425,923	157	101,129
38R	5,241	177	7,829	430,780	154	206,810
GROUP 2						
34X	13,595	82	11,344	286,859	249	19,175
40Q	3,731	211	8,461	292,792	174	36,310
48Y	25,084	155	8,099	564,761	152	57,224
5BM	18,803	193	14,831	450,116	237	69,047
UNGROUPED						
8Q	5,871	159	7,667	341,639	145	73,846
43K	71,546	272	14,858	202,228	2,089	74,053
45P	5,894	171	7,062	363,165	158	54,903
GROUP 3						
3R	3,601	134	3,778	237,659	89	41,168
4W	5,212	107	6,748	197,126	126	56,132
6L	2,401	107	2,515	143,999	67	37,869
9T	3,491	120	4,186	253,024	104	50,364
11R	3,066	86	5,119	132,404	201	42,213
19K	3,901	127	4,520	252,482	119	42,064
26X	3,846	83	4,137	246,121	110	55,123
27K	3,066	79	5,181	179,333	200	45,648
35K	4,001	140	3,925	247,885	92	50,941
37P	5,595	86	5,297	283,303	132	65,713
39W	9,633	81	5,786	157,283	174	26,074
46R	5,471	77	7,234	252,582	253	87,233
49L	4,387	110	6,346	151,404	248	47,112
GROUP 4						
3A BM	6,068	1,162	2,469	117,183	64	24,773
3B BM	6,122	935	3,379	155,441	98	39,632
4A BM	8,119	3,076	6,232	77,612	142	39,308
13Y	7,784	1,930	5,447	147,535	152	63,870
15N	26,410	1,836	10,768	88,273	292	33,772
16Q	5,885	1,059	2,650	121,424	66	32,076
17T	3,527	1,187	1,915	51,766	53	15,368
18X	9,075	10,042	2,471	154,360	75	32,956
21Y	6,264	2,212	2,978	152,686	87	38,954
24Q	24,628	1,400	8,187	67,997	141	23,376
25T	21,063	1,410	6,789	67,239	119	21,480
29P	12,603	3,387	6,512	101,281	271	58,968
30L	11,333	5,001	5,465	110,395	145	48,464
GROUP 4						
31N	15,187	2,030	5,977	103,294	120	25,963
32Q	4,939	777	2,547	70,176	71	35,063

Table 21 Raw values for NIMRUD colourant and related elements (ppb)

NIMRUD Colourant and Related Elements (ppb)									
Sample	Group	Co	Ni	Cu	Zn	Cd	Sn	Sb	Pb
blue T									
3R	3	2,791	6,552	12,710,166	21,194	21	110,825	206,735	420,857
8Q	ungrouped	2,029	26,210	13,623,895	52,861	23	6,278	83,571	89,828
9T	3	1,542	9,928	9,491,834	21,436	24	63,742	9,659	3,092
11R	3	3,571	9,948	12,359,642	14,650	16	174,194	65,460	15,604
19K	3	4,544	9,216	12,274,016	22,393	19	511,093	61,466	60,923
26X	3	2,169	7,854	11,215,703	21,113	21	52,057	12,375	2,625
27K	3	2,902	7,339	11,669,479	17,625	18	87,435	31,219	3,523
37P	3	8,707	12,181	14,194,789	23,808	17	318,878	2,149,787	9,237
46R	3	2,874	10,961	10,626,507	25,044	17	59,650	351,021	39,098
48Y	2	8,773	27,239	13,694,915	559,306	47	62,716	31,973	368,743
turq O									
4W	3	3,935	9,303	15,615,487	80,673	43	141,046	8,205,595	221,605
34X	2	1,212	8,716	20,530,159	179,815	176	20,818	10,618,195	862,399
40Q	2	8,652	39,839	37,438,186	60,384	37	2,454,521	11,510,930	492,348
43K	ungrouped	3,926	30,891	14,702,626	96,523	70	5,163	9,735,207	289,935
49L	3	9,576	23,341	36,277,462	20,225	42	558,651	21,806,061	207,917
colourless									
5Y	1	1,955	6,044	12,044	22,039	53	234	1,960,698	2,878
6L	3	451	3,224	10,431	14,188	18	104	5,129,009	11,187
7N	1	1,949	7,776	12,023	22,468	58	278	2,705,830	11,663
10P	1	1,893	6,077	15,270	20,780	65	237	5,211,531	13,704
12W	1	1,798	6,251	10,907	21,593	59	282	2,446,042	10,849
23N	1	1,973	6,499	13,183	26,466	62	588	1,582,300	1,000,114
35K	3	1,721	6,358	25,235	18,669	24	192	2,700,773	14,662
36N	1	2,576	7,245	14,499	22,869	70	489	2,322,180	201,998
39W	3	1,346	9,022	19,446	14,077	24	486	6,647,956	3,214
45P	ungrouped	1,834	11,001	23,995	32,791	37	236	4,141,840	8,853
purple T									
22L	1	4,669	17,977	14,296	24,700	100	429	104,571	8,419
38R	1	6,362	14,056	30,418	28,775	62	1,546	1,730,743	13,629
dark blue									
3A BM	4	963,381	439,489	100,874	472,393	61	746	16,067,902	49,325
3B BM	4	1,519,858	1,402,241	95,632	1,992,804	34	285	12,518,491	86,931
4A BM	4	847,251	868,350	52,312	1,427,005	227	670	14,508,602	12,979,131
13Y	4	572,543	362,534	106,348	414,011	60	954	22,016,072	83,059
15N	4	855,393	596,478	898,169	672,623	66	2,352	14,010,902	63,963
16Q	4	972,356	460,812	103,192	442,669	69	687	16,253,372	45,465
17T	4	1,150,308	1,190,552	37,284	1,207,489	26	717	10,649,881	6,825
18X	4	949,087	728,369	37,136	1,046,322	82	912	19,149,924	23,939
21Y	4	999,671	661,823	102,797	512,201	51	1,159	12,760,173	63,976
24Q	4	1,520,876	772,669	1,175,706	1,246,890	43	4,977	161,190	201,489
25T	4	1,617,923	801,343	1,019,368	1,263,837	37	5,373	158,637	343,168
29P	4	1,243,901	1,016,865	209,830	1,598,427	77	8,182	9,852,955	23,333,670
30L	4	910,635	606,874	125,101	830,664	50	883	12,794,023	39,320
31N	4	1,300,837	758,151	863,435	1,113,817	40	11,633	110,992	123,302
32Q	4	666,485	435,597	9,481	108,297	27	654	24,739,862	9,819

Appendix VI

EGYPT: Results of Compositional Analyses (Chapters 8 and 9)

Table 22 Raw values for EGYPT base glass oxides (wt%)

EGYPT Base Glass Oxides (wt%)																	
Sample	Colour	SiO_2	TiO_2	Al_2O_3	FeO	MgO	CaO	Na_2O	K_2O	MnO	CuO	Sb_2O_5	NiO	CoO	Cl	SO_2	Sum
Egypt 1	dark blue	58.35	3.60	3.54	0.42	3.60	6.69	22.98	0.55	0.38		0.95		0.27	1.26	0.42	103.01
Egypt 2	dark blue	62.60	0.14	0.70	0.83	3.13	6.56	20.98	1.57	0.24					1.27	0.34	98.38
Egypt 3	dark blue	62.03	0.12	3.07	0.72	4.32	8.64	17.09	0.74					0.24	0.95	0.28	98.18
Egypt 4	dark blue	64.65	0.26	1.53	0.77	2.35	7.49	18.65	0.73		0.65				1.32	0.31	98.70
Egypt 5	dark blue	64.54		1.48	0.44	2.77	5.75	21.00	0.53						1.25	0.28	98.05
Egypt 6	dark blue	64.71		1.94	0.47	3.13	6.56	20.98	1.57	0.24					1.26	0.34	101.20

Blank spaces indicate oxide not detected

Analyses courtesy of Professor Ian Freestone

Table 23 Raw values for EGYPT rare earth elements (ppb)

EGYPT Rare Earth Elements (ppb)														
Sample	La	Ce	Pr	Nd	Sm	Eu	Gd	Tb	Dy	Ho	Er	Tm	Yb	Lu
Egypt 1	3,480	7,237	955	4,091	960	236	1,057	167	982	219	651	106	708	112
Egypt 2	2,723	5,947	787	3,301	700	162	681	105	598	129	357	55	343	53
Egypt 3	7,671	18,644	2,549	11,611	2,689	739	3,696	508	2,708	555	1,419	195	1,105	171
Egypt 4	4,148	11,026	1,709	8,339	2,284	659	3,126	449	2,345	429	985	126	675	93
Egypt 5	3,135	7,024	851	3,466	735	187	788	122	673	140	384	57	356	54
Egypt 6	2,677	6,932	867	3,763	920	260	1,247	172	880	169	425	57	352	52

Table 24 Raw values for EGYPT sediment-related elements (ppb)

EGYPT Sediment-Related Elements (ppb)														
Sample	Sc	Ti	V	Cr	Ga	Y	Zr	Nb	Mo	Hf	Ta	W	Tl	Th	U
Egypt 1	953	428,873	6,791	5,423	1,132	6,671	35,448	1,525	146	841	96	22	5	734	1,284
Egypt 2	885	471,920	7,810	4,217	963	3,742	38,245	1,363	116	978	88	27	8	595	281
Egypt 3	1,733	823,622	11,160	8,855	1,738	16,280	63,455	2,808	120	1,505	182	25	6	1,408	606
Egypt 4	1,331	599,842	8,166	5,506	1,347	10,722	44,218	2,043	70	1,059	127	14	5	1,092	839
Egypt 5	1,049	602,427	8,093	11,229	1,088	4,144	49,317	1,753	238	1,140	110	41	2	740	1,027
Egypt 6	1,124	561,856	10,237	15,188	1,118	4,989	32,365	1,536	216	761	97	32	11	777	615

Table 25 Raw values for EGYPT alkali and alkaline earth elements (ppb)

EGYPT Alkali and Alkaline Earth Elements (ppb)						
Sample	Li	Be	Rb	Sr	Cs	Ba
Egypt 1	8,596	661	5,520	634,651	79	57,713
Egypt 2	9,475	354	3,931	499,790	49	46,884
Egypt 3	7,655	1,256	3,342	669,568	83	51,513
Egypt 4	7,365	660	2,988	467,997	43	52,541
Egypt 5	8,221	263	4,208	503,708	72	38,899
Egypt 6	11,854	302	6,903	253,392	171	54,214

Table 26 Raw values for EGYPT colourant and related elements (ppb)

EGYPT Colourant and Related Elements (ppb)								
Sample	Co	Ni	Cu	Zn	Cd	Sn	Sb	Pb
Egypt 1	715,787	475,967	2,333,295	139,478	152	225,448	1,349,900	67,503
Egypt 2	454,317	284,324	1,447,853	187,725	236	142,962	226,518	1,442,274
Egypt 3	1,356,315	772,669	393,082	1,817,200	81	39,869	606,970	13,318
Egypt 4	2,128,679	1,158,889	99,303	2,551,610	98	1,114	6,299,755	10,995
Egypt 5	783,270	514,650	2,337,715	534,972	89	96,717	20,672	215,923
Egypt 6	572,884	195,844	10,258,962	293,728	116	74,489	475,013	338,260